ROBUST MODEL-BASED FAULT DIAGNOSIS FOR DYNAMIC SYSTEMS

The Kluwer International Series on
ASIAN STUDIES IN COMPUTER AND INFORMATION SCIENCE

Series Editor

Kai-Yuan Cai

Beijing University of Aeronautics
Beijing, CHINA

Also in the Series:

SOFTWARE DEFECT AND OPERATIONAL PROFILE MODELING
by Kai-Yuan Cai
ISBN: 0-7923-8259-5

FUZZY LOGIC IN DATA MODELING: Semantics, Constraints, and Database Design
by Guoqing Chen
ISBN: 0-7923-8253-6

ROBUST MODEL-BASED FAULT DIAGNOSIS FOR DYNAMIC SYSTEMS

by

Jie Chen
Brunel University, Uxbridge, UK

Ron J. Patton
University of Hull, Hull, UK

KLUWER ACADEMIC PUBLISHERS
Boston / Dordrecht / London

Distributors for North, Central and South America:
Kluwer Academic Publishers
101 Philip Drive
Assinippi Park
Norwell, Massachusetts 02061 USA
Telephone (781) 871-6600
Fax (781) 871-6528
E-Mail <kluwer@wkap.com>

Distributors for all other countries:
Kluwer Academic Publishers Group
Distribution Centre
Post Office Box 322
3300 AH Dordrecht, THE NETHERLANDS
Telephone 31 78 6392 392
Fax 31 78 6546 474
E-Mail <orderdept@wkap.nl>

Electronic Services <http://www.wkap.nl>

Library of Congress Cataloging-in-Publication Data

A C.I.P. Catalogue record for this book is available
from the Library of Congress.

Printed on acid-free paper.

Printed in the United States of America

Contents

List of Figures

List of Tables

Preface

There is an increasing demand for dynamic systems to become more safe and reliable. This requirement extends beyond the normally accepted safety-critical systems of nuclear reactors and aircraft where safety is paramount important, to systems such as autonomous vehicles and fast railways where the system availability is vital. It is clear that fault diagnosis (including fault detection and isolation, FDI) has been becoming an important subject in modern control theory and practice. For example, the number of papers on FDI presented in many control-related conferences has been increasing steadily. The subject of fault detection and isolation continues to mature to an established field of research in control engineering.

A large amount of knowledge on model-based fault diagnosis has been accumulated through the literature since the beginning of the 1970s. However, publications are scattered over many papers and a few edited books. Up to the end of 1997, there is no any book which presents the subject in an unified framework. The consequence of this is the lack of "common language", different researchers use different terminology. This problem has obstructed the progress of model-based FDI techniques and has been causing great concern in research community. Many survey papers have been published to tackle this problem. However, a book which presents the materials in a unified format and provides a comprehensive foundation of model-based FDI is urgently needed. Such a book would promote the subject of model-based FDI and make the techniques more accessible for engineers and research students. This view has been shared by many researchers in this field. Such a book is also possible, because many important definitions have been made and the correspondences between different model-based FDI methods have been established.

This new book presents the subject of model-based FDI in a unified framework. It contains many important topics and methods, however perfect coverage and completeness is not the primary concern. The book focuses on fundamental issues such as basic definitions and the importance of robustness in FDI approaches. In this book, FDI concepts and methods are illustrated by either simple academic examples or practical applications. The first two chapters are of tutorial value and provide a starting point for new comers to this field. The rest of the book presents the state-of-the-art in model-based FDI by discussing many important robust approaches and their applications. This will certainly appeal to experts in this field.

The book targets both new comers, who want to get into this subject, and experts, who are concerned with fundamental issues and are also looking for inspiration for future research. The book is useful for both researchers in academia and professional engineers in industry because both theory and applications have been discussed. Although this is a research monograph, it will be an important text for MSc & PhD research students world-wide. The largest market, however will be academics, libraries and practising engineers and scientists throughout the world.

DR J. CHEN & PROF. R. J. PATTON

Acknowledgments

We have been enjoying many fruitful discussions and collaborations with other researchers world-wide. It is impossible to mention all of them. However, we would like to thank Prof. Paul M. Frank of University of Duisburg, Germany and Prof. Janos J. Gertler of George Mason University, USA with whom we had many discussions and debates. We are also grateful to Prof. Peter D. Roberts of City University, UK for his encouragement for starting this book. We have had some help in the writing of this book. Many thanks to Prof. J. Stoustrup of Alborg University, Denmark, Prof. M. Kinnaert of Université Libre de Bruxelles, Belgium, Prof. X. Ding of FH-Lausitz, Germany, Dr H. Wang of UMIST, UK, Dr D. N. Shields of Coventry University, UK and Prof. M. M. Polycarpou of University of Cincinnati, USA for sending us their recent publications which enable us to include updated information. Special thank goes to Dr Ming Hou of University of Hull, UK who provides us with his new research results in files.

We would like to express our sincere gratitude and thanks to our families for their support. Dr Chen is very grateful to his wife Ling Hao and daughter Nancy for their patience and understanding when he devotes many of his evening and weekends to this book. Prof. Patton would like to thank his wife Ann and children Libby, Kirstie and Stephen for their understanding when he carried around the draft manuscript of this book in his weekends and holidays.

1 INTRODUCTION

1.1 Background

1.1.1 Importance of fault diagnosis

Modern control systems are becoming more and more complex and control algorithms more and more sophisticated. Consequently, the issues of availability, cost efficiency, reliability, operating safety and environmental protection are of major importance, These issues are important to, not only normally accepted safety-critical systems such as nuclear reactors, chemical plants and aircraft, but also other advanced systems employed in cars, rapid transit trains, etc. For safety-critical systems, the consequences of faults can be extremely serious in terms of human mortality, environmental impact and economic loss. Therefore, there is a growing need for on-line supervision and fault diagnosis to increase the reliability of such safety-critical systems. Early indications concerning which faults are developing can help avoid system breakdown, mission abortion and catastrophes. For systems which are not safety-critical, on-line fault diagnosis techniques can be used to improve plant efficiency, maintainability, availability and reliability. Indeed, industry is starting to reconsider the implications of using predictive maintenance tools and is looking for alternative methods to ensure plant availability and safety, whilst at the same time obviating costly maintenance during plant down-time. To provide insight into the state of a system, which allows a true on-condition maintenance plan to be implemented, modern fault diagnosis methods can be considered.

Since the beginning of the 1970s, research in fault diagnosis has been gaining increasing consideration world-wide in both theory and application (Patton and Chen, 1998; Frank and Ding, 1997). This development was (and still is) mainly stimulated by the trend of automation towards more complexity and the growing demand for higher availability and security of control systems. However, a strong impetus also comes from the side of modern control theory that has brought forth powerful techniques of mathematical modeling, state estimation and parameter identification that have been made feasible by the spectacular progresses of computer technology.

1.1.2 Fault diagnosis terminology

By going through the literature, one recognizes immediately that the terminology in the field of fault diagnosis is not consistent. This makes it difficult to understand the goals of the particular contributions and to compare the different approaches. To address this problem, The IFAC Technical Committee: *SAFEPROCESS* (Fault Detection, Supervision and Safety for Technical Processes) has started an initiative to define common terminology. The definitions established by this initiative are listed in Appendix A. Further discussions regarding terminology can be found in references: Isermann and Ballé (1997) and van Schrick (1997). The terminology adapted in this book is largely consistent with *SAFEPROCESS* terminology.

A "fault" is to be understood as an unexpected change of system function[1], although it may not represent physical failure or breakdown. Such a fault or malfunction hampers or disturbs the normal operation of an automatic system, thus causing an unacceptable deterioration of the performance of the system or even leading to dangerous situations. We use the term "fault" rather than "failure" to denote a malfunction rather than a catastrophe. The term *failure* suggests complete breakdown of a system component or function, whilst the term *fault* may be used to indicate that a malfunction may be tolerable at its present stage. A fault must be diagnosed as early as possible even it is tolerable at its early stage, to prevent any serious consequences.

A monitoring system which is used to detect faults and diagnose their location and significance in a system is called a "fault diagnosis system". Such a system normally consists of the following tasks:

- **Fault detection:** to make a binary decision – either that something has gone wrong or that everything is fine.
- **Fault isolation:** to determine the location of the fault, e.g., which sensor or actuator has become faulty.
- **Fault identification:** to estimate the size and type or nature of the fault.

[1]An alternative definition given by Isermann (1984): a "fault" is defined as "a non permitted deviation of a characteristic property which leads to the inability to fulfil the intended purpose".

The relative importance of three tasks are obviously subjective, however the detection is an absolute must for any practical system and isolation is almost equally important. Fault identification, on the other hand, whilst undoubtedly helpful, may not be essential if no reconfiguration action is involved. Hence, fault diagnosis is very often considered as fault detection and isolation, abbreviated as FDI, in the literature.

1.1.3 Fault diagnosis in intelligent fault-tolerant control

There is an increasing need for controlled systems to continue operating acceptably to fulfil specified functions following faults in the system being controlled or in the controller. A control system with this kind of fault-tolerance capability is defined as a fault-tolerant control system. There may be some graceful performance degradation for a fault-tolerant system to operate under a faulty condition, however the primary objective is to maintain system operation and give the human operator (or automatic monitoring system) reasonable time to repair the system or to use alternative measures to avoid catastrophes. Fault-tolerant control has received increasing attention recently, motivated by the need to achieve high levels of reliability, maintainability and performance in situations where the controlled system can have potentially damaging effects on the environment if faults in its components take place (Patton, 1997a). For instance, in hazardous chemical and nuclear plants, the consequences of an improper control action following a control system component fault can be disastrous. In the case of flight control systems, safety is the greatest priority, which implies that even in the presence of failed components the aircraft must be able to land safely.

A fault-tolerant control system is designed to retain some portion of its control integrity in the event of a specified set of possible component faults or large changes in the system operating conditions that resemble these faults. This can only be done if the control system has built in an element of automatic reconfiguration, once a malfunction has been detected and isolated. Fault diagnosis plays an important role in the fault-tolerant control, as before any control law reconfiguration is possible the fault must be reliably detected, isolated, and the information should be passed to a supervision mechanism to make proper decision.

Fault-tolerance is considered as one of the characteristics of intelligent systems. According to Stengel (1991): "By design or implementation, failure-tolerant control systems are *intelligent* systems". Aström (1991) has also stated: "Fault diagnosis is an essential ingredient property of an intelligent control system". Many important issues in fault-tolerant control systems can be found in a recent survey paper by Patton (1997a).

1.1.4 Model-based fault diagnosis

In practice, the most frequently used diagnosis method is to monitor the level (or trend) of a particular signal, and taking action when the signal reached a given threshold. This method of limit checking, whilst simple to implement, has serious drawbacks. The first drawback is the possibility of false alarms in the event of noise, the input variations and the change of operating point. The second drawback is that a single fault could cause many system signals to exceed their limits and appear as multiple faults, and hence fault isolation is very difficult. The use of consistency checking for a number of system signals which can eliminate the above problems, is an important way of enhancing the detection and isolation or fault diagnosis capability of an automated system. However, a mathematical model which gives functional relationships among different system signals is needed.

A traditional approach to fault diagnosis in the wider application context is based on "hardware (or physical/parallel) redundancy" methods which use multiple lanes of sensors, actuators, computers and software to measure and/or control a particular variable. Typically, a voting scheme is applied to the hardware redundant system to decide if and when a fault has occurred and its likely location amongst redundant system components. The use of multiple redundancy in this way is common, for example with digital fly-by-wire flight control systems e.g. the AIRBUS 320 and its derivatives (Favre, 1994) and in other applications such as in nuclear reactors. The major problems encountered with hardware redundancy are the extra equipment and maintenance cost and, furthermore, the additional space required to accommodate the equipment.

In view of the conflict between the reliability and the cost of adding more hardware, it is sensible to attempt to use the dissimilar measured values together to *cross check*[2] each other, rather than replicating each hardware individually; this is the concept of "analytical (functional) redundancy" which uses redundant analytical (or functional) relationships between various measured variables of the monitored process (eg inputs/outputs; outputs/outputs; inputs/inputs). Fig.1.1 illustrated the hardware and analytical redundancy concepts. No additional hardware faults are introduced into an analytical redundant scheme, because no extra hardware is required, hence analytical redundancy is potentially more reliable than hardware redundancy (van Schrick, 1991; van Schrick, 1993).

In analytical redundancy schemes, the resulting difference generated from the consistency checking of different variables is called as a *residual* signal. The residual should be zero-valued when the system is normal, and should diverge from zero when a fault occurs in the system. This zero and non-zero property of the residual is used to determine whether or not faults have occurred. Analytical redundancy makes use of a mathematical model of the monitored process and is therefore often referred to as the "model-based approach" to fault diagnosis.

Consistency checking in analytical redundancy is normally achieved through a comparison between a measured signal with its estimation. The estimation is generated by the mathematical model of the system being considered. The

[2]This procedure is sometimes referred to as *data reconciliation*.

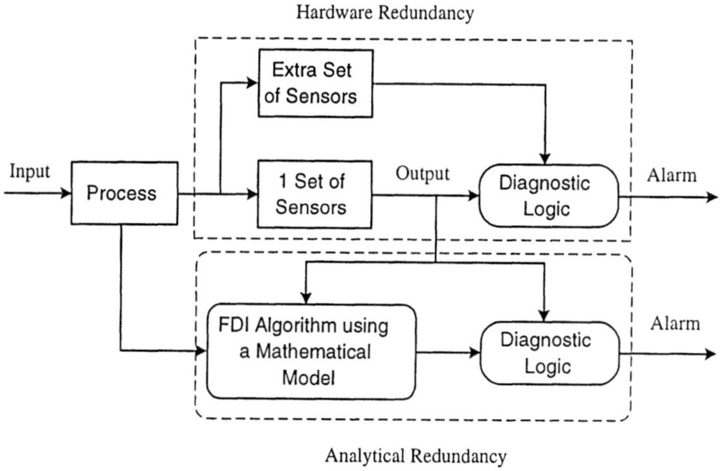

Figure 1.1. Hardware vs analytical redundancy

comparison is done using the residual quantities which give the difference between the measured signals and signals generated by the mathematical model. Hence, *model-based fault diagnosis* can be defined as *the determination of faults of a system from the comparison of available system measurements with a priori information represented by the system's mathematical model, through generation of residual quantities and their analysis.* A residual is a fault indicator or an accentuating signal which reflects the faulty situation of the monitored system.

The major advantage of the model-based approach is that no additional hardware components are needed in order to realize an FDI algorithm. A model-based FDI algorithm can be implemented in software on the process control computer. Furthermore, the measurements necessary to control the process are, in many cases, also sufficient for the FDI algorithm so that no additional sensors have to be installed. Under theses circumstances, only additional storage capacity and possibly greater computer power is needed for the implementation of a model-based FDI algorithm. Immense developments in computer technology have made such methods very feasible and practicable (Frank, 1990).

1.1.5 Robustness in model-based fault diagnosis

Model-based FDI makes use of mathematical models of the supervised system, however a perfectly *accurate* and *complete* mathematical model of a physical system is never available. Usually, the parameters of the system may vary with time in an uncertain manner, and the characteristics of the disturbances and noise are unknown so that they cannot be modeled accurately. Hence, there is always a mismatch between the actual process and its mathematical model even if there are no process faults. Apart from the modeling used for the purpose of control, such discrepancies cause fundamental methodology difficulties in FDI applications. They constitute a source of false and missed alarms which can corrupt the FDI system performance to such an extent that it may even become totally useless. The effect of modeling uncertainties is therefore the most crucial point in the model-based FDI concept, and the solution of this problem is the key for its practical applicability (Frank, 1991a; Patton and Chen, 1996b; Patton and Chen, 1997).

To overcome the difficulties introduced by modeling uncertainty, a model-based FDI has to be made robust, i.e. insensitive or even invariant to modeling uncertainty. Sometimes, a mere reduction of the sensitivity to modeling uncertainty does not solve the problem because such a sensitivity reduction may be associated with a reduction of the sensitivity to faults. A more meaningful formulation of the robust FDI problem is to increase the robustness to modeling uncertainty, whilst without losing (or even with an increase of) fault sensitivity. An FDI scheme designed to provide satisfactory sensitivity to faults, associated with the necessary robustness with respect to modeling uncertainty, is called *a robust FDI scheme* (Frank, 1991a; Patton and Chen, 1996b; Patton and Chen, 1997). The importance of robustness in model-based FDI has been widely recognized by both academia and industry. The development of robust model-based FDI methods has been a key research topic during the last 10 years. A number of methods have been proposed to tackle this problem, for example, the unknown input observer, eigenstructure assignment, optimally robust parity relation methods. However, the research is still under the way to develop the practically applicable methods.

An important task of the model-based FDI scheme is to be able to diagnose *incipient faults* [3] in a system before they are manifested as problems require either human operator or automatic system intervention. The diagnosis of hard and abrupt faults is relatively easy, because their effects on the FDI system are larger than modeling uncertainty and can be diagnosed by placing an appropriate threshold on the residual. However, incipient faults have a small effect on residuals, and can be hidden as a consequence of modeling uncertainty. This highlights the need of robustness in FDI. The effect of an incipient fault on the monitored system is very small and almost unnoticeable when it occurs. However, it may develop slowly to cause very serious consequences, although it may be tolerable in its early stage. It is important to note that a soft fault is a malfunction condition which is non-serious (in its present state) and which often develops in a continuous way (i.e. which does not contain discontinuous signal characteristics brought about as a consequence of abrupt changes). The presence of soft faults may not necessarily downgrade the performance of the plant significantly, however, such faults will indicate that the sensor (or other component) should be replaced, or that the system should be re-configured be-

[3] Small and slowly developing faults are normally defined as incipient faults, and sometimes called soft faults.

fore the probability of more serious malfunction increases. Prompt indication of incipient faults can give the operator (or an automatic monitoring system) enough information and time to take decisive actions to prevent any serious failure in the system. The successful detection and diagnosis of soft faults can therefore be considered as the hardest challenge for the design and evaluation of algorithms working in a safety-critical environment.

1.2 Brief history of model-based fault diagnosis

Beard-Jones Fault Detection Filter:

The development of model-based fault diagnosis began at various places in the early 1970s. The idea of replacing hardware redundancy by analytical redundancy was originated in MIT by Beard (1971), in which fault (failure) detection filters generating directional residuals for FDI was developed. This approach was then redefined in a geometric interpretation by Jones (1973) and Massoumnia (1986b). This particular line of inquiry has led to what is known as the *Beard-Jones Fault Detection Filter* or *Beard Fault Detection Filter.* The design problem was later resolved by White and Speyer (1987), Park and Rizzoni (1994a), Park, Halevi and Rizzoni (1994), Park, Rizzoni and Ribbens (1994) and Park and Rizzoni (1994b) and recently revisited by Douglas and Speyer (1996), Liu and Si (1997) and Chung and Speyer (1998).

Stochastic Systems FDI:

In parallel with the development of the Beard-Jones Fault Detection Filter, the statistical approaches were also developed starting in the early 1970s. Mehra and Peschon (1971) introduced a general procedure for FDI using innovations (or residuals) generated by a Kalman filter. The faults are diagnosed by statistical testing on whiteness, mean and covariance of residuals. Willsky and Jones (Willsky and Jones, 1974; Willsky and Jones, 1976) then developed an FDI strategy which uses Generalized Likelihood Ratio (GLR) testing on a residual generated by a Kalman filter to diagnose faults. In the well-known early survey paper, Willsky (1976) presented key concepts of analytical redundancy in model-based FDI with an emphasis on stochastic systems and jump detection. Following this line of inquiry, Basseville (1988) addressed the problems of detection, estimation and diagnosis of changes in dynamical properties of signals or systems, with particular emphasis on statistical methods for detection, to provide a general framework for change detection in signals and systems. The development of statistical approaches was later surveyed in an excellent paper by Tzafestas and Watanabe (1990). In this survey paper, other approaches such as knowledge-based techniques were discussed. The most distinguish feature of this paper was its excellence on the survey of stochastic techniques. The fundamental methods and recent development of statistical approaches can be found in Basseville and Nikiforov (1993), Da and Lin (1995) and Keller, Summerer, Boutayeb and Darouach (1996).

One of the statistical approaches is the multiple model adaptive filter approach which involves multiple hypothesis testing on residuals generated by a bank of Kalman filters (Willsky, Deyst and Crawford, 1974; Willsky, Deyst and Crawford, 1975; Montgomery and Caglayan, 1976). New developments and applications can be referred to Berec (1998), Menke and Maybeck (1995) and Eide and Maybeck (1996).

Recently, the statistical methods based on the use of principal component analysis has received favorable attention (Martin, Morris and Zhang, 1996; Zhang, Martin and Morris, 1996; Zhang, Martin and Morris, 1997).

Observer-based FDI approach:

Clark and co-workers first applied Luenberger observers for fault detection (Clark, Fosth and Walton, 1975) and various sensor fault isolation schemes were later developed (Clark, 1978a; Clark, 1978b; Clark, 1979). Frank's comprehensive survey paper (Frank, 1987) established the position of observer-based methods in model-based FDI. In this survey paper, many different schemes using both linear and non-linear observers are reviewed and some application examples were presented.

Parity Relation Approach for FDI:

The parity relation approach to generate the residual (or parity vector), based upon consistency checking on system input and output data over a time window, was originally proposed by Mironovski (1979; 1980) although he used a different terminology. Unfortunately, his papers have not received enough attention due to their limited availability. The approach was later, independently proposed by Chow and Willsky (1984), and has been expressed in several different versions. For example, Gertler (1988) gave a parity relation design method in the z-domain. Chen and Zhang (1990) developed a stochastic system FDI approach based upon a direct development of the parity vector concept used in hardware redundancy. The latest development regarding parity relation approaches can be found in Gertler (1997) and Gertler (1998).

Parameter Estimation Approach for FDI:

One of the important FDI approaches is the use of parameter estimation which is based directly on system identification techniques. This approach was first illustrated by Bakiotis, Raymond and Rault (1979) and Geiger (1982). Isermann and co-workers followed this line of research since early 1980s. In 1984, Isermann illustrated that process fault diagnosis can be achieved using the estimation of unmeasurable process parameters and/or state variables in his survey paper (Isermann, 1984). This paper gave a generalized structure of FDI based on process models and unmeasurable quantities. This structure has been referred to in many subsequent papers, e.g. Frank (1990). Isermann (1987) reported some experiences in the use of parameter estimation for process FDI. Isermann and Freyermuth (1990) studied on-line FDI systems using a combination of parameter estimation and heuristic process knowledge. This paper was followed by another survey paper (Isermann and Freyermuth, 1991a), and an application paper (Isermann and Freyermuth, 1991b). Isermann (1991a) gave an application-oriented review of parameter estimation FDI methods based a number of real or laboratory applications. The latest development and applications of parameter estimation FDI approaches can be found in Isermann (1997) and Isermann and Ballé (1997).

Two-stages Model-based FDI Structure:

Chow and Willsky (1980; 1984) first defined the model-based FDI as a two-stages process: (1) residual generation, (2) decision-making (including residual evaluation). This two-stages process is accepted as a standard procedure for model-based FDI nowadays.

Robustness Problem in FDI:

Leininger (1981) pointed out the impact of modeling errors on FDI performance. The first attempt of improving robustness of observer-based FDI approaches is attributed to Frank and Keller (1981).

Robust FDI using Unknown Input Observers:

To solved the robust FDI problem, Watanabe and Himmelblau (1982) introduced a robust sensor detection method using an unknown input observer (UIO). Robust FDI based on UIOs has been studied extensively by Frank's group at the University of Duisburg, Germany, and many contributions have been made by this group, for example, Frank and Wünnenberg (1987), Frank and Wünnenberg (1989), Wünnenberg (1990), Frank (1990), Frank (1991a), Frank and Seliger (1991) and Seliger and Frank (1991a). Chen and Zhang (1991) proposed a robust actuator fault isolation scheme and demonstrated this using a chemical process. Ge and Fang (1988; 1989) developed a robust component FDI approach using the so-called robust observation method which is similar to UIOs, in principle. Viswanadham et al. (Viswanadham and Srichander, 1987; Phatak and Viswanadham, 1988) proposed an actuator fault isolation scheme which is an important original contribution, however they did not consider robustness issues. A comprehensive treatment on robust FDI using UIOs is presented in Chapter 3 of this book.

Robust FDI via Eigenstructure Assignment:

Patton, Willcox and Winter (1986) proposed an FDI method based on eigenstructure assignment and this approach has been studied extensively by Prof. Patton's group. Many developments have been made, for example, Patton (1988), Patton and Kangethe (1989), Patton and Chen (1991g), Chen (1995) and Patton and Chen (1997). The details are described in Chapter 4 of this book.

Optimal Parity Relations for Robust FDI:

Lou, Willsky and Verghese (1986) developed a strategy to design "optimally robust parity relations" for diagnosing faults in systems represented by multiple models. Following the same philosophy, Wünnenberg and Frank (Wünnenberg and Frank, 1988; Frank, 1990; Wünnenberg and Frank, 1990; Wünnenberg, 1990) studied the design of optimal parity relations by adopting a modified criterion which is the ratio of the modeling uncertainty response effect to that of the fault effect. Similar to this approach, Gertler and colleagues proposed a scheme to design robust parity relations using the "orthogonal parity relations" concept (Gertler and Luo, 1989; Gertler, Fang and Luo, 1990; Gertler and Singer, 1990; Gertler, 1991; Gertler and Kunwer, 1993). The optimal parity relation approach is re-examined in Chapter 7 and the recent development can be found in Wu and Wang (1995), Kinnaert (1996) and Ding and Guo (1998).

Frequency Domain Design for Model-based FDI:

Viswanadham, Taylor and Luce (1987) introduced a new residual generation method based on a factorization of the system transfer matrix. This approach was later developed by Ding and Frank (1990) and revisited by Kinnaert and Peng (1995). This is now referred to as the frequency domain residual generation approach. To address the robust FDI problem in the frequency domain, Viswanadham and Minto (1988) proposed solutions for improving the robustness of frequency domain residual generation using H_∞ optimization techniques. Studies on this problem have been extended by Ding and Frank in a series of papers, e.g., Ding and Frank (1991), Frank and Ding (1993), Ding, Guo and Frank (1993) and Frank and Ding (1994). Qiu and Gertler (1993) also solved the same problem with a different approach. The μ synthesis, a robust design technique was also applied to frequency domain robust FDI problem (Mangoubi, Appleby and Farrell, 1992; Appleby, Dowdle and Vander Velde, 1991). Recently, there has been a surge of interest in the frequency domain FDI approach. A detailed treatment of this approach is described in Chapter 8 and recent developments can be found in Sadrnia, Chen and Patton (1997a; 1997b; 1997c), Sauter, Rambeaux and Hamelin (1997) and Edelmayer, Bokor and Keviczky (1994; 1997b).

Robust Residual Evaluation and Adaptive Threshold:

When residuals cannot be made robust against system uncertainty, the robust FDI can be achieved by robust decision making using adaptive thresholds. Emami-Naeini, Akhter and Rock (1988) introduced the *threshold selector* concept to generate adaptive thresholds. Clark (1989) proposed a method to produce adaptive thresholds. The adaptive threshold determination was studied extensively by Ding and Frank (Ding and Frank, 1991; Frank and Ding, 1993; Ding et al., 1993; Frank and Ding, 1994; Ding and Guo, 1998). The application examples was reported in Chang, Hsu and Lin (1995) and Höfling and Isermann (1996). The idea of achieving robustness in decision-making stage was further advanced by Frank *et al* by introducing a fuzzy logic approach to residual evaluation and decision-making, see Frank and Kiupel

(1993), Frank (1994a; 1996), Schneider and Frank (1996), Frank and Ding (1997), Frank and Köppen-Seliger (1997a; 1997b). Recently, Ding and Guo (1998) demonstrated that robust FDI can be achieved through the optimization of the residual evaluation function in the design of diagnostic observers.

Uncertainty modeling for Robust FDI:

To solve robust FDI problems, a mathematical representation for describing modeling uncertainty is needed. Patton and Chen (Patton and Chen, 1991f; Patton and Chen, 1991b; Patton, Chen and Zhang, 1992; Chen, 1995) proposed several schemes to represent modeling uncertainties from various sources as additive disturbances with an estimated distribution matrix. Robust FDI is thus achieved using disturbance de-coupling approaches. To date, this is one of the most important contributions in robust FDI. So far, most robust residual generation methods are based on the assumption that disturbance distribution matrices are known, however this assumption is not valid for most real systems. The contributions by Patton and Chen have paved a way for real application of robust FDI techniques. Application results can be found in Chen (1995), Dalton, Patton and Chen (1996) and Patton and Chen (1997).

Rapid Progress of Model-based FDI:

The most extensive period of development of model-based FDI took place between late 1980s and early 1990s. During this period, many fundamental definitions were established and a general model-based FDI system structure was proposed. In his early tutorial paper, Gertler (1988) described basic concepts and essential definitions. Some problems discussed in this paper, such as isolability conditions and sensitivity and robustness, are still of tutorial value today. In 1989, Patton, Frank and Clark (1989) published their important edited book. The book covered most import model-based FDI methods in the 1980s and many application examples have been listed. The book has been and is still widely referred to. Following the success of this book, Frank (1990) outlined the principles and most important techniques of model-based residual generation using parameter identification and state estimation methods with emphasis upon the latest attempts to achieve robustness with respect to modeling uncertainty. The possibility of combining model-based and knowledge-based techniques for FDI was also discussed. Patton (1991) emphasized aerospace applications of model-based FDI and analytical redundancy.

To define the model-based FDI in a consistent framework, the relationships between different approaches have to be established. Patton and Chen (1991d; 1991h; 1991c) first established the connection between parity relation and observer-based approaches. The correspondence between parity relations and parameter estimation approaches was formally established by Gertler and DiPierro (1997). With the connections between different approaches in mind, Patton and Chen (1991e) unified the observer-based and parity relations approaches under a common parity space format. The model-based FDI has been re-stated by them as the generation and analysis of residual signals in the parity space. Their paper presented a generalized framework of residual generators and provided some important definitions, as well as demonstrating robust fault diagnosis methods using two tutorial examples. The survey paper by Patton and Chen (1992a) followed the same philosophy given by Patton and Chen (1991e). The emphasis was on different synthesis methods for residual generators with a particular reference on aerospace applications.

The important paper by Patton and Chen (1991e) was presented in the IFAC Symposium: *SAFEPROCESS'91*. At the same conference, Frank (1991a) presented his view on enhancing robustness in observer-based FDI with disturbance de-coupling observers, optimal parity relations, H_∞ observers and adaptive thresholds. Gertler (1991) presented a tutorial on residual generator synthesis methods. The best known residual generation methods, including parity equations, diagnostic observers and Kalman filtering, were presented in a consistent framework. The discussion was organized along two residual enhancement concepts, namely structured and fixed direction residual sets. A numerical example was used to show how parity relation and observer based designs lead to equivalent residual generators, once the design objectives are specified. Isermann (1991a) presented a tutorial on parameter estimation approaches with an emphasis on applications.

In 1993, an important conference dedicated to fault diagnosis was organized in France: *TOOLDIAG'93* (Labarrère, 1993). In this conference, Frank (1993) reviewed the advanced methods of observer-based FDI. The paper discussed the issue of improving decision-making robustness using fuzzy logic. The paper, however was limited in its scope to research developments within Frank's group. The paper by Gertler and Kunwer (1993) studied both perfect and approximate disturbance de-coupled residual generator designs, with an emphasis on z-domain parity relation design methods and with a numerical example to demonstrate approximate de-coupling. Patton (1993) studied the robustness issues in fault-tolerant control systems, including diagnosis and reconfiguration issues. The paper pointed out that the best way forward in fault-tolerant control is to integrate together FDI and controller functions in analysis and design, so that joint stability and performance robustness properties can be optimized.

There have been so many FDI methods developed, a difficult problem one would face is how to choose the right method to solve a specific problem. To address this problem, Patton, Chen and Nielsen (1994) presented some guide-lines for engineers in the choice of different model-based FDI methods. Isermann (1993) discussed the *applicability* of different FDI methods based on their requirements and results from simulations. The work was followed by a more comprehensive paper (Isermann, 1994) in which the integration of different FDI methods was also studied.

Non-linear Dynamic Systems FDI:

Recently, the research has focused on non-linear systems FDI. Traditionally, the FDI problem for non-linear dynamic systems has been approached in two steps. Firstly, the model is linearized at an operating point, and then robust techniques are applied to generate residuals. This method only works well when the linearization does not cause a large mismatch between linear and non-linear models and the system operates close to the operating point specified. To deal with systems with high nonlinearity and wide operation range, the FDI problem has to be tackled directly using non-linear techniques. Early development of non-linear observer methods can be found in survey papers (Frank, 1994b; García and Frank, 1997). Krishnaswami, Luh and Rizzoni (1995) extended the parity relations method to non-linear systems.

The FDI problem for one special class of non-linear systems, bilinear systems, has been studied extensively over last a few years (Yu, Shields and Daley, 1996; Yu and Shields, 1996; Shields, 1997; Yu and Shields, 1997; Yang and Saif, 1997). Neural networks, as powerful tools for handling non-linear problems, have been successfully applied to non-linear systems FDI problems, for example, Watanabe, Matsuura, Abe, Kubota and Himmelblau (1989), Naidu, Zafiriou and McAvoy (1990), Himmelblau, Barker and Suewatanakul (1991), Sorsa, Koivo and Koivisto (1991), Willis, Massimo, Montague, Tham and Morris (1991), Sorsa and Koivo (1993), Kavuri and Venkatasubramanian (1994), Leonard and Kramer (1993), Napolitano, Neppach, Casdorph, Naylor, Innocenti and Silvestri (1995), Napolitano, Casdorph, Neppach, Naylor, Innocenti and Silvestri (1996), MakiLopa97, and Patton and Chen (1996a). Apart from the non-linear modeling ability of neural networks, the learning ability has also been utilized recently for non-linear robust FDI (Polycarpou and Vemuri, 1995; Polycarpou and Helmicki, 1995; Vemuri and Polycarpou, 1997b; Vemuri and Polycarpou, 1997a; Vemuri, Polycarpou and Diakourtis, 1998). Recently, the fuzzy logic's ability in handling non-linear problems has been applied to model-based FDI (Patton, Chen and Lopez-Toribio, 1998), sometimes, combined with neural networks (Zhang and Morris, 1996; Ayoubi and Isermann, 1997; Pfeufer and Ayoubi, 1997; Ballé, Fischer, Füssel, Nelles and Isermann, 1998). The problem of non-linear systems model-based FDI is discussed in Chapter 9.

Remarks:

There have been many developments in model-based fault diagnosis since the beginning of 1970s. The approaches and papers discussed above, however represent (in the authors' opinion) the majority of the key developments in model-based fault diagnosis. The list of papers discussed above is, nevertheless inconclusive. To advance and promote the research filed of model-based FDI, many excellent survey and tutorial papers have been published. Different papers discussed varied aspects of the problem from different perspectives. Apart from survey papers discussed above, many other survey papers emphasizing different aspects of the problem are worth bringing to reader's attention, e.g., Walker (1983); Himmelblau (1986); Tzafestas (1989); Frank (1991b); Ray and Luck (1991); Frank (1992b); Frank and Köppen (1993); Korbicz, Fathi and Ramirez (1993); Martin (1993); Patton and Chen (1993a); Stein (1993); Stengel (1993);

Patton and Chen (1993b); Patton (1994); Krishnaswami and Rizzoni (1994b); Wise and Gallagher (1996); Basseville (1997); Patton (1997b); Isermann (1998).

The model-based FDI techniques have been applied to many diagnostic problems in either real situation or laboratory simulation. The survey paper Frank (1987) and the book (Patton et al., 1989) listed many published application examples during 1970s and 1980s. A selection of application examples and diagnosis methods used between 1991 to 1995 can be found in Isermann and Ballé (1997).

There are three encyclopedia articles on model-based FDI techniques available, Frank (1992a) presented basic principles, Patton and Chen (1992b) discussed robustness issues and Labarrère and Patton (1993) emphasized aerospace applications.

Model-based FDI techniques have been summarized in the following books: Pau (1975); Himmelblau (1978); Basseville and Benveniste (1986); Singh, Hindi, Schmidt and Tzafestas (1987); Singh et al. (1987); Viswanadham, Sarma and Singh (1987); Patton et al. (1989); Brunet, Jaume, Labarrère, Rault and Vergé (1990); Basseville and Nikiforov (1993); Poulizeos and Stravlakakis (1994) and Gertler (1998). It should be pointed out that most of the books on model-based FDI are multi-authored books, this is mainly because this technique is still in a developmental stage.

Papers on model-based FDI techniques can be found in many engineering journals and IFAC, IMACS, IEEE, IEE and other conferences. The most important source of information is the three symposia organized by IFAC Technical Committee: Fault Detection, Safety and Supervision of Technical Processes (*SAFEPROCESS*): *SAFEPROCESS'91* (Isermann, 1991b), *SAFEPROCESS'94* (Ruokonen, 1994) and *SAFEPROCESS'97* (Patton and Chen, 1998).

1.3 Outline of the Book

To detect and isolate faults in a dynamic system, based on the use of an analytical model, a declarative or residual signal must be used, which is derived from a combination of real measurements and estimates (generated by the model). The robustness problem can be tackled by defining the independent sensitivities of the residual to uncertainties and faults. Following from the definition given above, a robust FDI scheme is one whose residual is insensitive to uncertainties whilst sensitive (in a certain way) to faults. The aim of robust design of the FDI scheme is to reduce the effects of uncertainties on the residuals, and (or) to enhance the effects of faults acting on the residuals. The success of fault diagnosis depends on the quality of the residuals. A preliminary requirement of residuals for successful diagnosis is the robustness with respect to modeling uncertainty. The *main theme* of this book is *to develop robust residual generation strategies for model-based fault diagnosis of dynamic uncertain systems.* The book consists of 9 chapters. Each chapter is devoted to a particular problem in robust residual generation, and hence the chapters are relatively independent although they are related in some ways. It would be of benefit to readers if

Chapter 2 is read first before other chapters. Chapter 5 is best served after the reading of Chapter 3 and/or Chapter 4.

Chapter 2 reviews the state of the art of model-based fault diagnosis techniques. The fault diagnosis problem is formalized in a uniform framework by presenting the mathematical description and definitions. This properly defined framework gives a clear picture of the principles and problems associated with model-based fault diagnosis. The fundamental issue of model-based methods is the generation of residual signals using the mathematical model of the monitored system. By analyzing the fault-indicating signal – residual, the nature of faults can be obtained. A generalized structure of the residual generator is presented in this chapter. This gives ideas of how to design and implement the residual generators. The residual generator can be purposely designed for achieving the required diagnosis performances, e.g, fault isolation, disturbance de-coupling and residual frequency response shaping.

In order to design a robust residual generator, we need to make some assumptions about the modeling uncertainty. The most frequently used assumption is that the modeling uncertainty is expressed as a disturbance term in the system dynamic equation. Although the magnitude of the disturbance is unknown, its distribution (or direction) is assumed known *a priori*. Based on this assumption, the disturbance de-coupling residual generator can be designed using unknown input observer theory or via the eigenstructure assignment technique. Robust fault diagnosis is then achievable using disturbance de-coupled residuals. Following this philosophy, Chapters 3 and 4 present some strategies for designing disturbance de-coupling residual generators.

Chapter 3 introduces the approach to robust residual generation with the aid of the unknown input observer (UIO). The principle of the UIO is to make the state estimation error de-coupled from the disturbance. Since the residual is defined as the *weighted output estimation error*, the residual is also de-coupled from disturbances. This chapter describes a full-order unknown input observer structure. The necessary and sufficient conditions for a UIO to exist presented in this chapter are very easy to verify and the design procedure is simple. Robust sensor and actuator fault isolation schemes based on UIOs are presented in this chapter and a chemical reactor is used to illustrate the robust actuator isolation principles. This chapter also introduces a method to make the residual have both disturbance de-coupling and directional properties, by combining the unknown input observer and fault detection filter theories. The directional property makes fault isolation achievable. Finally, this chapter introduces the optimal state estimation of stochastic systems with unknown inputs. It is proved that the design freedom left after disturbance de-coupling can be used to make the state estimation error have minimal variance. The use of this optimal disturbance de-coupled observer in fault detection is illustrated using a simplified flight control example.

Chapter 4 focuses on the disturbance de-coupled residual generator design via eigenstructure assignment. The most challenging problem in model-based fault diagnosis is the correct design of the residual. State estimation is not necessary in FDI, and hence the state estimation error does not need to be de-coupled from the disturbance. What is actually required is that the disturbance be de-coupled from the residual. The correct disturbance de-coupling can be achieved by assigning left observer eigenvectors orthogonal to disturbance directions or assigning right observer eigenvectors parallel to disturbance directions. The parametric eigenstructure assignment method is described to assign left observer eigenvectors. This chapter introduces a method for assigning right eigenvectors of the observer. This is equivalent to the assignment of left eigenvectors for a controlled system. The principles, existence conditions and the design procedure for the eigenstructure assignment approach to robust residual generation are presented in Chapter 4, where it is also shown that the remaining design freedom, after the disturbance de-coupling has been satisfied, can be utilized to optimize other performance indices (such as fault sensitivity). For a discrete-time design, a dead-beat disturbance de-coupling residual generator can be designed which has a direct correspondence with parity relations. Three numerical examples are presented in this chapter to illustrate the design procedure and de-coupling principles.

The theory of disturbance de-coupling for robust fault diagnosis has being developed for some years. However few investigators have shown how to apply this method to real applications. The difficulty is caused by the mis-match between the theoretical assumptions and practical reality. In most practical systems, the disturbance distribution matrix is not known. Disturbance de-coupling methods which require the disturbance distribution matrix cannot be applied directly to the system with unknown disturbance distribution matrix.

Chapter 5 demonstrates how to apply the disturbance de-coupling method to a system with modeling uncertainty. It is proved that an approximate disturbance term with an estimated distribution matrix can be used to represent the effect of modeling uncertainty on the system. Using this approximate distribution matrix in the disturbance de-coupling residual design, the *nearly robust* fault diagnosis is achievable. A number of methods for finding the approximate distribution matrix are given to deal with different uncertainty cases, based on either optimization or identification techniques. The methods described in Chapter 5 are applied to a jet engine simulation system to demonstrate the effectiveness of robust residual for detecting incipient faults. The simulation shows satisfactory results. The jet engine is a complex, highly non-linear and high order system and, any techniques applicable to such a system should also be applicable to other complex non-linear and uncertain dynamical systems.

The purpose of robust residual design is to make the residual maximally sensitive to faults and minimally sensitive to modeling uncertainty. **Chapter 6** introduces an approach to the design of optimal residuals for detecting incipient faults, based on multi-objective optimization and the genetic algorithm. In this approach the residual is generated via an observer. To reduce false and missed alarm rates in fault detection, a number of performance indices are introduced into the observer design. Some performance indices are expressed in the frequency domain to take account of the frequency distributions of faults, noise and modeling uncertainties. All objectives are then reformulated into a set of inequality constraints on the performance indices. The genetic algorithm is thus used to search an optimal solution to satisfy these inequality constraints on performance indices. The approach described is applied to a flight control system example and simulation results show that incipient sensor faults can be detected reliably in the presence of modeling uncertainty.

Chapter 7 discusses the robust residual generation using optimally robust parity relations. The system parameters are considered to vary within known bounds, and representative points in uncertainty regions are chosen to represent the uncertainty. The system dynamics are effectively describable using multiple linear models. A robust residual should be insensitive to changes in these models. This objective is achievable by minimizing a defined performance index. To avoid the reduction of fault sensitivity during the minimization of the sensitivity to uncertainty, the fault sensitivity is also used as a performance index to be maximized. Thus, the robust residual design is formulated as a multi-objective optimization problem. The chapter shows a number of ways of mixing these two performance indices together to form a single objective optimization problem. This problem is then solved using singular value decomposition and the computation of the generalized eigenstructure. Some methods in the design of robust parity relations are review in this chapter. A numerical example is used to demonstrate the method developed in this chapter.

Chapter 8 introduces frequency domain design and H_∞ optimization methods for robust fault diagnosis. The idea is to apply frequency domain and H_∞ control techniques to robust FDI problems. The main emphasis is on the formulation of robust FDI problems using frequency domain performance criteria. Once the problem is formulated, the solution can be readily found using techniques existing in H_∞ control. The chapter starts with the factorization approach for the design of residual generators and then perfect and optimal disturbance de-coupling problems are solved using factorization-based methods. To fully exploit the powerful H_∞ design techniques, this chapter formulates robust FDI problems into a standard H_∞ optimization problem which can be solved via the algebraic Riccati equation method. The problems considered and formulated in this chapter are residual generation with disturbance attenuation, optimal fault estimation and optimal fault estimation with disturbance attenuation. The formulation into a standard H_∞ problem is made on the assumption of there is no modeling errors or modeling errors have been transformed to disturbances. The possibility of including modeling errors in the standard problem framework is also discussed. This chapter also demonstrates that the robust FDI problems can also be solved through the use of linear matrix inequalities (LMI) due to the duality between estimation and control. With the LMI formulation, both disturbance robustness and fault sensitivity can be considered simultaneously in residual design. This is achieved by minimizing the disturbance sensitivity with a constraint on fault sensitivity.

Chapter 9 studies the fault detection and isolation problems for non-linear dynamic systems. In the beginning of the chapter, a review on non-linear dynamic systems FDI is given in this chapter. This is followed by the introduction of observer-based approaches. A number of observer-based approaches, which are an extension of linear observer-based approaches are discussed briefly and a detailed treatment is given to bilinear observers for FDI. After observer-based approaches, this chapter discusses how to exploit the ability of a neural network has to handle non-linear FDI problems. The ways of using neural networks for FDI are discussed. Then neural networks-based FDI schemes which utilize both modeling and classification abilities are presented and the schemes are demonstrated using a laboratory 3 tanks system. This chapter gives a particular emphasis on a novel fuzzy observer-based approach for non-linear systems FDI. The fuzzy observer, based on the idea of Takagi-Sugeno fuzzy models, comprises a number of locally linear observers and the final state estimate is a fuzzy fusion of all local observer outputs. To ensure good estimation performance, the eigenvalues of the fuzzy observer are assigned in a predefined region in the complex plane. The stability as well as eigenvalue constraint conditions for the fuzzy observer design are presented and solved in the framework of linear matrix inequalities. The application of fuzzy observers in detecting and isolating intermittent faults in the induction motor of a railway traction system is demonstrated. Finally, this chapter introduces a neuro-fuzzy approach for integrating quantitative and qualitative information in FDI. In this approach, residuals are generated and evaluated via a B-spline functions network. The configuration adopted allows the designer to extract symbolic knowledge from the trained network to provide reliable diagnostic information. The effectiveness of the FDI strategy is illustrated through a simulation study of a non-linear 2-tanks system.

2 BASIC PRINCIPLES OF MODEL-BASED FAULT DIAGNOSIS

2.1 Introduction

The model-based approach to fault diagnosis in automated processes has been receiving considerable attention since the beginning of 1970s, both in a research context and also in the domain of application studies on real processes. There are a great variety of methods in the literature, based on the use of mathematical models of the monitored processes and modern control theory.

The most important issue in model-based fault diagnosis is the robustness against modeling uncertainty which arises from incomplete knowledge and understanding of the monitored processes. Robust fault diagnosis has become a central research issue over recent years. As this book focuses on the development of robust model-based fault diagnosis techniques, this chapter introduces basic principles of model-based fault diagnosis. Attention is first turned to the modeling of the system with all possible faults. Residual generation is then identified as an essential problem in model-based FDI, as an information processing procedure which, if not designed correctly could lose some fault information. A general framework for the residual generator is described. Residual generators based on different methods, such as observers and parity relations, are just special cases in this general framework. This chapter also shows that, to fulfil FDI tasks successfully, the residual signal has to satisfy fault detectability and isolability conditions. Some most important residual generation methods are discussed in this chapter. This chapter also gives some general guide-lines about the applicability of different model-based FDI approaches.

The robust FDI issue is discussed in this chapter and some commonly used robust approaches are commented. This formalizes a basis for the studies described in later chapters. The use of adaptive thresholds in FDI is also discussed. Finally, a discussion of fuzzy logic, qualitative modeling and knowledge based approaches in FDI is given. Some perspectives in the future development of FDI, by combining quantitative and qualitative techniques are also discussed.

2.2 Model-based Fault Diagnosis Methods

Model-based fault diagnosis can be defined as the detection, isolation and characterization of faults in components of a system from the comparison of the system's available measurements, with *a priori* information represented by the system's mathematical model.

Faults are detected by setting a (fixed or variable) threshold on a residual quantity generated from the difference between real measurements and estimates of these measurements using the mathematical model. A number of residuals can be designed with each having special sensitivity to individual faults occurring in different locations in the system. The subsequent analysis of each residual, once a threshold is exceeded, then leads to fault isolation.

Fig.2.1 illustrates the general and conceptual structure of a model-based fault diagnosis system comprising two main stages of residual generation and decision making. This two-stages structure was first suggested by Chow and Willsky (1980) and now is widely accepted by the fault diagnosis community. These two main stages are described as follows:

(1) Residual Generation: Its purpose is to generate a fault indicating signal – residual, using available input and output information from the monitored system. This auxiliary signal is designed to reflect the onset of a possible fault in the analyzed system. The residual should be normally zero or close to zero when no fault is present, but is distinguishably different from zero when a fault occurs. This means that the residual is characteristically independent of system inputs and outputs, in ideal conditions. The algorithm (or processor) used to generate residuals is called a *residual generator*. Residual generation is thus a procedure for extracting fault symptoms from the system, with the fault symptom represented by the residual signal. The residual should ideally carry only fault information. To ensure reliable FDI, the loss of fault information in residual generation should be as small as possible.

(2) Decision-Making: The residuals are examined for the likelihood of faults, and a decision rule is then applied to determine if any faults have occurred. A decision process may consist of a simple threshold test on the instantaneous values or moving averages of the residuals, or it may consist of methods of statistical decision theory, e.g., generalized likelihood ratio (GLR) testing or sequential probability ratio testing (SPRT) (Willsky, 1976; Basseville, 1988; Tzafestas and Watanabe, 1990; Basseville and Nikiforov, 1993; Da, 1994).

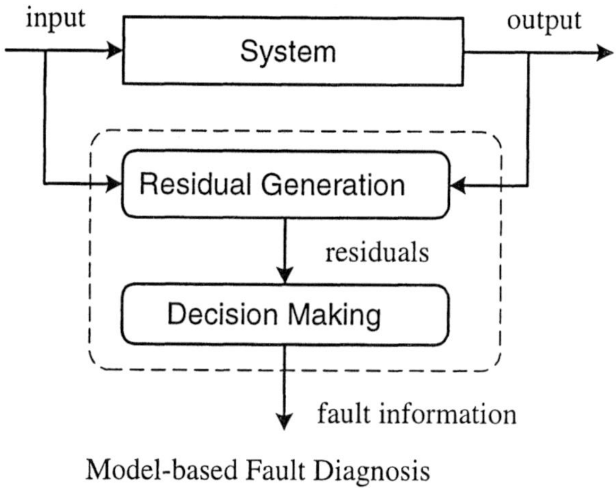

Figure 2.1. Conceptual structure of model-based fault diagnosis

Most of the work in the field of quantitative model-based fault diagnosis is focused on the residual-generation problem because the decision-making based on well designed residuals is relatively easy. However, this does not imply that the research on decision-making is not important (Frank, 1996; Schneider and Frank, 1996; Frank and Ding, 1997). The book will concentrate on the quantitative residual generation stage of fault diagnosis by introducing a number of strategies in the enhancement of residual robustness.

2.3 On-line Fault Diagnosis

Model-based FDI is concerned mainly with on-line fault diagnosis, in which the diagnosis is carried out during system operation. This is because the system input and output information required by model-based FDI is only available when the system is in operation. Opening of feedback loops in the system being tested or supplying test actions leading to incorrect functioning are considered inadmissible. The relationship between the fault diagnosis (or supervision) with the control loop is shown in Fig.2.2.

The information used for FDI is the measured output from sensors and the input to the actuators. The measured output is normally needed in the feedback control, whereas the input to the actuators is the required control action generated by the controller, which is normally implemented in the microprocessor. Hence, we do not normally require extra hardware resources to implement the fault diagnosis function with the exception of requiring some additional computing power.

From Fig.2.2, it can be seen that the system model required in model-based FDI is the open-loop system model although we consider that the system is in the control loop. This is because the input and output information required in

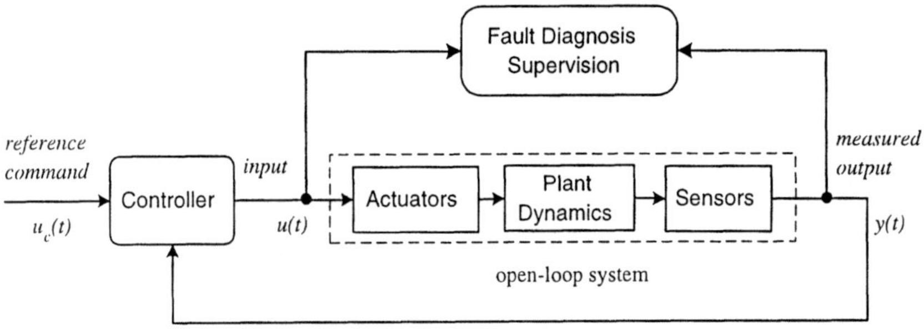

Figure 2.2. Fault diagnosis and control loop

model-based FDI is related to the open-loop system. Hence, it is not necessary to consider the controller in the design of a fault diagnosis scheme. This is consistent with the separation principle in control theory because fault diagnosis can be broadly treated as an observation problem. Once the input to the actuators is available, the fault diagnosis problem is the same no matter how the system is working in open-loop or in the closed-loop.

In the cases when the input to the actuator $u(t)$ is not available, we have to use the reference command $u_c(t)$ in FDI. Hence, the model involved is the relationship between the reference command $u_c(t)$ and the measured output $y(t)$, i.e., the closed-loop model. For those cases, the controller plays an important role in the design of diagnostic schemes. A robust controller may desensitize fault effects and make the diagnosis very difficult. This problem has been recognized by some researchers, e.g. Wu (1992), Patton (1997a), and the best solution is to design the fault diagnosis scheme and the controller simultaneously (Nett, Jacobson and Miller, 1988; Jacobson and Nett, 1991; Stoustrup, Grimble and Niemann, 1997). The interconnection between fault diagnosis and robust control is a topic for future research and is not considered further in this book.

2.4 Modeling of Faulty Systems

The first step in the model-based approach is to build a mathematical model of the system to be monitored. This book is concerned with multiple-input and multiple-output linear dynamic systems. In the case of a non-linear system, this implies a model linearization around an operating point.

As discussed in the previous section, we use the open-loop system model in model-based FDI. For the purposes of modeling, an open-loop system can be separated into three parts: actuators, system dynamics and sensors as illustrated in Fig.2.3.

The system dynamics shown in Fig.2.4 can be described by the state space model as:

$$\begin{cases} \dot{x}(t) & = & Ax(t) + Bu_R(t) \\ y_R(t) & = & Cx(t) + Du_R(t) \end{cases} \tag{2.1}$$

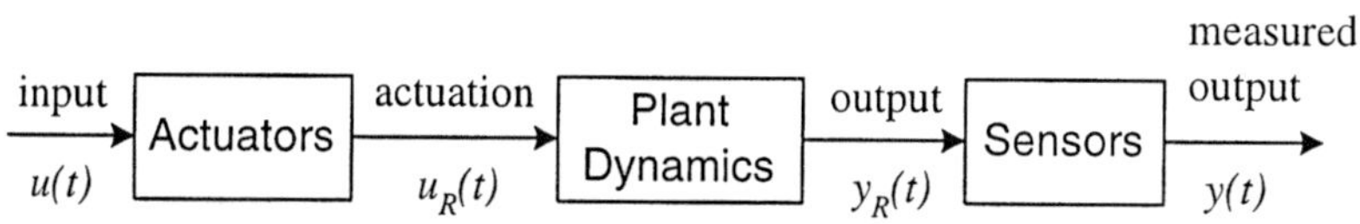

Figure 2.3. Open-loop system

where $x \in \mathbb{R}^n$ is the state vector, $u_R \in \mathbb{R}^r$ is the input vector to the actuator and $y_R \in \mathbb{R}^m$ is the real system output vector; A, B, C and D are known system matrices with appropriate dimensions.

When a component fault occurs in the system (see Fig.2.4), the dynamic model of the system can be described as:

$$\dot{x}(t) = Ax(t) + Bu_R(t) + f_c(t) \tag{2.2}$$

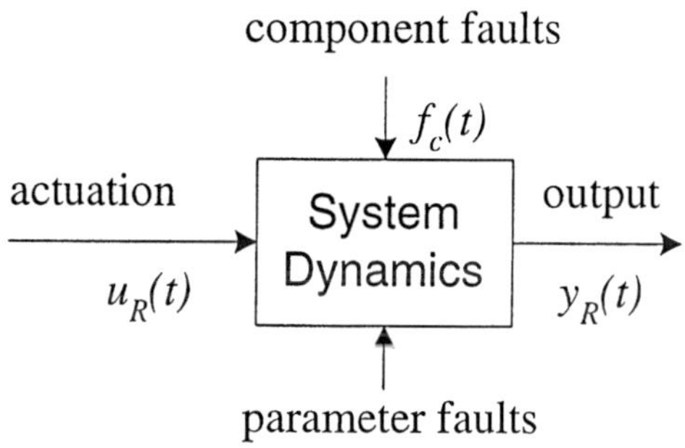

Figure 2.4. The system dynamics

The component fault is represented as the case when some condition changes in the system rendering the dynamic relation invalid, for example a leak in a water tank in the three tank system (Wünnenberg, 1990). In some cases, the fault could be expressed as a change in the system parameter, for example a change in the i_{th} row and j_{th} column element of the matrix A, the dynamic equation of the system can then be described as:

$$\dot{x}(t) = Ax(t) + Bu_R(t) + I_i \Delta a_{ij} x_j(t) \tag{2.3}$$

Here, $x_j(t)$ is the j_{th} element of the vector $x(t)$ and I_i is an n-dimensional vector with all zero elements except a 1 in the i_{th} element.

Generally speaking, the actual output $y_R(t)$ of the system is not directly accessible, and sensors are then used to measure the system output. This is shown in Fig.2.5 and can be described mathematically as (when the sensor dynamics are neglected):

$$y(t) = y_R(t) + f_s(t) \tag{2.4}$$

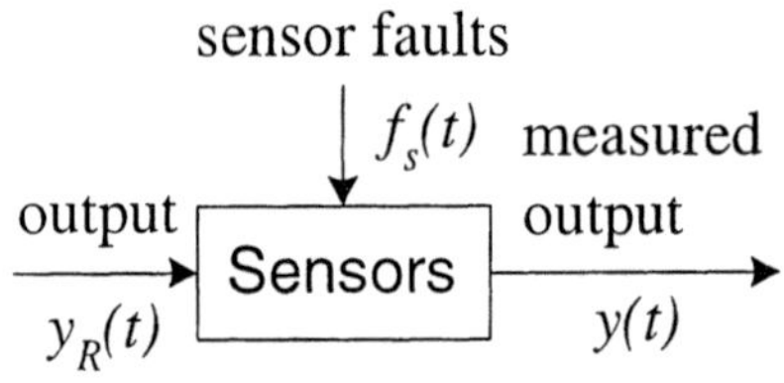

Figure 2.5. Sensors, output and measured output

where $f_s \in \mathbb{R}^m$ is the sensor fault vector.

By choosing the vector f_s correctly, we can then describe all sensor fault situations. When the sensors are "stuck at a particular value" (say at zero), the measurement vector is $y(t) = 0$ and the fault vector is $f_s(t) = -y_R(t)$. When there is a variation in the sensor scalar factors (multiplicative faults), the measurement becomes $y(t) = (1+\Delta)y_R(t)$ and the fault vector can be then written as $f_s(t) = \Delta y_R(t)$.

It is also true that the actual actuation (u_R) of the system is often not directly accessible. For a controlled system, u_R is the actuator response to an actuator command $u(t)$, this is shown in Fig.2.6 and can be described as (when the actuator dynamics are neglected):

$$u_R(t) = u(t) + f_a(t) \tag{2.5}$$

where $f_a \in \mathbb{R}^r$ is the actuator fault vector and $u(t)$ is the known control command. Similar to the sensor fault situation, different actuator fault situations can be represented by a proper fault function $f_a(t)$.

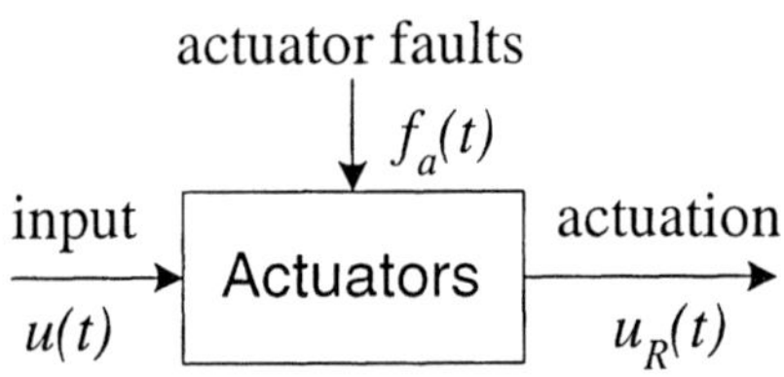

Figure 2.6. Actuator, input and actuation

If the system input is unknown (e.g., in an uncontrolled system), an input sensor can be used to measure the input to the actuator, this is shown in the Fig.2.7 and can be represented by the model:

$$u(t) = u_R(t) + f_{is}(t) \tag{2.6}$$

or

$$u_R(t) = u(t) + [-f_{is}(t)] \tag{2.7}$$

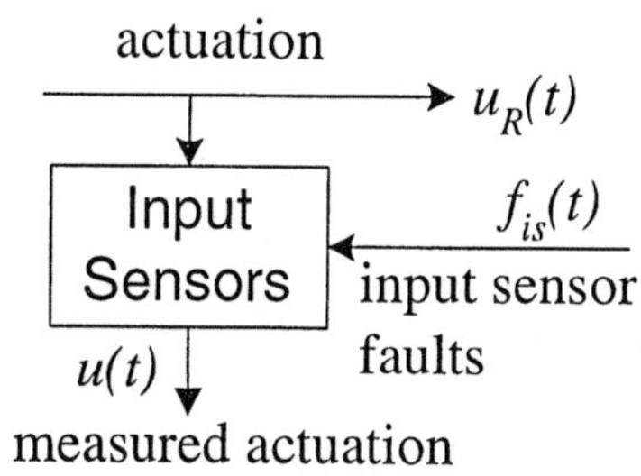

Figure 2.7. Input sensor, actuation and measured actuation

When the system has all possible sensor, component and actuator faults (this is the most common situation to be considered), the system model is described as:

$$\begin{cases} \dot{x}(t) = Ax(t) + Bu(t) + Bf_a(t) + f_c(t) \\ y(t) = Cx(t) + Du(t) + Df_a(t) + f_s(t) \end{cases} \quad (2.8)$$

Considering the general cases, a system with all possible faults can be described by the state space model as:

$$\begin{cases} \dot{x}(t) = Ax(t) + Bu(t) + R_1 f(t) \\ y(t) = Cx(t) + Du(t) + R_2 f(t) \end{cases} \quad (2.9)$$

where $f(t) \in \mathbb{R}^g$ is a fault vector, each element $f_i(t)$ $(i = 1, 2, \cdots, g)$ corresponds to a specific fault. From a practical point of view, it is unreasonable to make further assumptions about the fault characteristics but consider these as unknown time functions. The matrices R_1 and R_2 are known as fault entry matrices which represent the effect of faults on the system. The vector $u(t)$ is the input to the actuator or measured actuation, and the vector $y(t)$ is the measured output, and both vectors are known for FDI purpose.

In the FDI literature, the vectors $u(t)$ and $y(t)$ are simply called the input and output vectors of the monitored system. The terminology is not very precise, although no confusion arises and it is accepted widely in the FDI literature and in this book unless it is specifically stated.

An input-output transfer matrix representation for the system with possible faults is then described as:

$$y(s) = G_u(s)u(s) + G_f(s)f(s) \quad (2.10)$$

where

$$\begin{cases} G_u(s) = C(sI - A)^{-1}B + D \\ G_f(s) = C(sI - A)^{-1}R_1 + R_2 \end{cases} \quad (2.11)$$

The general model for a faulty system described by Eq.(2.9) in the time-domain and by Eq.(2.10) in the frequency-domain has been widely accepted in the fault diagnosis literature, e.g. in survey papers (Frank, 1990; Frank, 1991a; Patton and Chen, 1991e; Gertler, 1991; Frank, 1993; Patton, 1993; Gertler and Kunwer,

1993; Patton, 1994) (Patton and Chen, 1996b; Patton and Chen, 1997; Frank and Ding, 1997). However, most papers have just accepted them without any clue of how a particular individual fault fits into this model. Gertler and Luo (1989) and Gertler, Fang and Luo (1990) considered all possible fault sources in the monitored system and Chen and Patton (1994a) discussed briefly the modeling structure presented in this section.

2.5 A General Structure of Residual Generation in Model-based FDI

In practice, the most frequently used FDI approach uses information known *a priori* about the characteristics of certain signals (e.g. amplitude and frequency properties). As an example, we can take checking of the level or the dynamic range of the signal, the maximum rate of its variation and its spectrum. The main shortcomings of this group of methods can be listed as: (a) the necessity to have *a priori* information about the characteristics of the signals, (b) the unavoidable dependence of these characteristics on operating states of the system which are not known *a priori* and can change beforehand.

To eliminate the shortcomings of the traditional methods, the most significant contribution in modern model-based approaches is the introduction of residuals which are independent of the system operating state and respond to faults in characteristic manners. Residuals are quantities that represent the inconsistency between the actual system variables and the mathematical model. Based on the mathematical model, many invariant relations (dynamic or static) among different system variables can be derived, and any violation of these relations can be used as residuals.

The residual generation can be interpreted in terms of redundant signal structure as illustrated in Fig.2.8 (Mironovski, 1980; Basseville, 1988). In this structure, the system (processor or algorithm) $F_1(u, y)$ generates an auxiliary (redundant) signal z which, together with y generate the residual r which satisfy the following invariant relation:

$$r(t) = F_2(y(t), z(t)) = 0 \tag{2.12}$$

for the fault-free case. When any fault occurs in system, this invariant relation will be violated and the residual will be non-zero.

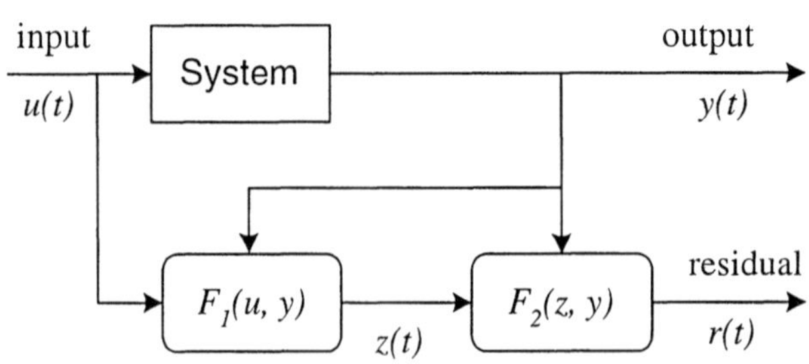

Figure 2.8. Redundant signal structure in residual generation

The simplest approach to residual generation is the use of system duplication, i.e. the system F_1 is made identical to the original system model and has the same input signal as the system. In this case, the signal y is not required in the system block F_1 which is then simply a system simulator. The signal z is the simulated output of the system, and the residual is the difference between z and y. The simplicity is the advantage of this method, but the disadvantage is that the stability of the simulator cannot be guaranteed when the system being monitored is unstable, as a consequence of the use of the open-loop system model in FDI (although it is under feedback control) (see Fig.2.2).

A direct extension to the simulator-based residual generation is to replace the simulator by an output estimator which requires both system input and output. In this case, the system $F_1(u, y)$ uses both signal u and y to generate an estimation of a linear function of the output y, say My, and the system F_2 can be defined as $F_2(y, z) = Q(z - My)$ with Q as a static (or dynamic) weighting matrix.

No matter what type of method is used, a residual generator is just a linear processor whose input consist of both input and output of the system being monitored. A general structure for a residual generator is shown in Fig.2.9 (Patton and Chen, 1991e).

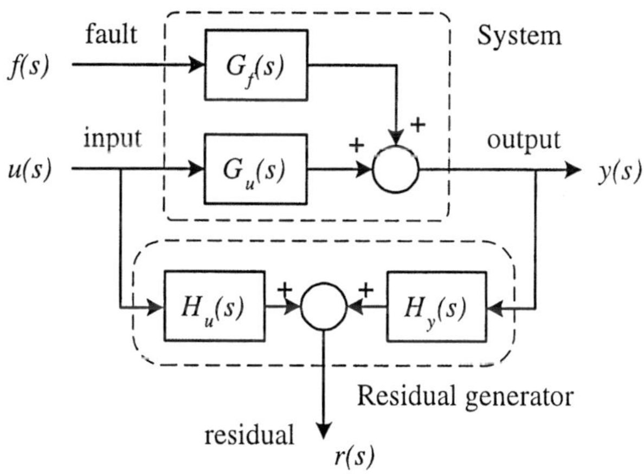

Figure 2.9. The general structure of a residual generator

This structure is expressed mathematically as:

$$r(s) = [H_u(s) \quad H_y(s)] \begin{bmatrix} u(s) \\ y(s) \end{bmatrix} = H_u(s)u(s) + H_y(s)y(s) \tag{2.13}$$

Here, $H_u(s)$ and $H_y(s)$ are transfer matrices which are realizable using stable linear systems. According to the definition, the residual is designed to become zero for the fault-free case and nonzero for faulty cases, i.e.

$$r(t) = 0 \qquad \text{if and only if} \qquad f(t) = 0 \tag{2.14}$$

To satisfy this condition, the transfer matrices $H_u(s)$ and $H_y(s)$ must satisfy the constraint condition:

$$H_u(s) + H_y(s)G_u(s) = 0 \tag{2.15}$$

Eq.(2.13) is a *generalized* representation of all residual generators (Patton and Chen, 1991e). The design of the residual generator results simply in the choice of the transfer matrices $H_u(s)$ and $H_y(s)$ which must satisfy Eq.(2.15). The various ways of generating residuals correspond to different parameterizations of $H_u(s)$ and $H_y(s)$. One can obtain different residual generators using different forms for $H_u(s)$ and $H_y(s)$. Using the design freedom, the desired performance of the residual can be achieved by suitable selection of $H_u(s)$ and $H_y(s)$.

A fault can be detected by comparing the residual evaluation function $J(r(t))$ with a threshold function $T(t)$ according to the test give in below:

$$\begin{cases} J((r(t)) \leq T(t) & \text{for} \quad f(t) = 0 \\ J((r(t)) > T(t) & \text{for} \quad f(t) \neq 0 \end{cases}$$

If this test is positive (i.e. the threshold is exceeded by the residual evaluation function), we can hypothesize that a fault is likely. There are many ways of defining evaluation functions and determining thresholds. As an example, the residual evaluation function is chosen as a norm of the residual vector and the threshold can be chosen as a constant positive value (fixed threshold).

2.6 Fault Detectability

When faults occur in the monitored process, the response of the residual vector is:

$$r(s) = H_y(s)G_f(s)f(s) = G_{rf}(s)f(s) = \sum_{i=1}^{g}[G_{rf}(s)]_i f_i(s) \tag{2.16}$$

where $G_{rf}(s) = H_y(s)G_f(s)$ is defined as a fault transfer matrix which represents the relation between the residual and faults, $[G_{rf}(s)]_i$ is the i_{th} column of the transfer matrix $G_{rf}(s)$ and $f_i(s)$ is the i_{th} component of $f(s)$. The above relationship is well illustrated by Fig.2.10.

2.6.1 *Fault detectability condition*

The fault transfer matrix plays an important role in FDI and must be examined in detail. In order to detect the i_{th} fault f_i in the residual $r(s)$, the i_{th} column $[G_{rf}(s)]_i$ of the transfer matrix $G_{rf}(s)$ should be non-zero:

$$[G_{rf}(s)]_i \neq 0 \tag{2.17}$$

If this condition holds true, the i_{th} fault f_i is *detectable* in the residual r. This is defined as the *fault detectability condition* of the residual r to the fault f_i. One must ask whether this condition is enough for detecting faults? This question will be answered using the following example.

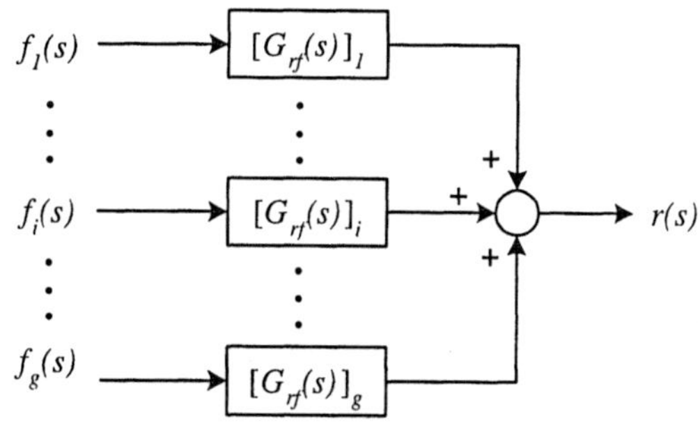

Figure 2.10. Faults and the residual

Example: The laboratory inverted pendulum system (described in Appendix B) is used as an example to illustrate the fault detectability (Chen and Patton, 1994b). The simulated fault detection results are shown in Figs.2.11–2.14. In the simulation, the same fault signal is applied to three sensors.

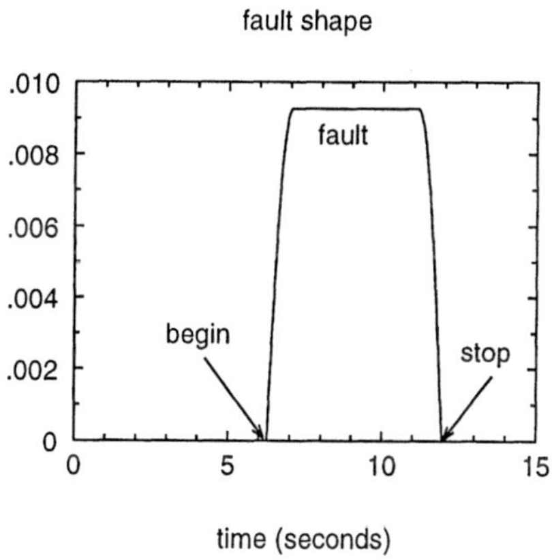

Figure 2.11. Inverted pendulum example (Fault signal)

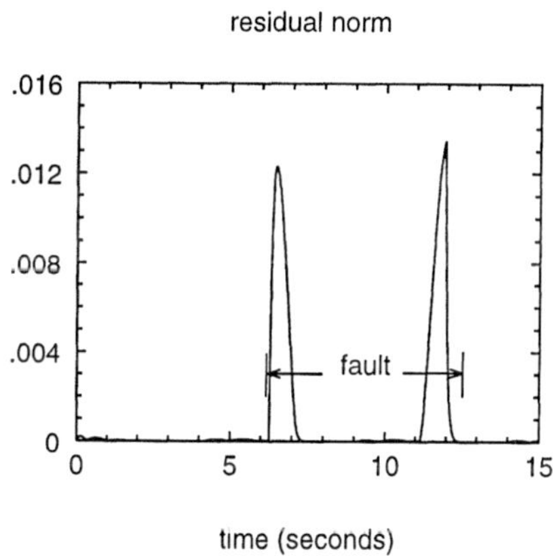

Figure 2.12. Inverted pendulum example (Fault in sensor 1)

However, the residual response for the fault in the first sensor is significantly different from the faults in the other sensors. The responses for the faults in sensor 2 and sensor 3 almost reproduce the shape variations of the fault signal. However, the response for the fault in the sensor 1 only reflects the change in the fault level. After a short transient, the residual returns back to zero, although the fault is still present in the system. It is possible to give a misinterpretation of faults if this observer-based residual generator is used to detect faults in the first sensor. To examine the fault detectability, we find that: $[G_{rf}(s)]_1 \neq 0$, $[G_{rf}(s)]_2 \neq 0$, $[G_{rf}(s)]_3 \neq 0$, i.e. the faults in *three* sensors are all detectable from the residual designed. This example illustrates that fault detectability alone is not enough to achieve reliable fault detection.

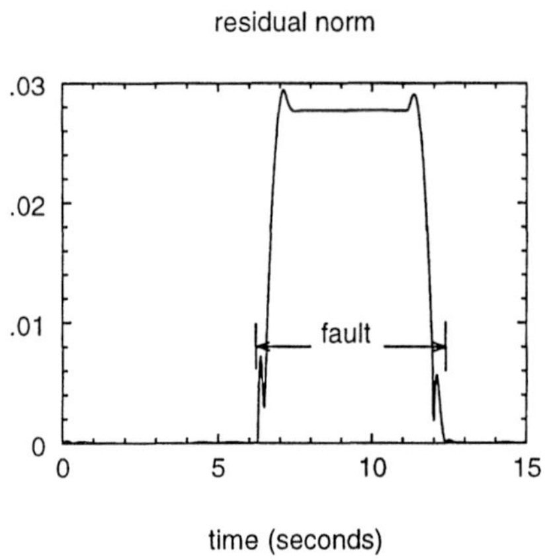

Figure 2.13. Inverted pendulum example (Fault in sensor 2)

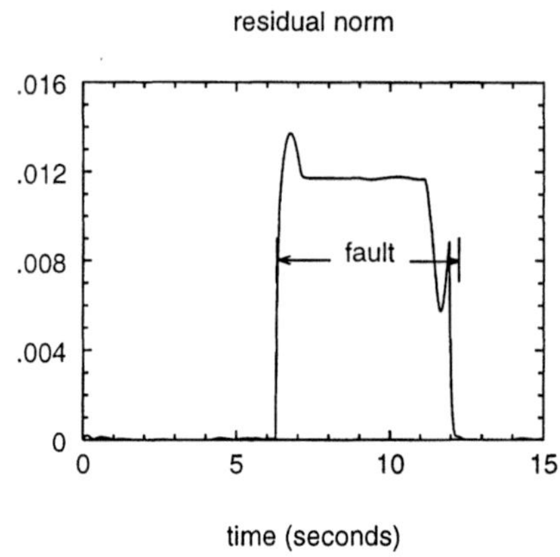

Figure 2.14. Inverted pendulum example (Fault in sensor 3)

On examining the *steady* state gains of the residual generators, we find that: $[G_{rf}(0)]_1 = 0$, $[G_{rf}(0)]_2 \neq 0$ and $[G_{rf}(0)]_3 \neq 0$. This easily shows why the effect of the fault in sensor 1 on the residual generator disappears after a short transient period.

2.6.2 Strong fault detectability condition

The example shows that fault detectability is not enough to achieve reliable fault detection. Hence, the strong fault detectability is introduced here as:

$$[G_{rf}(0)]_i \neq 0 \tag{2.18}$$

If this condition is satisfied, we define that the i_{th} fault f_i is *strongly detectable* in the residual r. This condition can also be defined as the *strong fault detectability condition* of the residual r to the fault f_i.

The misinterpretation problem due to the undesirable residual response has been noticed in the FDI research by a number of investigators (Patton and Kangethe, 1989; Frank, Ding and Köppen, 1993). There were some discussion in a benchmark testing session at the "International Conference of Fault Diagnosis at Toulouse (*TOOLDIAG'93*)" following a presentation by Frank et al. (1993). One explanation for this problem is that *the effect of a fault on the system* disappears, although the fault itself still exists. This is not a satisfactory explanation and the correct explanation is that *the effect of the fault on the residual* disappears, although the fault effect on the system still exists. That is to say, the residual generator which is used for FDI is not a good design.

We now examine the inverted pendulum system in more detail. Referring to the Appendix B, we find that the strong detectability for faults in the first sensor cannot be achieved no matter what observer gain matrix is used, if the residual generator is based on a full-order observer. It is also interesting to note that a residual generator based on a 1_{st} or 2_{nd} order parity relation also gives similar residual responses (the results are not shown in this book as they are very similar to the results shown in Figs.2.11–2.14.

The question arises as to how the above problem (for the inverted pendulum example) can be solved? One way is to design other residual generators which could satisfy the strong fault detectability, and this requires comprehensive research. The other possibility is to shape the frequency response of the residual according to the frequency distribution of the faults. For example, if the residual generated by the observer is filtered through a filter with transfer function $1/(s+0.01)$, the filtered residual can produce a satisfactory response for the fault given in Fig.2.11. This simple operation shows that the frequency response of the fault transfer function should also be studied in the residual design. If the frequency band of a certain fault is available, the residual can be designed maximally sensitive to this fault by frequency-shaping. This can be done by maximizing the following criterion:

$$J = \inf_{\omega \in [\omega_1, \omega_2]} \underline{\sigma}\{G_{rs}(j\omega)\} \tag{2.19}$$

where $\underline{\sigma}\{\cdot\}$ denotes the minimal singular values, and $[\omega_1, \omega_2]$ denotes the frequency range in which the fault is most likely to occur. This problem will be studied in Chapter 6. Other investigations are described in papers by Frank and Ding (Ding and Frank, 1989; Frank and Ding, 1993; Frank and Ding, 1994).

2.7 Fault Isolability

The successful detection of a fault is followed by the fault isolation procedure which will distinguish (isolate) a particular fault from others. Whilst a single residual signal is sufficient to detect faults, a set of residuals (or a vector of residuals) is usually required for fault isolation. If a fault is distinguishable from other faults using one residual set (or a residual vector), it can be said that this fault is *isolable* using this residual set (or this residual vector). If the residual set (or the residual vector) can isolate all faults, we can then say that the residual set (or the residual vector) has the required isolability property.

2.7.1 Structured residual set

One approach to fulfil the fault isolation task is to design a set of structured residuals. Each residual is designed to be sensitive to a subset of faults, whilst remaining insensitive to the remaining faults. The residual set which has the required sensitivity to specific faults and insensitivity to other faults is known as the *structured residual set* (Gertler, 1991). The design procedure consists of two steps, the first step is to specify the sensitivity and insensitivity relationships between residuals and faults according to the assigned isolation task, and the second is to design a set of residual generators according to the desired sensitivity and insensitivity relationships. The advantage of the structured residual set is that the diagnostic analysis is simplified to determining which of the residuals are non-zero. The threshold test may be performed separately for each residual, yielding a Boolean decision table, and the isolation task can be fulfilled using this table.

If all possible faults are to be isolated, a residual set can be designed according to the following fault sensitivity conditions:

$$r_i(t) = R(f_i(t)); \qquad i \in \{1, 2, \cdots, g\} \tag{2.20}$$

where $R(\cdot)$ denotes a functional relation. This is called as a *dedicated residual set* which is inspired by the *dedicated observer scheme* proposed by Clark (1978a). A simple threshold logic can be used to make decision about the appearance of a specific fault by the logic decision according to:

$$r_i(t) > T_i \implies f_i(t) \neq 0; \qquad i \in \{1, 2, \cdots, g\} \tag{2.21}$$

where T_i $(i = 1, \cdots, g)$ are thresholds. This isolable residual structure is very simple and all faults can be detected simultaneously, however it is difficult to design in practice. Even when this structured residual set can be designed, there is normally no design freedom left to achieve other desirable performances such as robustness against modeling errors (Wünnenberg, 1990). A most commonly used and better scheme in designing the residual set is to make each residual sensitive to all but one fault, i.e.

$$\left\{ \begin{array}{rcl} r_1(t) & = & R(f_2(t), \cdots, f_g(t)) \\ & \vdots & \\ r_i(t) & = & R(f_1(t), \cdots, f_{i-1}(t), f_{i+1}(t), \cdots, f_g(t)) \\ & \vdots & \\ r_g(t) & = & R(f_1(t), \cdots, f_{g-1}(t)) \end{array} \right. \tag{2.22}$$

This is defined as *a generalized residual set.* If all residuals of the generalized residual set are generated using a bank of observers (observer-based residual generators), the structure is known as the *generalized observer scheme* (Frank, 1987; Patton et al., 1989). The isolation can again be performed using simple threshold testing according to the following logic:

$$\left. \begin{array}{l} r_i(t) \leq T_i \\ r_j(t) > T_j \quad \forall j \in \{1, \cdots, i-1, i+1, \cdots, g\} \end{array} \right\} \implies f_i(t) \neq 0; \tag{2.23}$$

for $i = 1, 2, \cdots, g$

As a simple example of isolating *three* different faults $\{f_1, f_2, f_3\}$, the structured residual set can be designed in two different ways as shown in Figs. 2.15a & 2.15b. Faults can be uniquely isolated using either of two methods.

2.7.2 *Fixed direction residual vector*

An alternative way of achieving the isolability of faults is to design a *directional* residual vector which lies in a fixed and fault-specified direction (or subspace) in the residual space, in response to a particular fault. This is to make:

$$r(t \mid f_i(t)) = \alpha_i(t) l_i; \quad i \in \{1, 2, \cdots, g\} \tag{2.24}$$

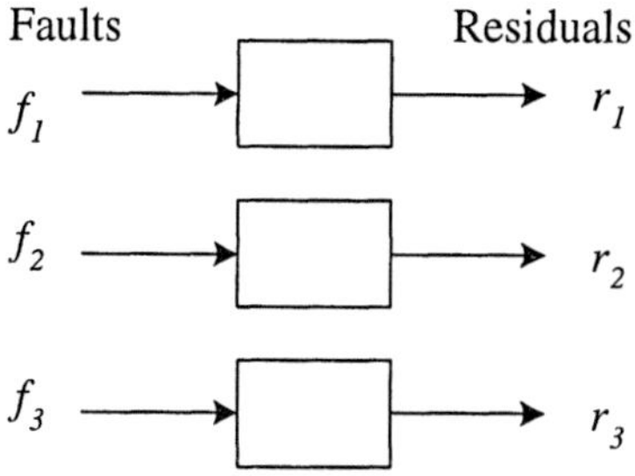

Figure 2.15a. Structured residual set (Dedicated scheme)

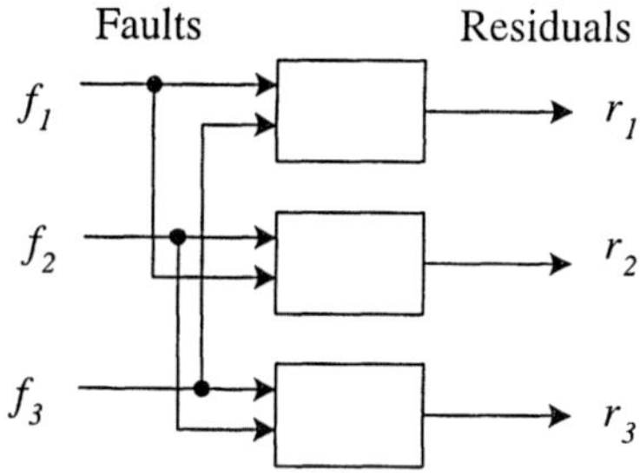

Figure 2.15b. Structured residual set (Generalized scheme)

where the constant vector l_i is the *signature direction* of the i_{th} fault in the residual space and α_i is a scalar that depends on the fault size and dynamics. With the fixed directional residual, the fault isolation problem is one of determining which of the known fault signature directions the generated residual vector lies the closest to. To isolate faults reliably, each fault signature has to be uniquely related to one fault. Fig. 2.16 illustrates this fault isolation approach using a *directional residual vector* in which the residual is closed to the signature direction of the fault f_2 and hence the fault f_2 is most likely the one in the system.

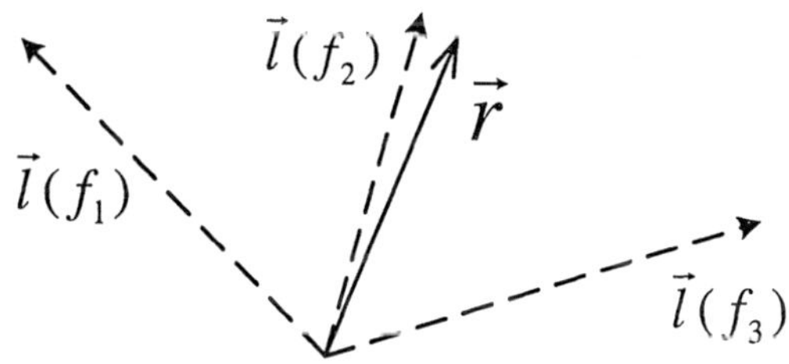

Figure 2.16. Directional residual vector for fault isolation

2.7.3 *Sensor and actuator faults isolation*

If we are only interested in sensor faults, the system output is given by:

$$y(s) = G_u(s)u(s) + f_s(s) \tag{2.25}$$

If one wants to design a residual signal which is sensitive to one group of faults in $f_s^1(s)$ and insensitive to another group of faults in $f_s^2(s)$, the above equation can be decomposed into:

$$\begin{bmatrix} y^1(s) \\ y^2(s) \end{bmatrix} = G_u(s)u(s) + \begin{bmatrix} f_s^1(s) \\ f_s^2(s) \end{bmatrix} \tag{2.26}$$

The residual generator then takes on the following format:

$$r^1(s) = H_u^1(s)u(s) + H_y^1(s)y^1(s) \tag{2.27}$$

On substituting $y^1(s)$ into the above equation, we have:

$$r^1(s) = [H_u^1(s) + H_y^1(s)G_u(s)]u(s) + H_y^1(s)f_s^1(s) \tag{2.28}$$

The residual will be only sensitive to the fault group $f_s^1(s)$, when the transfer function matrices of the residual generator satisfy:

$$\begin{cases} H_u^1(s) & = & -H_y^1(s)G_u(s) \\ H_y^1(s) & \neq & 0 \end{cases} \tag{2.29}$$

This is the normal requirement for a residual generator as shown in Eq.(2.15). That is to say that there is no additional requirement for the sensor fault isolation problem. The transfer matrix $H_y^1(s)$ can be chosen freely according to specific requirements. The only constraint on $H_y^1(s)$ is that it should be stable and realizable. Once it has been chosen, $H_u^1(s)$ can be determined by $H_u^1(s) = -H_y^1(s)G_u(s)$. As the transfer matrix $H_y^1(s)$ can be chosen freely, sensor fault isolation is always possible.

When actuator faults occur in the system, the system output is:

$$y(s) = G_u(s)[u(s) + f_a(s)] \tag{2.30}$$

If we want to design a residual signal which is sensitive to one group of faults $f_a^1(s)$ and insensitive to another group of faults $f_a^2(s)$, the above equation can be decomposed into:

$$y(s) = G_u^1(s)[u^1(s) + f_a^1(s)] + G_u^2(s)[u^2(s) + f_a^2(s)] \tag{2.31}$$

The residual generator is now:

$$r^1(s) = H_u^1(s)u^1(s) + H_y^1(s)y(s) \tag{2.32}$$

On substituting $y(s)$ into Eq.(2.32), we have:

$$\begin{aligned} r^1(s) &= [H_u^1(s) + H_y^1(s)G_u^1(s)]u^1(s) + H_y^1(s)G_u^1(s)f_a^1(s) \\ &\quad + H^1(s)G_u^2(s)[u^2(s) + f_a^2(s)] \end{aligned} \tag{2.33}$$

To make the residual only sensitive to the fault group $f_a^1(s)$, we need the following conditions:

$$\begin{cases} H_u^1(s) & = & -H_y^1(s)G_u^1(s) \\ H_y^1(s)G_u^2(s) & = & 0 \\ H_y^1(s)G_u^1(s) & \neq & 0 \end{cases} \tag{2.34}$$

These equations illustrate that an extra constraint ($H_y^1(s)G_u^2(s) = 0$) is required for the actuator isolation problem. A stable and implementable transfer matrix $H_y^1(s)$, which satisfies this constraint does not always exist. That is to say, we do not have full freedom to achieve the required actuator fault isolation requirement. Hence, actuator fault isolation is not always possible.

2.8 Residual Generation Techniques

The generation of residual signals is a central issue in model-based fault diagnosis. A rich variety of methods are available for residual generation and this Section discusses briefly some of the most common approaches. It must be pointed out that most residual generation approaches are applicable for both continuous and discrete system models, however some approaches can only work for discrete models. In this book, if the continuous model is used, it implies that the technique can be applied to both continuous and discrete models, otherwise the technique is only applicable for discrete models. For example, the parity relation approach is developed specially for discrete models although there have been some studies into the use of the parity relation approach for continuous models (Mironovski, 1979; Magni and Mouyon, 1991; Medvedev, 1995; Medvedev, 1996).

2.8.1 Observer-based approaches

The basic idea behind the observer or filter-based approaches is to estimate the outputs of the system from the measurements (or a subset of measurements) by using either Luenberger observer(s) in a deterministic setting (Beard, 1971; Clark et al., 1975; Clark, 1979; Frank, 1987; Frank, 1990; Patton and Kangethe, 1989; Patton and Chen, 1997) or Kalman filter(s) in a stochastic setting (Mehra and Peschon, 1971; Willsky, 1976; Frank, 1987; Basseville, 1988; Tzafestas and Watanabe, 1990; Da and Lin, 1996; Zolghadri, 1996; Sohlberg, 1998). Then, the (weighted) output estimation error (or innovations in the stochastic case), is used as a residual. The flexibility in selecting observer gains has been fully exploited in the literature yielding a rich variety of FDI schemes, the most recently development can be found in various survey papers: e.g. Frank (1994b; 1996), Frank and Ding (1997), García and Frank (1997), Patton (1994; 1997b), Patton and Chen (1994; 1996b; 1997), and conference proceedings such as, Isermann (1991b), Labarrère (1993), Ruokonen (1994) and Patton and Chen (1998).

What we are interested in FDI is the estimation of outputs using an observer, whilst the estimation of the state vector is unnecessary. As a matter of fact, a functional observer is suitable for this task. In practice, the order of the functional observer is less than the order of a state observer. It is desired to estimate a linear function of the state, i.e. $Lx(t)$, using a functional (or generalized) Luenberger observer with the following structure:

$$\left\{ \begin{array}{lcl} \dot{z}(t) & = & Fz(t) + Ky(t) + Ju(t) \\ w(t) & = & Gz(t) + Ry(t) + Su(t) \end{array} \right. \tag{2.35}$$

where $z(t) \in \mathbb{R}^q$ is the state vector of this functional observer, F, K, J, R, G and S are matrices with appropriate dimensions. The output $w(t)$ of this observer is said to be an estimate of $Lx(t)$, for the system described in Eq.(2.9), in an asymptotic sense if in the absence of faults:

$$\lim_{t \to \infty} [w(t) - Lx(t)] = 0 \tag{2.36}$$

To introduce a transformation matrix T, the observer shown in Eq.(2.35) will generate the estimate $Lx(t)$ in the *asymptotic sense* if and only if the following conditions hold (O'Reilly, 1983):

$$\begin{cases} F \text{ has stable eigenvalues} \\ TA - FT = KC \\ J = TB - KD \\ RC + GT = L \\ S + RD = 0 \end{cases} \tag{2.37}$$

The necessary and sufficient condition for the existence of the observer given by Eq.(2.35) for the system Eq.(2.9) is that the pair (C, A) is observable (O'Reilly, 1983). In order to generate residuals, we need to estimate the system output. If we assign:

$$L = C \tag{2.38}$$

We have the output estimation as:

$$\hat{y}(t) = w(t) + Du(t) \tag{2.39}$$

The residual vector $r(t)$ is defined as:

$$r(t) = Q[y(t) - \hat{y}(t)] = L_1 z(t) + L_2 y(t) + L_3 u(t) \tag{2.40}$$

where:

$$\begin{cases} L_1 &= -QG \\ L_2 &= Q - QR \\ L_3 &= -Q(S + D) \end{cases}$$

Now, the residual generator based on a generalized Luenberger is illustrated in Fig. 2.17 and given by the following equation:

$$\begin{cases} \dot{z}(t) &= Fz(t) + Ky(t) + Ju(t) \\ r(t) &= L_1 z(t) + L_2 y(t) + L_3 u(t) \end{cases} \tag{2.41}$$

And the matrices in this equation should satisfy the following conditions:

$$\begin{cases} F \text{ has stable eigenvalues} \\ TA - FT = KC \\ J = TB - KD \\ L_1 T + L_2 C = 0 \\ L_3 + L_2 D = 0 \end{cases} \tag{2.42}$$

The Laplace transformation of the residual is thus:

$$r(s) = [L_1(sI - F)^{-1}K + L_2]y(s) + [L_1(sI - F)^{-1}J + L_3]u(s) \tag{2.43}$$

The residual generator based on a generalized Luenberger observer is shown in Fig. 2.17. It can be seen that there is a feedback structure imbedded within it. The feedback can be used to improve the dynamic behavior of residuals.

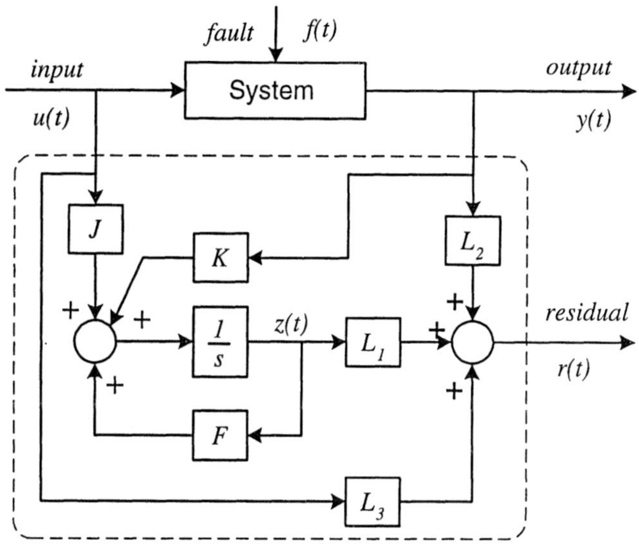

Figure 2.17. Residual generation via a generalized Luenberger observer

When we apply the residual generator described by Eq.(2.41) to the system described by Eq.(2.9), the residual will be:

$$\begin{cases} \dot{e}(t) = Fe(t) - TR_1 f(t) + KR_2 f(t) \\ r(t) = L_1 e(t) + L_2 R_2 f(t) \end{cases} \tag{2.44}$$

where $e(t) = z(t) - Tx(t)$. It can be seen that the residual depends solely and totally on faults.

The simplest method in observer-based residual generation is to use a full order observer, in this case the observer dimension q equals n and we have:

$$\begin{cases} T = I \\ F = A - KC \\ J = B - KD \end{cases} \qquad \begin{cases} L_1 = QC \\ L_2 = -Q \\ L_3 = QD \end{cases}$$

Hence, the transfer function matrices for a full-order observer based residual generator are given by:

$$\begin{cases} H_y(s) = Q\{C[sI - (A - KC)]^{-1}K - I\} \\ H_u(s) = Q\{C[sI - (A - KC)]^{-1}(B - KD) + D\} \end{cases} \tag{2.45}$$

To alter the frequency response of the residual, the residual weighting matrix Q can be changed into a dynamic weighting $Q(s)$.

For any dynamic system, the observer-based residual generator always exists. This is because any input-output transfer function matrix has the observable realization. That is to say, the output estimator always exists although a suitable state observer cannot always be designed. The minimal order q_0 of a functional observer satisfies the inequality (O'Reilly, 1983; Mironovski, 1979; Mironovski, 1980):

$$q_0 \leq \mu - 1 \tag{2.46}$$

where μ is the observability index of the system which is defined as the minimum number for which:

$$rank[C^T, \ (CA)^T, \ \cdots, \ (CA^{\mu})^T] = n$$

For observable systems the observability index lies within the limits:

$$\frac{n}{m} \leq \mu \leq n - m + 1$$

Inequality (2.46) gives only the *minimum* possible order of a functional observer. In order to provide additional freedom for achieving required diagnostic performance, the observer order is normally larger than this minimum possible order. For example, the residual must be made sensitive to faults.

To isolate faults, the observer-based approaches can be used to design structured residual sets or fixed residual vectors. For sensor faults, the design of a structured residual set is very straightforward. If we require that a residual is sensitive to faults in all but one of the sensors, the observer used to generate this residual should be driven by outputs excluding that single sensor measurement. To be more specific, if we replace the output vector $y = (y_1, \cdots, y_m)$ by $(y_1, \cdots, y_{i-1}, y_{i+1}, \cdots, y_m)$, the residual will be insensitive to the fault in the i_{th} sensor. However, the design of a structured residual set for actuator fault isolation is more difficult. This problem can be solved via unknown input observers (Viswanadham and Srichander, 1987; Phatak and Viswanadham, 1988; Frank, 1990) and eigenstructure assignment (Patton et al., 1986; Patton, 1988; Patton and Chen, 1991g; Chen, 1995), however the isolation of actuator faults is not always possible. The problem of designing structured residual set via unknown input observers is discussed in Chapter 3. The schemes used in designing observer-based structured residual set have being called the dedicated observer scheme and the generalized observer scheme etc. (Frank, 1987; Frank, 1990; Patton et al., 1989; van Schrick, 1994a). The fixed residual vector can be designed by the so-called "fault (or failure) detection filter" originated by Beard (1971) and revisited by White and Speyer (1987) and Park and Rizzoni (1994b).

2.8.2 Parity vector (relation) methods

In the early development of fault diagnosis, the *parity vector (relation)* approach was applied to *static or parallel redundancy* schemes (Potter and Suman, 1977; Gai, Harrison and Daly, 1978; Daly, Gai and Harrison, 1979; Desai and Ray, 1984) which may be obtained directly from measurements or from analytical relations. A survey of these schemes can be found in Ray and Luck (1991). There are typically two cases for arranging hardware redundancy, one is the use of sensors having identical or similar functions to measure the same variable, another is the use of dissimilar sensors to measure different variables but with their outputs being relative to each other. The basic idea of the parity relation approach is to provide a proper check of the parity (consistency) of the measurements of the monitored system. To begin with this problem, let us consider a general problem of the measurement of an n-dimensional vector using m sensors. The measurement (algebraic) equation is:

$$y(k) = Cx(k) + f(k) + \xi(k) \tag{2.47}$$

where $y(k) \in \mathbb{R}^m$ is measurement vector, $x(k) \in \mathbb{R}^n$ is the state vector, and $f(k)$ is the vector of sensor faults, $\xi(k)$ is a noise vector and C is an $m \times n$ measurement matrix.

With hardware (direct) redundancy there are more than the minimum number of sensors (eg., two or more for a scalar state variable, and four or more for a three-dimensional state vector). That is to say that the dimension of $y(k)$ is larger than the dimension of $x(k)$.

$$m > n; \quad \text{and} \quad rank(C) = n$$

For such system configurations, the number of measurements is greater than the number of variables to be sensed. Inconsistency in the measurement data is then a metric that can be used initially for detecting faults and, subsequently for fault isolation. This technique has been successfully applied to fault diagnosis schemes for inertial navigation (Potter and Suman, 1977; Gai, Harrison and Daly, 1978; Daly et al., 1979; Desai and Ray, 1984) where relationships between gyroscope readings and/or accelerometer assemblies provide analytical forms of redundancy.

For FDI purposes, the vector $y(k)$ can be combined into a set of linearly independent parity equations to generate the parity vector (residual):

$$r(k) = Vy(k) \tag{2.48}$$

The residual generation scheme based on direct redundant measurements is shown in Fig.2.18.

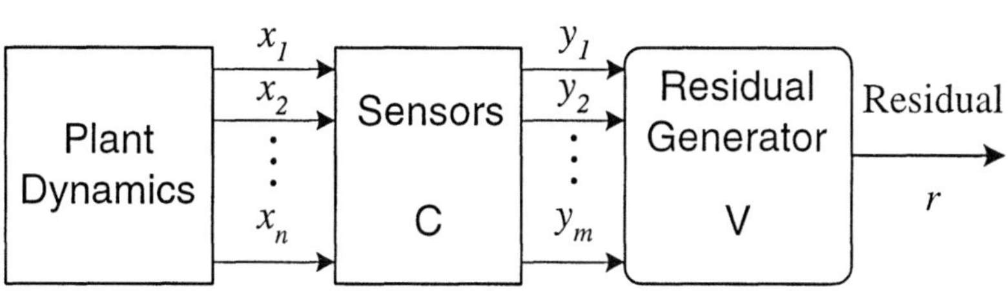

Figure 2.18. Residual generation via parallel redundancy

In order to make $r(k)$ satisfy the usual requirement for a residual (zero-valued for the fault-free case), the matrix V must satisfy the condition:

$$VC = 0 \tag{2.49}$$

When this condition holds true, the residual (parity vector) only contains information on the faults and noise:

$$r(k) = v_1[f_1(k) + \xi_1(k)] + \cdots + v_m[f_m(k) + \xi_m(k)] \tag{2.50}$$

where v_i is the i_{th} column of V, $f_i(k)$ is the i_{th} element of $f(k)$ which denotes the fault in the i_{th} sensor.

Eq.(2.50) reveals that the parity vector only contains information due to faults and noise (uncertainty), and is independent of the unmeasured state $x(k)$. Eq.(2.50) also shows that the parity space (or residual space) is spanned by the columns of V, i.e. the columns of V form a basis for this space. Moreover, the following attractive property can also be exploited: a fault in the i_{th} sensor implies a growth of the residual $r(k)$ in the direction v_i. This ensures that a fault in the i_{th} sensor, implies a magnification of the norm of $r(k)$ in the direction v_i. The space $span\{V\}$ is called a "parity space". The term "parity" was first used in connection with digital logic systems and computer software reliability to enable "parity checks" to be performed for error checking. In the fault diagnosis field, it has similar meaning in the context of providing an indicator for the presence of a fault (or error) in system components.

Using the notation of Daly et al. (1979), a *fault detection decision function* is defined as:

$$DFD(k) = r(k)^T r(k) \tag{2.51}$$

If a fault occurs in the sensors, $DFD(k)$ will be greater than an predetermined threshold.

The *fault isolation decision function* is then:

$$DFI_i(k) = v_i^T r(k); \quad i \in \{1, 2, \cdots, m\} \tag{2.52}$$

For a given $r(k)$, a malfunctioning sensor is identified by computing the m values of $DFI_i(k)$. If $DFI_j(k)$ is the largest one of these values, the sensor corresponding to $DFI_j(k)$ is the one which is most likely to have become faulty.

In the parity space point of view, the columns of V define m distinct *fault signature directions* (v_i, $i = 1, 2, \cdots, m$). After a fault has been declared, it can be isolated by comparing the orientation of the parity vector to each these signature directions. Indeed, the fault isolation function $DFI_i(k)$ is a measure of the correlation of the residual vector with fault signature directions. In order to isolate faults reliably, the generalized angles between fault signature directions should be "as large as possible", i.e., to make $v_i^T v_j$ $(i \neq j)$ "as small as possible". Thus, optimal fault isolation performance will be achieved when v_i determined by:

$$\begin{cases} \min\{v_i^T v_j\}; & i \neq j, \;\; i, j \in \{1, 2, \cdots, m\} \\ \max\{v_i^T v_i\}; & i \in \{1, 2, \cdots, m\} \end{cases} \tag{2.53}$$

The traditional sub-optimal solution of the matrix V is to make (Ray and Luck, 1991):

$$VV^T = I_{m-n} \tag{2.54}$$

A further consequence of conditions (2.49) and (2.54) is that:

$$V^T V = I_m - C(C^T C)^{-1} C^T \tag{2.55}$$

The condition for the existence of a solution V for Eq.(2.49) is that $rank(C) = n < m$. This implies that the rows of C are linearly dependent, i.e., the outputs of the sensors are related by a static relation.

In order to completely determine the matrix V, Potter and Suman (1977) suggested making V an upper triangular matrix with positive diagonal elements so that the Gram-Schmidt orthogonalization scheme (Golub and Van Loan, 1989) can be used to determine V. The Potter's algorithm is as follows:

$$\left\{\begin{array}{ll} 1. & \Theta = I - C(C^T C)^{-1} C^T \\ 2. & v_{11}^2 = \theta_{11} \\ & v_{1j} = \theta_{1j}/v_{11},\ j = 2, \cdots, m \\ & v_{ij} = 0,\ i = 2, \cdots, m-n;\ j = 1, \cdots, i-1 \\ 3. & v_{ii}^2 = \theta_{ii} - \sum_{l=1}^{i-1} v_{li}^2,\ i = 2, \cdots, m-n \\ 4. & v_{ij} = (\theta_{ij} - \sum_{l=1}^{i-1} v_{li} v_{lj})/v_{ii},\ i = 2, \cdots, m-n;\ j = i+1, \cdots, m \end{array}\right. \tag{2.56}$$

For the case $rank(C) = m < n$, the *direct redundancy* relation does not exist however, we may construct redundancy relations by collecting sensor outputs over a time interval (data window) (say, $\{y(k-s), y(k-s+1), \cdots, y(k)\}$). This is known as "temporal redundancy" or "serial redundancy". The dynamic model must be known and used in this case, as the redundancy is related to time. Here, we consider that the system is given by the linear discrete state space equations as follows:

$$\left\{\begin{array}{rcl} x(k+1) & = & Ax(k) + Bu(k) + R_1 f(k) \\ y(k) & = & Cx(k) + Du(k) + R_2 f(k) \end{array}\right. \tag{2.57}$$

where $x \in \mathbb{R}^n$ is state vector, $y \in \mathbb{R}^m$ is output vector, $u \in \mathbb{R}^r$ is input vector, $f \in \mathbb{R}^g$ fault vector, and A, B, C, D, R_1 and R_2 are real matrices of compatible dimensions.

As a direct extension of the case of parallel redundancy, the parity relation concept was first generalized by (Chow and Willsky, 1984) using the *temporal redundancy* relations of the dynamic system. Extended researches have been done by various other authors as, Lou et al. (1986), Massoumnia and Van-der Velde (1988), Frank and Wünnenberg (1989), Wünnenberg (1990), Gertler and Singer (1990), Patton and Chen (1991e). It is important however, to note that essentially the same scheme has been suggested by the Russian expert Mironovski (1979) (see also: Mironovski (1980) and Basseville (1988)). Although he did not use the term "parity relation", the essential ideas are the same as those of the remaining authors.

The redundancy relations are now specified mathematically as follows. Combining together Eq.(2.57) from time instant $k-s$ to time instant k yields the following redundant relations:

$$\underbrace{\begin{bmatrix} y(k-s) \\ y(k-s+1) \\ \vdots \\ y(k) \end{bmatrix}}_{Y(k)} - H \underbrace{\begin{bmatrix} u(k-s) \\ u(k-s+1) \\ \vdots \\ u(k) \end{bmatrix}}_{U(k)} = Wx(k-s) + M \underbrace{\begin{bmatrix} f(k-s) \\ f(k-s+1) \\ \vdots \\ f(k) \end{bmatrix}}_{F(k)} \tag{2.58}$$

where

$$H = \begin{bmatrix} D & 0 & \dots & 0 \\ CB & D & \dots & 0 \\ \vdots & \vdots & \ddots & \vdots \\ CA^{s-1}B & CA^{s-2}B & \dots & D \end{bmatrix} \in \mathbb{R}^{(s+1)m \times (s+1)r}$$

$$W = \begin{bmatrix} C \\ CA \\ \vdots \\ CA^s \end{bmatrix} \in \mathbb{R}^{(s+1)m \times n}$$

and the matrix M is constructed by replacing $\{D, B\}$ with $\{R_2, R_1\}$ in the matrix H.

To simplify the notation, Eq.(2.58) can be rewritten as:

$$Y(k) - HU(k) = Wx(k-s) + MF(k) \tag{2.59}$$

According to Chow and Willsky (1984) and Lou et al. (1986), a residual signal can be defined as:

$$r(k) = V[Y(k) - HU(k)] \tag{2.60}$$

where $V \in \mathbb{R}^{p \times (s+1)m}$ and p is the residual vector dimension. Eq.(2.60) is termed an s_{th} order parity equation or parity relation. It is the *computational form* of a residual generator which shows the residual signal as a function of measured inputs and outputs of the monitored system. Substituting Eq.(2.59) into Eq.(2.60), we have:

$$r(k) = VWx(k-s) + VMF(k) \tag{2.61}$$

This is the *evaluation* format of the residual. In order to make the parity vector useful for FDI, one should make it insensitive to system inputs and states, i.e.

$$VW = 0 \tag{2.62}$$

To satisfy the fault detectability condition, the matrix V should also satisfy the following condition:

$$VM \neq 0 \tag{2.63}$$

Once we have the matrix V, the residual signal can be generated using Eq. (2.60). The residual generator design depends on solutions of Eq.(2.62). For an appropriately large s (for example $s = n$), it follows from the Cayley-Hamilton theorem that the solution V of Eq. (2.62) always exists. This means that a parity relation-based residual generator for fault detection always exists. An appropriate value for s can be found by the designer by a systematic increase in s.

Of particular interest are those parity relations for which the order s (length $(s+1)$ of the data window) is minimal. The minimum order s_0 of the parity relations satisfies the two-sided inequality (Mironovski, 1979; Mironovski, 1980):

$$\frac{rank(W_o)}{rank(C)} \leq s_0 \leq rank(W_o) - rank(C) + 1$$

where W_o is the observability matrix of the pair (C, A). If the system is observable and the rows of the matrix C are linearly-independent, then the inequality takes the form:

$$\frac{n}{m} \leq s_0 \leq n - m + 1$$

The parity relation approach for residual generation of dynamic system is shown in Fig.2.19.

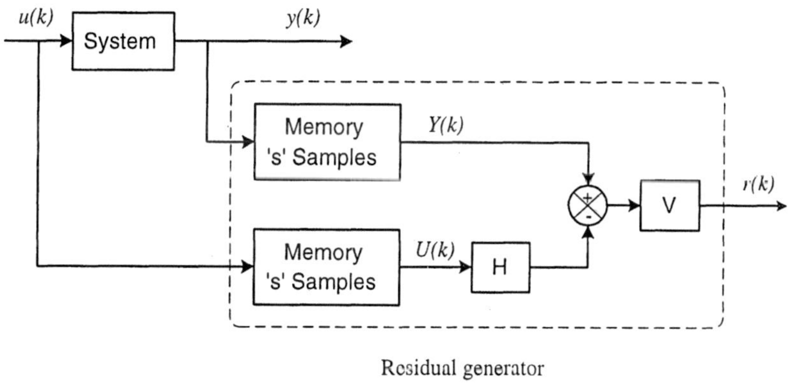

Figure 2.19. Residual generation via temporal redundancy

Here, we discussed the construction of parity relations based a state space model which is suggested by Chow and Willsky (1984) and Mironovski (1979). However, it must be pointed out that the parity relation can also be constructed using a z-transformed input-output model (or discrete transfer matrix representation). Gertler *et al.* (Gertler and Luo, 1989; Gertler, Fang and Luo, 1990; Gertler, Luo, Anderson and Fang, 1990; Gertler and Singer, 1990; Gertler, 1998) first introduced this design approach, preferring to call it the "parity equation" method. Gertler (1991) presented an excellent tutorial on this approach. Note that this input-output model based design method has also been studied by Mironovski (1980) and Massoumnia and Vander Velde (1988).

The parity relation approach can be used to design structured residual set for fault isolation (Massoumnia and Vander Velde, 1988; Gertler, 1991; Patton and Chen, 1991e; Gertler, 1997). The design for isolating sensor faults is very straightforward. If we use c_i^T (the i_{th} row of C) and y_i (the i_{th} component of y) instead of C and y, the parity relation will only contain the i_{th} sensor's output together with all inputs. This form of parity relation has been called a *single-sensor parity relation (SSPR)* (Massoumnia and Vander Velde, 1988; Tsai and Chou, 1993; Peng, Youssouf, Arte and Kinnaert, 1997) and the residual generated by this relation is only sensitive to the fault in the i_{th} sensor. For the actuator isolation problem, the structured residual set is more difficult to design. Massoumnia and Vander Velde (1988) studied this problem and pointed out that the isolation of actuator faults is not always possible since a *single-actuator parity relation (SAPR)* (Massoumnia and Vander Velde, 1988; Tsai and Chou, 1993; Peng et al., 1997). This conclusion is consistent with the observer-based approaches. Gertler *et al.* (Gertler and Luo, 1989; Gertler, Fang and Luo, 1990; Gertler and Singer, 1990; Gertler, 1997) suggested a so-called "orthogonal parity equation" approach in designing structured residual sets. The idea is to make the parity equation (and residual) orthogonal to a particular fault direction if we want the residual insensitive to this fault. The design of directional residual vectors using parity relations is not straightforward. Gertler (1991) studied this problem and illustrated the possibility based on examples. The systematic approaches of designing parity equations with directional properties are presented in Gertler and Monajemy (1995), Gertler (1997), and Gertler (1998).

It is clear therefore that some correspondence exists between observer-based and parity relation approaches. Massoumnia (1986a) first pointed out this correspondence, and this was later demonstrated by Frank and Wünnenberg (1989), Wünnenberg (1990) and Magni and Mouyon (1991). A full derivation of this equivalence has been given by Patton and Chen (1991e). Patton and Chen (1991d) have re-examined this problem in detail and the equivalence under different conditions and in different meanings have been discussed. It has been shown by Frank and Wünnenberg (1989), and more fully by Patton and Chen (1991e), that the parity relation approach is equivalent to the use of a dead-beat observer. A residual signal generated by a non dead-beat observer is equivalent to a post-filtered residual which generated by a dead-beat observer. This implies that the parity relation method provides less design flexibility when compared with methods which are based on observers without any restriction.

2.8.3 Factorization methods for residual generation

A residual generator can also be synthesized in the frequency domain via factorization of the transfer function matrix of the monitored system. This method was initiated by Viswanadham, Taylor and Luce (1987). A more comprehensive study and extension was made by Ding and Frank (1990), in which the FDI problem was systematically formulated and solved via factorization techniques. This approach is also studied by other investigators such as Marquez and Diduch (1992), Yao, Schaefers and Darouach (1994) and Kinnaert and Peng (1995). The recent developments including robustness issues can be found in (Frank and Ding, 1993; Ding et al., 1993; Frank and Ding, 1994) and a detailed discussions is presented in Section 8.2.

This approach is based on a fact that any $m \times r$ proper rational transfer function matrix $G_u(s)$ can be factorized as (Vidyasagar, 1985):

$$G_u(s) = \tilde{M}^{-1}(s)\tilde{N}(s) \tag{2.64}$$

where $\tilde{M}(s)$ and $\tilde{N}(s)$ are stable, rational and realizable transfer function matrices. Based on this factorization, a residual generator can be designed as:

$$r(s) = Q(s)[\tilde{M}(s)y(s) - \tilde{N}(s)u(s)] \tag{2.65}$$

where $Q(s)$ denotes a dynamic residual weighting. It was pointed out quite early on Section 2.4 that the system output is:

$$y(s) = G_u(s)u(s) + G_f(s)f(s) \tag{2.66}$$

On substituting Eq.(2.66) into Eq.(2.65) together with Eq.(2.64), the residual is:

$$r(s) = Q(s)\tilde{M}(s)G_f(s)f(s) \tag{2.67}$$

which is only affected by the fault. The weighting factor $Q(s)$ can be used to improve the residual performance responding to faults in a particular frequency region.

Note that Eq.(2.65) is a special representation of residual generators which can also fitted into the general framework given by Eq.(2.13) and Fig.2.9. The design of a residual generator is to construct the transfer function matrices $\tilde{M}(s)$ and $\tilde{N}(s)$ which can be given by (Nett, Jacobson and Balas, 1984):

$$\begin{cases} \tilde{M}(s) & = & -C[sI - (A - KC)]^{-1}L + I \\ \tilde{N}(s) & = & C[sI - (A - KC)]^{-1}(B - KD) + D \end{cases} \tag{2.68}$$

On comparing the above equations with the transfer function matrices for a full-order observer-based residual generator given in Eq.(2.45), one can see that they are almost identical. This demonstrates the correspondence between observer-based and factorization approaches. Ding, Guo and Frank (1994) presented a study on the design of linear observers, based on the transfer matrix factorization.

2.9 Model-based FDI via Parameter Estimation

Model-based FDI can also be achieved by the use of system identification techniques (Isermann, 1984; Isermann, 1987; Isermann and Freyermuth, 1990; Isermann, 1991a; Isermann, 1997). This approach is based on the assumption that the faults are reflected in the physical system parameters such as friction, mass, viscosity, resistance, inductance, capacitance, etc. The basic idea of the detection method is that the parameters of the actual process are repeatedly estimated on-line using well known parameter estimation methods and the results are compared with the parameters the reference model obtained initially under the faulty-free condition. Any substantial discrepancy indicated as a fault. This approach normally uses the input-output mathematical model of a system in the following form:

$$y(t) = f(P, u(t)) \tag{2.69}$$

where, P is the model coefficient vector which is directly related to physical parameters of the system. The function $f(\cdot,\cdot)$ can take *both* linear or non-linear formats.

The basic procedure for carrying out FDI using parameter estimation is:

- Establish the process model using physical relations.
- Determine the relationship between model coefficients and process physical parameters.
- Estimate the normal model coefficients.
- Calculate the normal process physical parameters.
- Determine the parameter changes which occur for the various fault cases.

By carrying out the last step for known faults, a database of faults and their symptoms can be built up. During the system operation, the coefficients of the system model are periodically identified from the measurable inputs and outputs, and compared with the normal and faulty model parameters.

To generate residuals using this approach, an on-line parameter identification algorithm should be used. If one has the estimation of the model coefficient at time step $k-1$ as $\hat{P}_{k-1}$, the residual can be defined in either of the following ways:

$$\begin{cases} r(k) &= \hat{P}_k - P_0 \\ r(k) &= y(k) - f(\hat{P}_{k-1}, u(k)) \end{cases} \tag{2.70}$$

where P_0 is the normal model coefficient.

It is not easy to achieve fault isolation using the parameter estimation method. This is because the parameters being identified are model parameters which cannot always be converted back to the system physical parameters (Isermann, 1984). However, the faults are represented by variations in physical parameters. Recently, Doraiswami and Stevenson (1996) proposed an influence matrix approach to overcome the isolation difficult. The idea is to identify the influence of each physical parameter on the residual.

2.10 Fault Diagnosis for Stochastic Systems

For stochastic systems, the FDI is based on statistical testing of the residuals (Willsky, 1976; Basseville and Benveniste, 1986; Basseville, 1988; Tzafestas and Watanabe, 1990; Basseville and Nikiforov, 1993; Da and Lin, 1996; Zolghadri, 1996; Basseville, 1997; Sohlberg, 1998), for example:

- The weighted sum-squared residual (WSSR) testing (Willsky et al., 1975; Tzafestas and Watanabe, 1990).

- χ^2 testing (Willsky, 1976; Da, 1994; Da and Lin, 1996; Chen and Patton, 1994b; Chen and Patton, 1996).

- Sequential probability ratio testing (SPRT) (Wald, 1947; Willsky, 1976; Tzafestas and Watanabe, 1990; Grainger, Holst, Isaksson and Ninness, 1995) and modified SPRT (Gai and Gurry, 1977; Speyer and White, 1984; Tzafestas and Watanabe, 1990).

- Generalized likelihood ratio (GLR) testing (Willsky and Jones, 1974; Willsky and Jones, 1976; Tanaka and Müller, 1990; Tanaka and Müller, 1993; Keller et al., 1996; Peng et al., 1997).

- Cumulative sum algorithm (Nikiforov, Varavva and Kireichikov, 1993).

- Multiple hypothesis testing (Bogh, 1995).

In order to suppress the effect of noise on the residuals, the residual generator has to be specifically designed to deal with the noise. A common approach is the use of Kalman filter-based residual generators. Whilst using a similar structure to the observer, approaches based on the Kalman filter comprise a residual generation mechanism derived by means of a stochastic model of the dynamic system. In normal operation the Kalman filter residual (or innovation) vector (the difference between the measurements and their Kalman filter estimates), is a zero-mean white noise process with known covariance matrix. Mehra and Peschon (1971) first proposed the use of different statistical tests on the innovation to detect faults in the system. Many variants of the idea of hypothesis testing for FDI have been published since (Willsky, 1976; Basseville, 1988; Tzafestas and Watanabe, 1990; Nikoukhah, 1994; Chang and Chen, 1995; Zolghadri, 1996; Basseville, 1997; Sohlberg, 1998). The idea which is common to all these approaches is to test, amongst all possible hypotheses, that the system has a fault or is fault-free. As each fault type has its own signature, a set of hypotheses can be used and checked for the likelihood that a particular fault has occurred.

Some Kalman filter-like state estimators are developed especially for FDI of stochastic systems, e.g:

- Multiple model adaptive filters (MMAFs) (Willsky et al., 1974; Willsky et al., 1975; Montgomery and Caglayan, 1976; Loparo, Buchner and Vasudeva, 1991; Menke and Maybeck, 1995; Eide and Maybeck, 1996; Berec, 1998).

- Two-stage bias-correction filters (Friedland and Grabousky, 1982; Chen, Zhang and Zhang, 1990).

2.11 Robust Residual Generation Problems

The reliability of fault diagnosis must be higher than the monitored system. The model-based fault diagnosis is based upon the use of mathematical models of the supervised system. The better the model used to represent the dynamic behavior of the system, the better will be the chance of improving the reliability and performance in diagnosing faults. However, modeling errors and disturbances in complex engineering systems are inevitable, and hence there is a need to develop robust fault diagnosis algorithms. The robustness of a fault diagnosis system means that it must be only sensitive to faults, even in the presence of model-reality differences (e.g. parameter variations, turbulence, and the effects of manoeuvres). Usually, parameter variations and disturbances act upon a real process in an uncertain way, so that it may be difficult to design a fault diagnosis system which is highly sensitive to faults, whilst insensitive to uncertainty and unmodeled disturbances.

The heart of model-based fault diagnosis is the generation of residuals. Both faults and uncertainty affect the residual, and discrimination between their effects is difficult. The task in the design of a robust FDI system is thus to generate residuals which are insensitive to uncertainty, whilst at the same time sensitive to faults, and therefore *robust*: Frank and Wünnenberg (1987), Frank (1990; 1991a; 1993; 1994b; 1996), Frank and Ding (1997), Patton et al. (1989), Gertler (1991), Gertler and Kunwer (1993), Patton and Chen (1991e; 1992b; 1992c; 1993b; 1994; 1996b; 1997), Patton (1993; 1994; 1997b). The robustness is of course only proved if the residual of interest remains insensitive to uncertainty over the whole range of operation of the system being monitored.

To approach the problem from the general point view, one must start with a mathematical description of the monitored system that includes all kinds of modeling uncertainty that can occur in practice and affect the behavior of the system. Therefore, the state space model of the system is given by:

$$\left\{ \begin{array}{rcl} \dot{x}(t) & = & (A+\Delta A)x(t) + (B+\Delta B)u(t) + E_1 d(t) + R_1 f(t) \\ y(t) & = & (C+\Delta C)x(t) + (D+\Delta D)u(t) + E_2 d(t) + R_2 f(t) \end{array} \right. \tag{2.71}$$

here $d(t) \in \mathbb{R}^q$ is an *unknown* input (disturbance) vector, however the unknown input distribution matrices E_1 and E_2 are assumed to be known. The matrices ΔA, ΔB, ΔC and ΔD are the parameter errors or variations which represent the *modeling errors*. The transfer function description of the system is then:

$$y(s) = (G_u(s) + \Delta G_u(s))u(s) + G_d(s)d(s) + G_f(s)f(s) \tag{2.72}$$

Here $G_d(s)d(s)$ represent the disturbance effect and

$$G_d(s) = E_2 + C(sI - A)^{-1}E_1$$

$\Delta G_u(s)$ is used to described modeling errors. The terms $G_d(s)d(s)$ and $\Delta G_u(s)u(s)$ together represent modeling uncertainty. If we substitute the system output $y(s)$ into the residual generator Eq.(2.13), the s-domain residual vector is:

$$r(s) = H_y(s)G_f(s)f(s) + H_y(s)\Delta G_u(s) + H_y(s)G_d(s)d(s) \tag{2.73}$$

Both faults and modeling uncertainty (disturbance and modeling error) affect the residual, and hence discrimination between these two effects is difficult. This is the heart of the robustness problem in FDI.

2.11.1 Robustness to disturbances

If the residual generator has been designed to satisfy:

$$H_y(s)G_d(s) = 0 \tag{2.74}$$

i.e., the disturbance is totally de-coupled from the residual $r(t)$, the residual is robust to the disturbance. This is the *principle of disturbance de-coupling* for robust residual generation.

Disturbance de-coupling designs can be achieved by using the unknown input observer (Watanabe and Himmelblau, 1982; Frank and Wünnenberg, 1987; Frank and Wünnenberg, 1989; Chen and Zhang, 1991; Saif and Guan, 1993; Hou and Müller, 1994b; Chang and Hsu, 1995; Yu and Shields, 1996; Chen, Patton and Zhang, 1996) or alternatively, eigenstructure assignment approaches (Patton et al., 1986; Patton, 1988; Patton and Chen, 1991f; Patton and Chen, 1991e; Patton and Chen, 1991g; Jorgensen, Patton and Chen, 1994; Patton and Chen, 1997; Shen, Chang and Hsu, 1998). These two approaches are studied in greater detail in Chapters 3 and 4. As far as the design of robust residuals is concerned, these methods are formally equivalent whilst using different mathematical tools to achieve the same goal in robustness (Gertler, 1991). Gertler *et al.* (Gertler and Luo, 1989; Gertler, 1991; Gertler and Kunwer, 1993; Gertler, 1997) proposed the disturbance de-coupling design based on the so-called orthogonal parity equations. Disturbance de-coupling can also be achieved using frequency domain design techniques (e.g H_∞-norm optimization) (Ding and Frank, 1991; Frank, 1991a; Frank and Ding, 1993; Frank and Ding, 1994).

If the condition (2.74) does not hold, perfect (accurate) de-coupling is not achievable. One can consider an optimal or approximate de-coupling by minimizing a performance index containing a measure of the effects of both disturbances and faults. One suitable choice of performance index can be defined in the frequency domain as (Ding and Frank, 1991):

$$J = \frac{\|H_y(j\omega)G_d(j\omega)\|}{\|H_y(j\omega)G_f(j\omega)\|} \tag{2.75}$$

By minimizing the performance index J over a specified frequency range, an approximate de-coupling design can be achieved (Ding and Frank, 1991; Frank, 1991a; Frank and Ding, 1993; Qiu and Gertler, 1993; Frank and Ding, 1994). The frequency domain design for optimal disturbance de-coupling is introduced in Chapter 8. The optimal approximate disturbance de-coupling design can also be defined and solved in the time domain (Frank and Wünnenberg, 1989; Wünnenberg, 1990; Chen, Patton and Zhang, 1993) and this is studied in Chapter 7.

2.11.2 Robustness to modeling errors

For modeling errors represented by $\Delta G_u(s)$, the robust problem is more difficult to solve. Two main approaches have been proposed. One, based on an attempt to account for uncertainty in designing the residual is known as *active robustness in FDI* (Patton and Chen (1991f; 1991e; 1991b; 1992b; 1993b), Frank (1996)). The second approach is called *passive robustness in FDI* (Patton and Chen, 1992b; Frank, 1996), which makes use of adaptive threshold at the decision-making stage and this is discussed in Section 2.12.

The active way of achieving a robust solution is to obtain an approximate structure for the uncertainty, i.e. to represent approximately modeling errors as disturbances:

$$\Delta G_u(s)u(s) \approx G_{d_1}(s)d_1(s) \tag{2.76}$$

where $d_1(s)$ is an unknown vector and $G_{d_1}(s)$ is a estimated transfer function matrix. When this approximate structure is used to design disturbance de-coupling residual generators, a suitably robust FDI is achievable. As the attempt is made to render the actual residuals robust with respect to uncertainty, we call this active robustness in FDI (Patton and Chen, 1991e; Patton and Chen, 1992b; Patton and Chen, 1993b; Gertler, 1994; Edelmayer, Bokor, Szigeti and Keviczky, 1997; Patton and Chen, 1997). As an example, let's assume that the parameter errors can be approximate as:

$$\Delta A \approx \sum_{i=1}^{N} a_i A_i \qquad \Delta B \approx \sum_{i=1}^{N} b_i B_i$$

$$\Delta C \approx \sum_{i=1}^{N} c_i C_i \qquad \Delta D \approx \sum_{i=1}^{N} d_i D_i$$

where A_i, B_i, C_i and D_i are known matrices and have the same dimension as matrices A, B, C, D respectively, a_i, b_i, c_i and d_i are scalar factors. In this case, the modeling error can be approximated by the disturbance term as:

$$E_1 d_1(t) = \Delta A x(t) + \Delta B u(t) = [A_1 \ \cdots \ A_N \ B_1 \ \cdots \ B_N] \begin{bmatrix} a_1 x(t) \\ \vdots \\ a_N x(t) \\ b_1 x(t) \\ \vdots \\ b_N x(t) \end{bmatrix}$$

$$E_2 d_2(t) = \Delta C x(t) + \Delta D u(t) = [C_1 \ \cdots \ C_N \ D_1 \ \cdots \ D_N] \begin{bmatrix} c_1 x(t) \\ \vdots \\ c_N x(t) \\ d_1 x(t) \\ \vdots \\ d_N x(t) \end{bmatrix}$$

The Laplace transformed representation is:

$$G_d(s)d(s) = E_2 d_2(s) + C(sI - A)^{-1} E_1 d_1(s)$$

2.11.3 *Discussion on robust FDI*

The disturbance de-coupling method for robust FDI has been studied extensively, however its effectiveness has not been fully demonstrated in real problems. The main difficulty arises as most of the disturbances only account for a small percentage of the uncertainty. The disturbance de-coupling method cannot be directly applied to the system with other uncertainties such as modeling errors. The approximate representation of modeling errors and other uncertain factors as the disturbance term provides a practical way to tackle the robustness issue for real systems. Chapter 5 studies different approaches for representing modeling errors and other uncertain factors via the disturbance term with an approximate or estimated distribution matrix. With this estimated distribution matrix, the disturbance de-coupled residual can be designed and the robust FDI problem is solvable. The study given in Chapter 5 covers most possible uncertain situations and the methods are assessed using realistic system simulation models. To extend the application domain of robust model-based FDI, the modeling uncertainty should have a very general format without structural constraints. Chapter 6 studies this problem, in which the robust design is reformulated into a multi-objective optimization problem and solved by a combination of the method of inequalities and genetic algorithms. Another way to tackle robustness problem against modeling errors is via the use of multiple models to cover all possible system operating ranges. This approach, which was originated by Lou et al. (1986) is further developed in Chapter 7.

2.12 Adaptive Thresholds in Robust FDI

Efforts to enhance the robustness of FDI can be made at the decision-making stage (Emami-Naeini, Akhter and Rock, 1986; Emami-Naeini et al., 1988; Ding and Frank, 1991; Frank, 1991a; Frank and Ding, 1993; Ding et al., 1993; Frank and Ding, 1994; Chang et al., 1995; Ding and Guo, 1998). Due to inevitable parameter uncertainty, disturbance and noise encountered in a practical application, one will rarely find a situation where the conditions for a perfectly robust residual generation are met. This is especially true for modeling errors. It is therefore necessary to provide sufficient robustness not only in the residual generation stage but also in decision-making. When the decision-making is made robust against uncertainty, we can speak of passive robustness in FDI (Patton and Chen (1991f; 1991e; 1991b; 1992b; 1993b), Frank (1996)) in which case it may not be necessary (or it may be difficult) to make the residual robust. Passive robustness is thus an alternative to active robustness which should be used when there is very limited system information available.

The goal of robust decision-making is thus to minimize the false and missing alarm rates due to the effects that modeling uncertainty and unknown disturbances will have on the residuals. This can be achieved in several ways, e.g. by statistical data processing, averaging, or by finding and using the most effective threshold.

In practical situations, the residual is never zero, even when no faults occur. A threshold must then be used in the residual evaluation stage. Normally, the threshold is set slightly larger than the largest magnitude of the residual evaluation function for the fault-free case. The smallest detectable fault is a fault which drives the residual evaluation function to just exceed the threshold. Any fault which produces a residual response smaller than this magnitude is not detectable. From our point of view, the purpose of the robust design is to decrease the magnitude of the fault free residual and maintain (even increase) the magnitude of faulty residuals. From this setting, "adaptive threshold" methods are not really robust FDI methods. They can however be grouped into the class of passive methods for robust FDI.

The decision-making stage normally involves a thresholding process, the choice of the threshold is not at all a straightforward issue, as first pointed out by Gai, Adams and Walker (1976). When fixed thresholds are used, the sensitivity to faults will be intolerably reduced if the threshold is chosen too high, whereas the false alarm rate will be too large when the threshold is chosen too low. The proper choice of the threshold is a delicate problem. Clearly, there should be an optimum choice of threshold level and Walker *et al.* (Gai et al., 1976; Gai, Adams, Walker and Smestad, 1978; Walker and Gai, 1979; Walker, 1989) showed how this can be achieved using the theory of Markov processes. Ding and Frank (Frank and Ding, 1993; Ding et al., 1993; Frank, 1993; Ding and Guo, 1998) proposed a way to calculate the minimum detectable fault in the frequency domain, with the threshold set just slightly higher than the residual evaluation function in response to the minimum detectable fault.

In the case of large manoeuvres, these changes might be large enough so that there is no fixed threshold that allows satisfactory FDI at a tolerable false and missed alarm rates. The solution for such problems is to use adaptive thresholds (Clark, 1989), where thresholds are varied according to the control activity and the noise and the fault signal properties of the monitored system. This concept is illustrated in Fig.2.20 which shows the typical shape of an adaptive threshold for direct residual evaluation.

An interesting question is how should we determine the functional form of the adaptive threshold law? Clark (1989) used an empirical adaptive law, whilst Emami-Naeini et al. (Emami-Naeini et al., 1986; Emami-Naeini et al., 1988) proposed the threshold selector (or threshold adaptor) method and Ding and Frank (1991) developed this concept further in connection with frequency domain approaches. This problem was also studied by Isaksson (1993). The determination of the threshold in the time-domain is studied by Seliger and Frank (1993) and has also discussed by Frank (1993) and Ding and Guo (1998). Faitakis, Thapliyal and Kantor (1994) discussed the computation of thresholds using vector and matrix norm operations, whilst Chang et al. (1995) derived a method to determine an adequate threshold based on robust control theory.

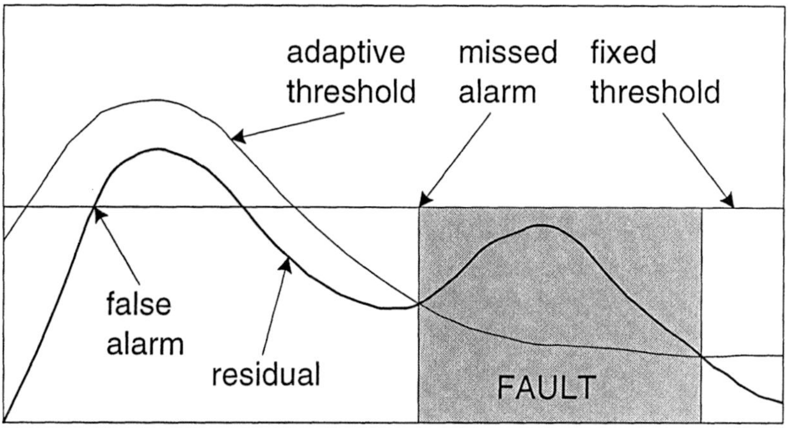

Figure 2.20. Application of an adaptive threshold

All the recent studies have shown that the adaptive threshold can be obtained in a systematic way and it presents a new and innovative tool for analysis and synthesis of FDI systems.

As a simple example for determining adaptive thresholds, assuming that the disturbance de-coupling condition for the uncertainty arising from disturbances is achieved (see Eq.(2.74)), and the residual uncertainty is only caused by modeling errors, i.e. the fault-free residual is:

$$r(s) = H_y(s)\Delta G_u(s)u(s) \tag{2.77}$$

Assuming that the modeling errors are bounded by a limiting value δ, i.e.

$$||\Delta G_u(j\omega)|| \leq \delta \tag{2.78}$$

In this situation, the frequency response of the fault-free residual will be bounded as:

$$\begin{aligned} ||r(j\omega)|| &= ||H_y(j\omega)\Delta G_u(j\omega)u(j\omega)|| \\ &\leq ||H_y(j\omega)u(j\omega)||||\Delta G_u(j\omega)|| \leq \delta||H_y(j\omega)u(j\omega)|| \end{aligned} \tag{2.79}$$

Therefore, an adaptive threshold $T(t)$ can be generated by a linear system as follows:

$$T(s) = \delta H_y(s)u(s) \tag{2.80}$$

It is readily seen that the threshold $T(t)$ is no longer fixed but depends on the input $u(t)$, thus being adaptive to the system operation. A fault is declared if $||r(t)|| > ||T(t)||$. A robust FDI scheme with the threshold adaptor or selector is shown in Fig.2.21.

As discussed above, the use of adaptive thresholds is a passive approach to robust FDI. By this we mean that no effort is made to design a robust residual.

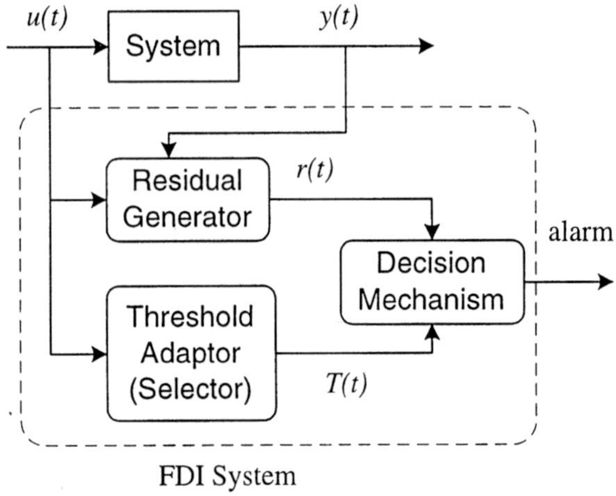

Figure 2.21. A FDI system with the threshold adaptor (selector)

The robust problem is tackled by reliable decision-making under the situation of uncertain residuals. A combination of active and passive approaches can offer potential for real robustness, especially when considering practical applications. It is believed that the success of an FDI scheme depends on the accuracy and choice of modeling of the monitored process. Hence, some attention in the field of robustness studies must be paid to ensure that sufficient modeling of the monitored process is achieved.

2.13 Applicability of Model-based FDI Methods

Many model-based FDI approaches have been developed so far. An engineer may find himself/herself in a dilemma when he/she wants to chose an approach to suit his problem. Some research attempts have been made in identifying the applicability of model-based FDI methods (Isermann, 1993; van Schrick, 1994b). The authors have also studied this problem and some general guidelines on the choice of FDI methods were published in Patton, Chen and Nielsen (1994). Some of the authors' opinions on the applicability of model-based FDI are presented in this Section. It must be stressed that the statements are only of a preliminary nature and there is no claim for their completeness.

A fault diagnosis technique should be able to complete the following important tasks:

- Detect and isolate faults in sensors, actuators and components.
- Detect and isolate incipient faults as well as abrupt faults.

In the design of fault diagnosis system, the following tasks and questions should be considered:

- How to handle noise in the system?
- How to handle multiple faults?
- How to handle disturbances (additive uncertainty)?
- How to handle modeling errors (multiplicative uncertainty)?
- How to handle nonlinearity?
- How to cope with detection delay?
- How to overcome complexity in the FDI algorithm design?
- How to minimize the complexity in FDI algorithm implementation (or execution)?
- What are the requirement for *a priori* modeling information?
- How good are self learning and adaptive capabilities?

The applicability of different model-based FDI approached are listed as follows.

2.13.1 Observer-based approaches

- The isolation task can be fulfilled via
  - ◇ a structured residual set designed by a dedicated or a generalized observer scheme.
  - ◇ a directional residual vector designed via a fault detection filter.
- Reaction to incipient faults is very fast.
- Very suitable for detecting and isolating faults in actuators and sensors.
- Possible of detecting and isolating faults in parameters, although complicated to achieve.
- Design procedure is systematic and simple.
- Easy to implement and execution algorithm is simple.
- Easy to handle multiple faults if the measurement number is sufficient.
- Handling noise in the system:
  - ◇ Statistical properties are unknown: An additional filter can be applied to the residual, based on assumptions on fault and noise frequency bands.
  - ◇ Statistical properties are known: A Kalman filter can be used to produce the fault-free residual with minimum variance and, consequently reducing false and missed alarms.
- Nonlinearity:

- ◇ The application of linear observers to a linearized model is simple but difficulties may be encountered for complex and highly non-linear systems.
- ◇ Non-linear observers are direct and accurate, however they are only applicable for particular classes of nonlinearities. The approach is not yet mature.

- Robustness: there are many mature techniques available, e.g.
  - ◇ Unknown input observer.
  - ◇ Eigenstructure assignment.
- Requirements for *a priori* modeling:
  - ◇ A reasonably accurate model is required.
- Adaptive and self-learning capability:
  - ◇ Adaptive observers can be employed for systems with unknown or time varying parameters.

The applicability of factorization methods is almost the same as the observer-based methods, however it can only be applied to *linear* or *linearized* models.

2.13.2 Parity relation approaches

As pointed out in Section 2.8.2, the parity relation approach is equivalent to the observer-based approach in certain conditions. Hence, most of their applicability conditions are the same. In the following, only the differences are listed.

- Fault isolation:
  - ◇ Structured residual set designed by orthogonal parity relations.
  - ◇ Directional residual is possible but difficult.
- Handling noise in the system:
  - ◇ An additional filter can be applied to a residual, based on assumptions on fault and noise frequency bands. It is not easy to incorporate noise statistics into the design.
- Nonlinearity:
  - ◇ Only linearized models can be used, simple but difficulties may be encountered for complex and highly non-linear systems.
- Robustness: there are many mature techniques available, e.g.
  - ◇ Orthogonal parity relations for additive uncertainty.

  - ◇ Optimally robust parity relations which can be used for systems with parameters within known error bounds.

- Adaptive and self-learning capability:
  - ◇ No available research on this aspect yet.

The observer-based and parity relation approaches can be designed not only in the time domain, but also in the frequency domain by factorizing the transfer matrix of the monitored system. The latter approach can make full use of the advantages in the frequency domain. The robust design can be achieved by enhancing fault responses and reducing noise and modeling uncertainty responses, based on the information on frequency distribution of faults, noise and modeling uncertainty. The residual response for a particular fault can also be shaped in the frequency domain according to performance requirements. The frequency domain design normally requires less accurate models than the time domain.

2.13.3 Parameter estimation approaches

- The isolation is normally achieved by analyzing the sensitivity matrix corresponding to the prediction errors with physical parameters. Fault isolation is not very easy because the physical parameters do not uniquely correspond to model parameters. The directional residual is not normally possible to design.

- Reaction for incipient faults is slow.

- The detection and isolation of faults in actuators and sensors are possible but complicated.

- The detection and isolation of faults in parameters are very straightforward.

- The design procedure is systematic but not simple.

- Implementation complexity:
  - ◇ Requires a large amount of computation.

- The detection and isolation of multiple faults is an not easy task unless a large number of sensors are installed.

- Noise is easy to handle in the parameter estimation procedure.

- Nonlinearity:
  - ◇ Possible to handle using identification techniques for non-linear systems.

- Robustness:
  - ◇ Dependent on the robustness properties of the estimation method.

- Requirements for *a priori* modeling:
  - ◇ Model structure, do not require model parameters.
- Adaptive and self-learning capability:
  - ◇ Excellent, if the parameter estimation method is adaptive.

2.13.4 Discussion on applicability

Some guide-lines about applicability of different methods have been given. However, the choice of FDI methods is still a complicated problem. The main factor to be considered is the availability of system information. Some information about the normal system operation (normal behavior) is necessary. This will serve as a reference base to be compared with. The information of the normal system behavior is usually expressed in terms of models, i.e. a model is necessary for FDI. The "black-box" assumption is not very suitable for advanced fault diagnosis and analysis and control reconfiguration. Some investigators argue that the observer-based methods require models, but parameter estimation methods do not require models. This is not really true because the principle of the parameter estimation approach is to compare estimated with known parameters of the system. Moreover, the modeling procedure is necessary to establish relationships between physical and model parameters. The system model can be in different formats, e.g., state space, parametric, frequency domain, qualitative model, etc. Hence, different methods require different model formats, and the first criterion in choosing model-based methods is the availability of the model type. As pointed by Gertler and Costin (1994), most of the time spent in developing a fault diagnosis scheme is spent in understanding the process to be diagnosed. It is hard to say whether a particular method is better than another method because one may be good in one aspect but bad in others. Hence, the second criterion in the choice of FDI method is dependent on the problem to be solved.

When we don't have *a priori* modeling information, what kind of model should we build for fault diagnosis purposes? This question is very difficult to answer, as the designer's experience and background play an important role. Sometimes, it depends just entirely on the designer's personal preference. If a particular criterion is needed, the more accurate and detailed the model, the better will be the fault diagnosis performance. If possible, a detailed state space model derived from physical laws is the best choice. However, accurate modeling would involve a large amount of work and, sometimes this is impossible. A cost effective way is to identify a parametric model using identification techniques, based on input and output data of the system under normal condition. However, fault diagnosis performance could be degraded if the identified model is not very accurate. Moreover, in-depth analysis of fault location and cause is not very easy if input-output models are used. If the quantitative (analytical) model is very difficult to obtain or, if uncertain factors are dominant in the system, one can consider building qualitative (heuristic) models which only require crude description. Some human knowledge about the system can also be expressed in heuristic format and included in the qualitative model. This would lead to the use of qualitative model-based approaches or even the use of a knowledge base.

When using a real life application of a FDI scheme, whose feasibility has been demonstrated (including the use of a laboratory demonstrator), many practical and unforeseen difficulties present themselves. To overcome these difficulties, one must understand the detailed design of the fault diagnosis scheme, as well as the nature of the practical problems. This usually requires the fault diagnosis designer to follow his work into the specific engineering field, either doing the implementation himself or working closely with those who do it. For this reason we should also include the application domain as far as possible into our research in this field.

There have been a significant number of application studies of fault diagnosis techniques, including some actual application to either process plant or laboratory experiments using real-time equipment. The book by Patton et al. (1989) provides some useful pre-1989 application examples. A selection of application examples and diagnosis methods used between 1991 to 1995 are analyzed in Isermann and Ballé (1997) More recent application examples can be found in the recent conference proceeding (Patton and Chen, 1998). However, there is still a great need for academia and industry to work together very closely to put fault diagnosis into the more useful setting of real application.

2.14 Integration of Fault Diagnosis Techniques

Many FDI methods have been developed and they show different properties with regard to the diagnosis of different faults in a process. To facilitate reliable FDI, taking advantages of different methods, a proper integration of several methods is a good solution (Isermann, 1994). Furthermore, a comprehensive fault diagnosis require a knowledge based treatment of all available, analytical and heuristic information. This can be performed by an integrated approach to knowledge-based fault diagnosis.

2.14.1 Fuzzy logic in fault diagnosis

The second stage of model-based FDI, decision making, is a logic decision process that transforms quantitative knowledge (residual signals) into qualitative statements (faulty or normal, etc). Therefore, the problem of decision-making can be treated in a novel way with the aid of fuzzy logic. To outline briefly the basic idea let us again consider the case that the residual due to faults is also contaminated by noise and the effect of uncertainty due to incomplete de-coupling, so that the residual will be non-zero even in the absence of faults. Typically, these effects will be time varying, i.e., the residual will fluctuate depending on the unknown time functions of the disturbances, noise and inputs of the process. This is a common situation, and hence fuzzy logic seems to be a natural tool to handle the decision making in a complicated and uncertain situation; based on incomplete information. The appealing feature of fuzzy logic is that it constitutes a powerful tool for modeling vague and imprecise facts and is therefore highly suited for applications where complete information about the system is not available to the designer.

Much effort has been spent on trying to decrease the uncertainty associated with quantitative residual generation. However, it is impossible to fully eliminate the effect of uncertainty. Based upon this limitation, the problem encountered in residual evaluation is to make the correct decisions on the basis of uncertain information. Non-Boolean reasoning (e.g. fuzzy logic) can be a suitable tool for this task. Contrary to classical logic which only allows a definite classification of fixed values, the fuzzy logic offers a form for the description of tolerances, i.e. fuzzy values, heuristic rules and their combination. There are, for instance, a lot of processes and experiences which can be grasped by humans heuristically, but which cannot be described analytically. The question of how this expert knowledge can be put into the form of a rule-based knowledge format has been answered partly through the use of fuzzy logic. Fuzzy logic endows machine intelligence with such human traits as the ability to make decisions based on shades of grey, instead of black-and-white information. Fuzzy processing can be divided into essentially the following stages. In the first, the residuals are compared with membership functions (or degree-of-belief curves) which are often assumed to be of triangular shape. In the second stage the lower of the two antecedent outputs is selected. Then the output of all rules is combined. Finally, the center of gravity (or another averaging methods) is used to defuzzify the output and lead to the possibility of definite decision-making. The introduction of fuzzy logic can improve the decision-making, and in turn will provide reliable and sufficient FDI which are applicable for real industrial systems. However, difficulty arises in the training of the algorithm in the inference mechanism.

Frank and his co-workers (Frank and Kiupel, 1993; Frank, 1993; Frank, 1994a; Frank, 1996; Schneider and Frank, 1996; Frank and Köppen-Seliger, 1997a) use fuzzy logic for residual evaluation. The aim is to release weighted alarms instead of yes-no decision. Such information can, if necessary, be given to a human operator to make the final yes-no decision or even train a specialist to perform the task. A similar approach was proposed by Ulieru and Isermann (Ulieru, 1993; Ulieru and Isermann, 1993), where analytical fault detection was integrated with fuzzy diagnostic decision-making. The approach solves the problem at two levels: first analytical redundancy is used to generate symptoms and then fault detection and isolation is achieved using heuristic techniques based on fuzzy logic. It should also be pointed out that fuzzy logic its own has been used extensively in fault diagnosis (Dexter, 1995; Leonhardt and Ayoubi, 1997; Dexter and Benouarets, 1997; Isermann, 1998)

2.14.2 Qualitative fault diagnosis

It may often be difficult and time consuming to develop a good mathematical model, there have been many attempts to use cruder descriptions (Lunze,

1994; Frank, 1996). Fault diagnosis of dynamic systems can also be based upon declarative knowledge of the system which is available in qualitative (rather than quantitative) form, e.g. Dvorak (1992), Leitch, Kraft and Luntz (1991), Leitch (1993), Leitch and Quek (1992), Shen and Leitch (1993), Lunze (1991), Lunze and Schiller (1992), Zhang (1991), Fathi, Ramirez and Korbicz (1993), Chang, Yu and Liou (1994), Howell (1994), Howell (1994), Leyval, Montmain and Gentil (1994), Betta, Dapuzzo and Pietrosanto (1995), Zhuang and Frank (1997). The qualitative approach is based upon the concept of a qualitative model which unlike the quantitative counterpart only requires declarative (heuristic) information e.g. the sign of variables, the tendencies of variables (increasing, decreasing or constant), order and/or relative magnitude, and hence can be robust with respect to uncertainty in a well defined sense. Clearly, this can be a significant advantage and qualitative methods can serve to confirm hypotheses already tested using the quantitative methods. The qualitative approach to fault diagnosis is motivated by the following circumstances encountered in practical applications:

- Faults cannot be reasonably described by analytical models, e.g. a valve is blocked or a pipe is broken.

- The on-line information available is not given by quantitative measurements of the system output but by qualitative assessments of the current operating conditions. For example, the information "the water level is high" cannot be unambiguously transformed into quantitative measurement data. Likewise, alarm messages are qualitative in nature because they do not provide precise state information. No analytical model can be used to process this kind of on-line information.

- If the system structure or parameters are not precisely known and diagnosis has to be based primarily on heuristic information, no quantitative model can be set up.

In these cases a qualitative approach to fault diagnosis is necessary. There have been several approaches in qualitative fault diagnosis, e.g. fault tree diagnosis and association-based diagnosis. The fault tree approach uses the evolution of the fault through the dynamic system which is described by a fault tree, event trees or causal networks. The association-based approach uses the relations among faults and the faulty system observations which are described by rules. The current attention is mainly focused on the qualitative model-based approach which uses the qualitative model derived directly from the physical laws of the system under consideration.

One of the disadvantages of the qualitative approach to fault diagnosis is the possibility of ambiguity which can arise when manipulating two or more declarative variables, for example the sum of a positive variable and a negative variable can either be negative or positive! This is clearly a situation to avoid when using these methods. Another disadvantage is that qualitative methods are relatively crude and usually cannot, on their own, be used to detect soft faults as the diagnosis is symptom-based.

Zhuang and Frank (1997) developed a qualitative observer FDI approach which is closely resemble to quantitative observer based approach. In their approach, Zhuang and Frank (1997) proposed a qualitative observer, which is based on qualitative simulation and observation filtering. This provides a framework in which the measured information can be utilized to reduce accumulated ambiguity and to avoid spurious solutions.

Quantitative and qualitative approaches have a lot of complementary features and can be suitably combined together to capitalize on their advantages by increasing the robustness of quantitative methods (Handelman and Stengel, 1989). This combination can also minimize the disadvantages of the two approaches; in particular it is important that ambiguity arising in qualitative reasoning is reduced or eliminated. Hence, one of the main aims of future research on model-based fault diagnosis is to find the way to combine these two approaches together to provide highly reliable diagnostic information.

2.14.3 Integrated fault diagnosis systems

Quantitative model-based fault diagnosis generates symptoms based on the analytical knowledge of the process. In most cases this is, however, not enough information to perform efficient FDI, i.e. to indicate the location, and the size of the fault. In such cases, fault diagnosis requires the use of a knowledge-based treatment (Milne, 1987; Isermann and Freyermuth, 1991a; Isermann and Freyermuth, 1991b; Tzafestas, 1989; Tzafestas and Watanabe, 1990). The intention is to transfer the existing knowledge of engineers, operators and maintenance staff into the supervision methodology and to develop on-line integrated expert systems for fault diagnosis (Frank, 1990; Frank, 1996; Isermann, 1991a; Isermann, 1994; Isermann, 1997; Ballé et al., 1998).

Fig.2.22 shows a typical integrated fault diagnosis system. Both analytical and heuristic knowledge are used in the system. Analytical knowledge includes: a quantitative model, normal process behavior, process history and fault statistics (if quantifiable), state estimation, parameter estimation, parity relations, etc. Heuristic knowledge (available from physical law and experience) includes: fault tree (connection of symptoms and faults), process history and fault statistics (if only qualitatively known), etc. The knowledge will be processed in terms of residual generation and feature extraction. The processed knowledge is then given to an inference mechanism which comprises residual evaluation, symptom observation and pattern recognition. For the last part of the problem solving, a certain amount of human expertise and judgement, expressed in rules and facts can be used. This can be formulated, for example, by different levels of diagnostic reasoning and different kinds of models.

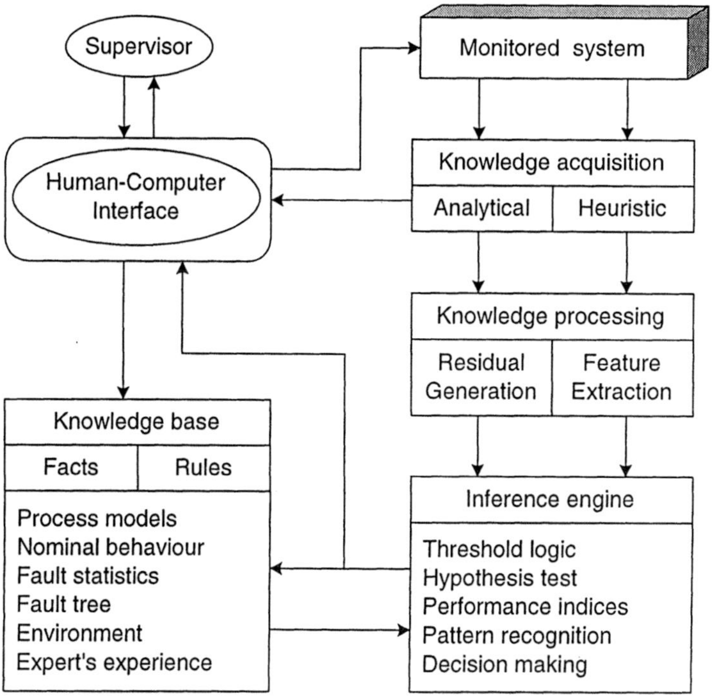

Figure 2.22. An integrated knowledge-based system for fault diagnosis

2.15 Summary

This chapter has presented a tutorial treatment on the basic principles of model-based FDI. The FDI problem has been formalized in a uniform framework by presenting mathematical descriptions and definitions. Within this framework, the residual generation has been identified as a central issue in model-based FDI. By analyzing a properly designed residual signal, FDI tasks can be performed. The residual generator has been summarized in a generalized structure which can cover all residual generation methods. The concept of fault detectability to guarantee reliable fault detection has been defined in this chapter. The ways of designing residuals for isolation have also been discussed. The most commonly used residual generation methods have been introduced in a tutorial setting and the applicability of model-based FDI techniques have been discussed. The success of fault diagnosis depends on the quality of the residuals. A prerequisite of residuals for successful diagnosis is the robustness with respect to modeling uncertainty. The robust FDI problem has been discussed in this chapter and a foundation has been laid down for further studies in the following chapters of the book. Other FDI methods such as fuzzy logic and qualitative modeling have been discussed briefly and some perspectives in forming an integrated knowledge-based fault diagnosis, utilizing all available analytical and heuristic information have been discussed.

3 ROBUST RESIDUAL GENERATION USING UNKNOWN INPUT OBSERVERS

3.1 Introduction

The generation of robust residuals is the most important task in model-based fault diagnosis techniques. As pointed out in Section 2.11, the disturbance de-coupling based approaches are the dominant approaches for robust residual generation. For those approaches, uncertain factors in system modeling are considered to act via an unknown input (or disturbance) on a linear system model. Although the unknown input vector is unknown, its distribution matrix is assumed known. Based on the information given by the distribution matrix, the unknown input (disturbance) can be de-coupled from the residual. Robust FDI is thus achievable using the disturbance de-coupled residual. This chapter focuses on the robust residual generation problem via unknown input observers. The principle of the unknown input observer (UIO) is to make the state estimation error de-coupled from the unknown inputs (disturbances). In this way, the residual can also be de-coupled from each disturbance, as the residual is defined as a weighted output estimation error. This approach was originally proposed by Watanabe and Himmelblau (1982) who considered the robust sensor fault detection and isolation problem for the system with modeling uncertainty. Later, Wünnenberg & Frank (Wünnenberg and Frank, 1987; Frank and Wünnenberg, 1989; Wünnenberg, 1990) generalized this approach for detecting and isolating both sensor and actuator faults by considering the case when unknown inputs also appear in the output equation. In parallel with this development, a robust scheme for diagnosing actuator faults via UIOs is proposed by Chen and Zhang (1991). A very important contribution of the paper by Chen and Zhang (1991) was to demonstrate the robust FDI approach via to a realistic chemical process system example. Note that Viswanadham and Srichander (1987) and Phatak and Viswanadham (1988) also studied the actuator fault detection and isolation problem via UIOs, however they failed to consider robustness issues. Many other investigators have considered the use of UIOs for robust FDI: e.g. Hou and Müller (1991), Hou and Müller (1994b), Frank and Seliger (1991), Seliger and Frank (1991a), Keller, Nowakowski and Darouach (1992), Chang and Hsu (1993a), Ragot, Maquin and Kratz (1993), Saif and Guan (1993), Wang and Daley (1993), Chen and Patton (1994b), Shields (1994), Yu, Shields and Mahtani (1994b), Yu and Shields (1996) and Hwang, Chang and Hsu (1997).

The first step to generate robust (in the sense of disturbance de-coupling) residuals is to design a UIO. The problem of UIO design dates back to 1975 (Wang, Davison and Dorato, 1975). Darouach, Zasadzinski and Xu (1994) and Hou and Müller (1994a) reported the recent developments, and different methods for designing UIOs are discussed in Section 3.2. This chapter introduces a new full-order UIO structure which is based on the paper by Chen, Patton and Zhang (1996). A rigorous mathematical foundation in designing a full-order UIO are laid down and the necessary and sufficient existence conditions are presented and thoroughly proved. When compared with other UIO design methods, the existence conditions are very easy to verify and the design procedure is simple. This avoids some of the unnecessary and complex computation that is otherwise required for UIO design. An example of a typical complexity is the Kronecker canonical form transformation method (Frank and Wünnenberg, 1989) which can also suffer from numerical conditioning problems.

Robust FDI schemes based on UIOs are discussed further in Section 3.3 where an application example of isolating actuator faults in a non-linear process is presented. Unlike some other work in which the reduced order structure has been used, this chapter is based exclusively on the use of the full-order UIO. The unknown input de-coupling conditions for a full-order UIO are not very different from those of the reduced order counterpart. However, for a full-order UIO, there is more design freedom available to achieve other required performances, after the disturbance de-coupling conditions have been satisfied. This is easy to understand, since the number of free parameters will increase if the observer order is increased. The exploitation of the remaining design freedom for designing *directional* residuals has been studied by Chen, Patton and Zhang (1996) and this is discussed in Section 3.4. The generation of the *minimal variance* state estimation using freedom left after unknown input de-coupling was investigated by Chen and Patton (1996) and this is described in Section 3.5.

As pointed in Section 2.7, one of the approaches for fault isolation is to design a directional residual vector, i.e. to make the residual vector lie in a fixed and fault-specific direction in the residual space in response to each fault. With directional residual vectors, the fault isolation problem is one of determining which of the known fault signature directions the residual vector lies the closest to. The most effective way to generate directional residual vectors is the use of the Beard fault detection filters (BFDF) (Beard, 1971; Jones, 1973; Massoumnia, 1986b; White and Speyer, 1987; Park and Rizzoni, 1993; Park and Rizzoni, 1994b; Douglas and Speyer, 1996; Liu and Si, 1997; Chung and Speyer, 1998). It should be pointed out that this class of observers has been known as the "failure detection filter" (Beard, 1971; Jones, 1973; White and Speyer, 1987) in the early development of fault (failure) diagnosis. Fault detection filters are a special class of full-order Luenberger observers with a specially designed feedback gain matrix, which can make the output estimation error (residual vector) have uni-directional characteristics associated with some known fault directions. This is the main and most appealing feature of fault detection filters. However, the robustness issues have not been considered in the context of BFDFs until recently. Hence, this approach does not account for the effects of disturbances, non-linearities, modeling errors, parameter variations and other uncertain factors in the system. There would be false or missed alarms when this approach is directly applied to industrial systems, in which the uncertain factors are unavoidable in modeling (specially for systems such as mechanical, eletromechanical, thermofluid and aircraft systems). The application of BFDFs has been obstructed by the lack of robustness. Section 3.4 describes a strategy (Chen, Patton and Zhang, 1996) for the design of robust fault detection filters which ensures that the residuals have both disturbance de-coupling and uni-directional properties. This is done by combing the UIO and the BFDF principles. By the use of the UIO principle, the residual has been made robust against unknown inputs (disturbances). The uni-directional property is achieved based on BFDF techniques using the design freedom available after the disturbance de-coupling conditions have been satisfied. A filter which can produce disturbance de-coupled and directional residuals is called a "robust (disturbance de-coupled) fault detection filter". The robust fault detection filter developed in this section is also demonstrated via a realistic example.

Section 3.5 considers the optimal filtering and robust fault diagnosis problems for stochastic systems with unknown disturbances. The optimal observer proposed by Chen and Patton (1996), which can produce disturbance de-coupled state estimation with minimum variance for time-varying systems with both noise and unknown disturbances, is introduced. The output estimation error with disturbance de-coupling and minimum variance properties is used as a residual signal. A statistical testing procedure is then applied to examine the residual and hence to diagnose faults. The method described is applied to an illustrative example and simulation results show that the optimal observer can give good state estimation; the fault detection approach taken is able to detect faults reliably in the presence of both modeling errors and noise. The scope of applications of the optimal observer described in Section 3.5 extends to a wide range of stochastic uncertain systems and is not confined to the fault diagnosis problem domain.

The primary requirement for a UIO or other disturbance de-coupling based robust residual generation approaches is that the unknown input distribution matrix must be known *a priori*, although the actual unknown input itself does not need to be known. If the uncertainty is caused by the disturbance, this requirement is easy to meet and hence the robustness in FDI with respect to unknown disturbances can be easily solved. However, the disturbance de-coupling approach cannot be directly applied to systems for which the uncertainty is caused by modeling errors, linearization errors, parameter variations etc. This is because the distribution matrix for such uncertain factors is normally unknown. This problem has obstructed the application of these robust FDI approaches in real industrial systems. To solve this problem, some investigators led by Patton & Chen (Patton and Chen, 1991f; Patton and Chen, 1991b; Patton, Chen and Zhang, 1992; Patton, Zhang and Chen, 1992; Gertler and Kunwer, 1993; Gertler, 1994; Chen, 1995; Edelmayer, Bokor, Szigeti and Keviczky, 1997; Patton and Chen, 1997) have suggested an approach in which modeling errors and other uncertain factors are represented approximately as unknown disturbances, with an *estimated* distribution matrix. In this way, an optimally robust solution is achievable. This approximate strategy has extended the application domain of disturbance de-coupling based robust residual generation approaches. All three application examples presented in this chapter illustrate how different kinds of uncertain factors can be represented approximately as unknown input terms. These uncertain factors are, for example, the non-linear terms in the dynamic equation of a non-linear process (Section 3.3), the linearization error in a system as complex as a jet engine (Section 3.4) and parameter variations in a flight control system (Section 3.5). The simulation results in all three examples show the power of these proposed methods. The problem of representing modeling errors as an unknown input term is examined in more detail in Chapter 5.

3.2 Theory and Design of Unknown Input Observers

This section deals with the observer design for a class of systems, in which the system uncertainty can be summarized as an *additive* unknown disturbance term in the dynamic equation described as follows:

$$\begin{cases} \dot{x}(t) &= Ax(t) + Bu(t) + Ed(t) \\ y(t) &= Cx(t) \end{cases} \tag{3.1}$$

where $x(t) \in \mathbb{R}^n$ is the state vector, $y(t) \in \mathbb{R}^m$ is the output vector, $u(t) \in \mathbb{R}^r$ is the known input vector and $d(t) \in \mathbb{R}^q$ is the unknown input (or disturbance) vector. A, B, C and E are known matrices with appropriate dimensions.

Remarks:

(a) There is no loss of generality in assuming that the unknown input distribution matrix E should be full column rank. When this is not the case, the following rank decomposition can be applied to the matrix E (see Appendix C):

$$Ed(t) = E_1 E_2 d(t)$$

where E_1 is a full column rank matrix and $E_2 d(t)$ can now be considered as a new unknown input.

(b) The term $Ed(t)$ can be used to describe an additive disturbance as well as a number of other different kinds of modeling uncertainties. Examples are: noise, interconnecting terms in large scale systems, non-linear terms in system dynamics, terms arise from time-varying system dynamics, linearization and model reduction errors, parameter variations. Some examples of this problem are presented in the following sections of this chapter and a detailed study can be found in Chapter 5.

(c) The disturbance term may also appear in the output equation, i.e.,

$$y(t) = Cx(t) + E_y d(t)$$

This case is not considered here because the disturbance term $E_y d(t)$ in the output equation can be nulled by simply using a transformation of the output signal $y(t)$, i.e.

$$y_E(t) = T_y y(t) = T_y Cx(t) + T_y E_y d(t) = T_y Cx(t)$$

where $T_y E_y = 0$, if one replaces $y(t)$ and C with $y_E(t)$ and $T_y C$, the problem will be equivalent to one without output disturbances.

(d) For some systems, there is a term relating the control input $u(t)$ in the system output equation, i.e.

$$y(t) = Cx(t) + Du(t)$$

As the control input $u(t)$ is known, a new output can be constructed as:

$$\overline{y}(t) = y(t) - Du(t) = Cx(t)$$

If the output $y(t)$ is replaced by $\overline{y}(t)$, the problem will be equivalent to the one without the term $Du(t)$. For brevity, the term $Du(t)$ is omitted in this chapter as this does not affect the generality of the discussion on the observer design.

Definition 3.1 (Unknown Input Observer (UIO)) *An observer is defined as an unknown input observer for the system described by Eq.(3.1), if its state estimation error vector $e(t)$ approaches zero asymptotically, regardless of the presence of the unknown input (disturbance) in the system.*

The problem of designing an observer for a linear system with *both* known and unknown inputs has been studied for over two decades (Wang et al., 1975). The problem is of considerable importance as, in practice there are many situations where disturbances are present. Alternatively, some of the system inputs are inaccessible (or unmeasurable), and therefore a conventional observer which uses all input signals cannot be used. It is more useful to assume no *a priori* knowledge about unknown inputs. Wang et al. (1975) proposed a minimal-order UIO for the system (3.1). The existence conditions for such an $(n-m)th$-order observer were shown by Kudva, Viswanadham and Ramakrishna (1980). After the work of Wang et al. (1975), many approaches for designing unknown input observers have been proposed, for example, the geometric method by Bhattacharyya (1978), the inversion algorithm by Kobayashi and Nakamizo (1982), the matrix algebra method by Watanabe and Himmelblau (1982), the generalized matrix inversion approach by Miller and Mukundan (1982), and the singular value decomposition technique by Fairman, Mahil and Luk (1984). Park and Stein (1988) studied the simultaneous estimation problem for both states and unknown inputs. The problem of designing reduced order UIOs has been revisited by Hou and Müller (1992) and Guan and Saif (1991) using algebraic approaches. In their studies, the state vector is divided into two parts, via a linear transformation onto the state equation, a part can be directly obtained from the measurements, and another part has to be estimated using a reduced order disturbance de-coupled observer. More recently, Hou and Müller (1994a) presented a unified viewpoint in designing UIOs.

Unlike all the above mentioned work in which the reduced order observer structure has been used, Kurek (1982) proposed a full-order unknown input observer structure. Yang and Richard (1988) gave a direct design procedure for full-order UIOs and have showed, through an example, that the reduced-order observer may restrict the convergence rate in estimation. However, the design procedure they presented is very complicated and involves some trial-and-error exercises. Furthermore, the existence conditions are not very easy to verify. This full-order UIO structure is re-examined by Darouach et al. (1994). It has been shown that the minimal order of a UIO is $(n-m)$, any order between $(n-m)$ to n is possible for a UIO to be exist. The disturbance de-coupling conditions for a full-order UIO are not very different from those of a reduced-order UIO. That is to say, there are no significant differences between two UIO structures, as far as unknown input (disturbance) de-coupling is concerned. However, there is more design freedom available for a full-order UIO to achieve other required performances such as the rate of convergence and minimal variance. This is easy to understand since the number of free parameters will increase if the observer order is increased.

Based on the scheme proposed by Chen, Patton and Zhang (1996), a full-order UIO structure is used in this study. This is because extra design freedom is required for generating directional residuals in fault isolation. A rigorous mathematical foundation in designing full-order UIOs is presented. The necessary and sufficient conditions for this observer to exist are given and thoroughly proved in this chapter. These conditions are easy to verify and the design procedure is systematic and easy to implement. Moreover, this chapter shows that the remaining freedom can be used to make the residual have directional properties (or make the state estimation error have minimal variance), after unknown input (or disturbance) de-coupling has been achieved.

3.2.1 Theory of UIOs

The structure for a full-order observer is described as:

$$\begin{cases} \dot{z}(t) &= Fz(t) + TBu(t) + Ky(t) \\ \hat{x}(t) &= z(t) + Hy(t) \end{cases} \tag{3.2}$$

where $\hat{x} \in \mathbb{R}^n$ is the estimated state vector and $z \in \mathbb{R}^n$ is the state of this full-order observer, and F, T, K, H are matrices to be designed for achieving unknown input de-coupling and other design requirements. The observer described by Eq.(3.2) is illustrated in Fig.3.1.

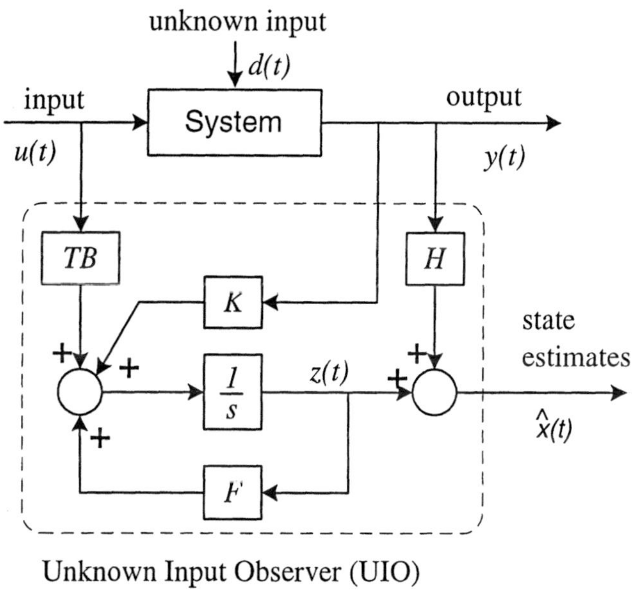

Figure 3.1. The structure of a full-order unknown input observer

When the observer (3.2) is applied to the system (3.1), the estimation error $(e(t) = x(t) - \hat{x}(t))$ is governed by the equation:

$$\begin{aligned} \dot{e}(t) &= (A - HCA - K_1C)e(t) + [F - (A - HCA - K_1C)]z(t) \\ &+ [K_2 - (A - HCA - K_1C)H]y(t) \\ &+ [T - (I - HC)]Bu(t) + (HC - I)Ed(t) \end{aligned} \tag{3.3}$$

where

$$K = K_1 + K_2 \tag{3.4}$$

If one can make the following relations hold true:

$$(HC - I)E = 0 \quad (3.5)$$

$$T = I - HC \quad (3.6)$$

$$F = A - HCA - K_1C \quad (3.7)$$

$$K_2 = FH \quad (3.8)$$

The state estimation error will then be:

$$\dot{e}(t) = Fe(t) \quad (3.9)$$

If all eigenvalues of F are stable, $e(t)$ will approach zero asymptotically, i.e. $\hat{x} \longrightarrow x$. This means that the observer (3.2) is an unknown input observer for the system (3.1) according to Definition 3.1. The design of this UIO is to solve Eqs.(3.4) – (3.8) and making all eigenvalues of the system matrix F be stable. Before we give the necessary and sufficient conditions for the existence of a UIO, two Lemmas are introduced.

Lemma 3.1 *Eq.(3.5) is solvable* iff*:*

$$rank(CE) = rank(E) \quad (3.10)$$

and a special solution is:

$$H^* = E[(CE)^T CE]^{-1}(CE)^T \quad (3.11)$$

Proof: Necessity: When Eq.(3.5) has a solution H, one has $HCE = E$ or

$$(CE)^T H^T = E^T$$

i.e., E^T belongs to the range space of the matrix $(CE)^T$ and this leads to:

$$rank(E^T) \leq rank((CE)^T)$$

i.e.

$$rank(E) \leq rank(CE)$$

However,

$$rank(CE) \leq \min\{rank(C), rank(E)\} \leq rank(E)$$

Hence, $rank(CE) = rank(E)$ and the necessary condition is proved.

Sufficiency: When $rank(CE) = rank(E)$ holds true, CE is a full column rank matrix (because E is assumed to be full column rank), and a left inverse of CE exists:

$$(CE)^+ = [(CE)^T CE]^{-1}(CE)^T$$

Clearly, $H = E(CE)^+$ is a solution to Eq.(3.5).

◇ **QED**

Lemma 3.2 : *Let:*

$$C_1 = \begin{bmatrix} C \\ CA \end{bmatrix}$$

then the detectability for the pair (C_1, A) *is equivalent to that for the pair* (C, A).

Proof: If $s_1 \in \mathbb{C}$ is an unobservable mode of the pair (C_1, A), we have:

$$rank\{\begin{bmatrix} s_1 I - A \\ C_1 \end{bmatrix}\} = rank\{\begin{bmatrix} s_1 I - A \\ C \\ CA \end{bmatrix}\} < n$$

This means that a vector $\alpha \in \mathbb{C}^n$ will exist such that:

$$\begin{bmatrix} s_1 I - A \\ C \\ CA \end{bmatrix} \alpha = 0$$

This leads to:

$$\begin{bmatrix} s_1 I - A \\ C \end{bmatrix} \alpha = 0 \qquad \text{or} \qquad rank\{\begin{bmatrix} s_1 I - A \\ C \end{bmatrix}\} < n$$

That is to say that s_1 is also an *unobservable mode* of the pair (C, A).

If $s_2 \in \mathbb{C}$ is an unobservable mode of the pair (C, A), we have:

$$rank\{\begin{bmatrix} s_2 I - A \\ C \end{bmatrix}\} < n$$

This means that a vector $\beta \in \mathbb{C}^n$ can always be found, such that:

$$\begin{bmatrix} s_2 I - A \\ C \end{bmatrix} \beta = 0$$

This leads to:

$$(s_2 I - A)\beta = 0 \qquad C\beta = 0$$
$$CA\beta = Cs_2\beta = s_2 C\beta = 0$$

Hence:

$$\begin{bmatrix} s_2 I - A \\ C \\ CA \end{bmatrix} \beta = \begin{bmatrix} s_2 I - A \\ C_1 \end{bmatrix} \beta = 0$$

i.e., s_2 is also an *unobservable mode* of the pair (C_1, A).

As the pairs (C_1, A) and (C, A) have the same unobservable modes, their detectability is formally equivalent.

◊ **QED.**

An alternative way to prove the Lemma 3.2 can be found in Appendix D. Note that the detectability (Chen, 1984) is a weaker condition than observability. A pair (C, A) is detectable when all unobservable modes for this pair are stable.

Theorem 3.1 *Necessary and sufficient conditions for (3.2) to be a UIO for the system defined by (3.1) are:*

(i) $rank(CE) = rank(E)$

(ii) (C, A_1) *is detectable pair, where*

$$A_1 = A - E[(CE)^T CE]^{-1}(CE)^T CA \tag{3.12}$$

Proof: Sufficiency: According to Lemma 3.1, the Eq. (3.5) is solvable when condition (i) holds true. A special solution for H is $H^* = E[(CE)^T CE]^{-1}(CE)^T$. In this case, the system dynamics matrix is:

$$F = A - HCA - K_1C = A_1 - K_1C$$

which can be stabilized by selecting the gain matrix K_1 due to the condition (ii). Finally, the remaining UIO matrices described in (3.2) can be calculated using Eqs.(3.4) – (3.8). Thus, the observer (3.2) is a UIO for the system (3.1).

Necessity: Since (3.2) is a UIO for (3.1), Eq. (3.5) is solvable. This leads to the fact that condition (i) hold true according to Lemma 3.1. The general solution of the matrix H for Eq.(3.5) can be calculated as:

$$H = E(CE)^+ + H_0[I_m - CE(CE)^+]$$

where $H_0 \in \mathbb{R}^{n \times m}$ is an arbitrary matrix and $(CE)^+$ is the left inverse of CE which is:

$$(CE)^+ = [(CE)^T CE]^{-1}(CE)^T$$

Substituting the solution for H into Eq.(3.7), the system dynamics matrix F is:

$$\begin{aligned} F &= A - HCA - K_1C \\ &= [I_n - E(CE)^+C]A - [K_1 \quad H_0]\begin{bmatrix} C \\ [I_m - CE(CE)^+]CA \end{bmatrix} \\ &= A_1 - [K_1 \quad H_0]\begin{bmatrix} C \\ CA_1 \end{bmatrix} \\ &= A_1 - \overline{K}_1\overline{C}_1 \end{aligned}$$

where

$$\overline{K}_1 = [K_1 \quad H_0] \quad \text{and} \quad \overline{C}_1 = \begin{bmatrix} C \\ CA_1 \end{bmatrix}$$

Since the matrix F is stable, the pair $(\overline{C}_1, A_1)$ is detectable, and the pair (C, A_1) also is detectable according to Lemma 3-2.

◊ **QED**

One should note that the number of independent row of the matrix C must not be less than the number of the independent columns of the matrix E to satisfy condition (i). That is to say, the maximum number of disturbances which can be de-coupled cannot be larger than the number of the independent measurements. It is very interesting to note that observer (3.2) will be a simple full-order Luenberger observer by setting $T = I$ and $H = 0$, when $E = 0$ (i.e. no unknown inputs in the system). In this situation, condition (i) in Theorem 3.1 is clearly hold true and condition (ii) is simply changed to that of (C, A) being detectable. This is a well known result in the design of a full-order Luenberger observer.

Condition (ii) can be verified in terms of the structural properties of the original system. In fact, this condition is equivalent to the condition that the transmission zeros from the unknown inputs to the measurements must be stable, i.e.

$$\begin{bmatrix} sI_n - A & E \\ C & 0 \end{bmatrix}$$

is of full column rank for all s with $Re(s) \geq 0$. This can be proved as follows:

It can be verified that:

$$\begin{bmatrix} I_n - E(CE)^+C & sE(CE)^+ \\ 0 & I_m \\ E(CE)^+C & -sE(CE)^+ \end{bmatrix} \begin{bmatrix} sI_n - A & E \\ C & 0 \end{bmatrix} = \begin{bmatrix} sI_n - A_1 & 0 \\ C & 0 \\ -E(CE)^+CA & E \end{bmatrix}$$

As the first matrix in the left side of the above equation is a full column rank matrix, we have:

$$\begin{aligned} rank \begin{bmatrix} sI_n - A & E \\ C & 0 \end{bmatrix} &= rank \begin{bmatrix} sI_n - A_1 & 0 \\ C & 0 \\ -E(CE)^+CA & E \end{bmatrix} \\ &= rank \begin{bmatrix} sI_n - A_1 \\ C \\ -E(CE)^+CA \end{bmatrix} + rank(E) \end{aligned}$$

We have assumed that E is a full column rank matrix. Hence, condition (ii) is equivalent to the case when the matrix of the left side of the above equation is full column rank for all s with $Re(s) \geq 0$. This is because the condition for pair (C, A_1) to be detectable is equivalent to the following matrix

$$\begin{bmatrix} sI - A_1 \\ C \end{bmatrix}$$

having full column rank for all s with $Re(s) \geq 0$.

From the above analysis, it can be seen that K_1 is a free matrix of parameters in the design of a UIO. After K_1 is determined, other parameter matrices in the UIO can be computed by Eqs.(3.4) – (3.8). The only restriction on the matrix K_1 is that it must stabilize the system dynamics matrix F. The matrix K_1 which stabilizes the matrix F is not unique due to the multivariable nature of the problem. That is to say there is still some design freedom left in the choice of K_1, after unknown input disturbance conditions have been satisfied. In the following sections, this freedom is exploited further to make the diagnostic residual has directional characteristics or minimum variance properties.

3.2.2 Design procedure for UIOs

One of the most important steps in designing a UIO is to stabilize $F = A_1 - K_1 C$ by choosing the matrix K_1, when the pair (C, A_1) is detectable. If (C, A_1) is observable, this can be achieved easily by using a pole placement routine which is widely available in any control system design packages such as Control System Toolbox for MATLAB. If (C, A_1) is not observable, an observable canonical decomposition procedure (Chen, 1984) should be applied to (C, A_1), which is:

$$PA_1P^{-1} = \begin{bmatrix} A_{11} & 0 \\ A_{12} & A_{22} \end{bmatrix} \qquad A_{11} \in \mathbb{R}^{n_1 \times n_1}$$

$$CP^{-1} = [C^* \quad 0] \qquad C^* \in \mathbb{R}^{m \times n_1}$$

where n_1 is the rank of the observability matrix for (C, A_1), and (C^*, A_{11}) is observable. The choice of the transformation matrix can be found in Appendix D and Chen (1984). If all eigenvalues of A_{22} are stable, (C, A_1) is detectable and the matrix F can be stabilized.

$$\begin{aligned} F &= A_1 - K_1 C = P^{-1}[PAP^{-1} - PK_1CP^{-1}]P \\ &= P^{-1}\left\{\begin{bmatrix} A_{11} & 0 \\ A_{12} & A_{22} \end{bmatrix} - \begin{bmatrix} K_p^1 \\ K_p^2 \end{bmatrix} [C^* \quad 0]\right\} P \\ &= P^{-1}\begin{bmatrix} A_{11} - K_p^1 C^* & 0 \\ A_{12} - K_p^2 C^* & A_{22} \end{bmatrix} P \end{aligned}$$

where:

$$K_p = PK_1 = \begin{bmatrix} K_p^1 \\ K_p^2 \end{bmatrix}$$

$$\{\text{Eigenvalues of } F\} = \{\text{Eigenvalues of } A_{22}\} \bigcup \{\text{Eigenvalues of } (A_{11} - K_p^1 C^*)\}$$

As (C^*, A_{11}) is observable, K_p^1 can be determined via the pole placement. The matrix K_p^2 can be any matrix, because it does not affect the eigenvalues of F. The design procedure of a UIO is thus given in Table 3.1.

Example: Consider the example used in (Wang et al., 1975; Miller and Mukundan, 1982; Yang and Richard, 1988; Hou and Müller, 1992) with the following parameter matrices:

$$A = \begin{bmatrix} -1 & 1 & 0 \\ -1 & 0 & 0 \\ 0 & -1 & -1 \end{bmatrix} \quad C = \begin{bmatrix} 1 & 0 & 0 \\ 0 & 0 & 1 \end{bmatrix} \quad E = \begin{bmatrix} -1 \\ 0 \\ 0 \end{bmatrix}$$

1°: It can easily be checked that $rank(CE) = rank(E) = 1$.

Table 3.1. Unknown input observer (UIO) design procedure

1° Check the rank condition for E and CE: If $rank(CE) \neq rank(E)$, a UIO does not exist, go to 10°.

2° Compute H, T and A_1:

$$H = E[(CE)^T CE]^{-1}(CE)^T \qquad T = I - HC \qquad A_1 = TA$$

3° Check the observability: If (C, A_1) observable, a UIO exists and K_1 can be computed using pole placement, go to 9°.

4° Construct a transformation matrix P for the observable canonical decomposition: To select independent $n_1 = rank(W_O)$ (W_O is the observability matrix of (C, A_1)) row vector $p_1^T, \cdots, p_{n_1}^T$ from W_O, together other $n - n_1$ row vector $p_{n_1+1}^T, \cdots, p_n^T$ to construct an non-singular matrix as:

$$P = [p_1, \cdots, p_{no}; p_{no+1}, \cdots, p_n]^T$$

5° Perform an observable canonical decomposition on (C, A_1)):

$$PA_1P^{-1} = \begin{bmatrix} A_{11} & 0 \\ A_{12} & A_{22} \end{bmatrix} \qquad CP^{-1} = [C^* \quad 0]$$

6° Check the detectability of (C, A_1): If any one of the eigenvalues of A_{22} is unstable, a UIO does not exist and go to 10°.

7° Select n_1 desirable eigenvalues and assign them to $A_{11} - K_p^1 C^*$ using pole placement.

8° Compute $K_1 = P^{-1}K_p = P^{-1}[(K_p^1)^T \quad (K_p^2)^T]^T$, where K_p^2 can be any $(n - n_1) \times m$ matrix.

9° Compute F and K: $F = A_1 - K_1C$, $K = K_1 + K_2 = K_1 + FH$.

10° STOP.

2°: The matrices H, T and A_1 are calculated as:

$$H = \begin{bmatrix} 1 & 0 \\ 0 & 0 \\ 0 & 0 \end{bmatrix} \quad T = \begin{bmatrix} 0 & 0 & 0 \\ 0 & 1 & 0 \\ 0 & 0 & 1 \end{bmatrix} \quad A_1 = \begin{bmatrix} 0 & 0 & 0 \\ -1 & 0 & 0 \\ 0 & -1 & -1 \end{bmatrix}$$

3°: The pair (C, A_1) is observable, a UIO exists, and the matrix K_1 can be determined via the pole placement procedure.

$$K_1 = \begin{bmatrix} 1 & 2 \\ -1 & -6 \\ 0 & 4 \end{bmatrix} \qquad \text{which assigns eigenvalues at: } \{-1, -2, -3\}$$

Note that the gain matrix K_1 is not unique for assigning the same set of eigenvalues.

9°: The matrices F and K are calculated as:

$$F = \begin{bmatrix} -1 & 0 & -2 \\ 0 & 0 & 6 \\ 0 & -1 & -5 \end{bmatrix} \qquad K = \begin{bmatrix} 0 & 2 \\ -1 & -6 \\ 0 & 4 \end{bmatrix}$$

Remarks: Due to the multivariable nature of the observer design problem, the choice of the gain matrix $K_1 \in \mathbb{R}^{3\times 2}$ is not unique. That is to say there is some design freedom left after the unknown input de-coupling conditions have been satisfied. This example was also studied by Hou and Müller (1992) in which a first order UIO was designed. The gain for their reduced order UIO is a scalar, and there is no design freedom left after the satisfaction of unknown input de-coupling and the assignment of the single eigenvalue. This demonstrates the advantage of the full-order UIOs in terms of design freedom.

3.3 Robust Fault Detection and Isolation Schemes based on UIOs

3.3.1 Robust fault detection schemes based on UIOs

The main task of robust fault detection is to generate a residual signal which is robust to the system uncertainty. To detect a particular fault, the residual has to be sensitive to this fault. The detailed discussion about fault detectability has been presented in Section 2.6. According to the study in Section 2.4, a system with possible sensor and actuator faults can be described as:

$$\left\{ \begin{array}{lcl} \dot{x}(t) & = & Ax(t) + Bu(t) + Ed(t) + Bf_a(t) \\ y(t) & = & Cx(t) + f_s(t) \end{array} \right. \tag{3.13}$$

where $f_a \in \mathbb{R}^r$ denotes the presence of actuator faults and $f_s \in \mathbb{R}^m$ denotes sensor faults. To generate a robust (in the sense of disturbance de-coupling) residual, a UIO described by Eq.(3.2) in Section 3.2 is required. When the state estimation is available, the residual can be generated as:

$$r(t) = y(t) - C\hat{x}(t) = (I - CH)y(t) - Cz(t) \tag{3.14}$$

When this UIO-based residual generator applied to the system described in Eq.(3.13), the residual and the state estimation error $(e(t))$ will be:

$$\left\{ \begin{array}{lcl} \dot{e}(t) & = & (A_1 - K_1C)e(t) + TBf_a(t) - K_1f_s(t) - H\dot{f}_s(t) \\ r(t) & = & Ce(t) + f_s(t) \end{array} \right. \tag{3.15}$$

From Eq.(3.15), it can be seen that the disturbance effects have been de-coupled from the residual. To detect actuator faults, one has to make:

$$TB \neq 0$$

More specifically, the fault in the i_{th} actuator will affect the residual *iff*:

$$Tb_i \neq 0$$

where b_i is the i_{th} column of the matrix B. Similarly, the residual has to be made sensitive to $f_s(t)$ if sensor faults are to be detected. This condition is normally satisfied, as the sensor fault vector $f_s(t)$ has a direct effect on the residual $r(t)$. The robust residual can be used to detect faults according to a simple threshold logic:

$$\left\{ \begin{array}{llll} \|r(t)\| & < & Threshold & \text{for} \quad \text{fault-free case} \\ \|r(t)\| & \geq & Threshold & \text{for} \quad \text{faulty cases} \end{array} \right. \tag{3.16}$$

3.3.2 Robust fault isolation schemes based on UIOs

The fault isolation problem is to locate the fault, i.e., to determine in which sensor (or actuator) the fault has occurred. As pointed out in Section 2.7, one of the approaches to facilitate fault isolation is to design a *structured* residual set. The term "structured" here means that each residual is designed to be sensitive to a certain group of faults and insensitive to others. The sensitivity and insensitivity properties makes isolation possible. The ideal situation is to make each residual only sensitive to a particular fault and insensitive to all other faults. However, this ideal situation is normally difficult to achieve (Patton et al., 1989). Even when the ideal situation can be achieved, the design freedom will be used up and no freedom will be left for achieving robustness. This problem was encountered by Wünnenberg (1990). To exploit the maximum design freedom for robustness, a commonly accepted scheme (Patton et al., 1989) in fault isolation is to make each residual to be sensitive to faults in all but one sensors (or actuators).

3.3.2.1 Robust sensor fault isolation schemes. To design robust sensor fault isolation schemes, all actuators are assumed to be fault-free and the system equations can be expressed as:

$$\left\{ \begin{array}{lll} \dot{x}(t) & = & Ax(t) + Bu(t) + Ed(t) \\ y^j(t) & = & C^j x(t) + f_s^j(t) \\ y_j(t) & = & c_j x(t) + f_{sj}(t) \end{array} \right. \quad \text{for} \quad j = 1, 2, \cdots, m \tag{3.17}$$

where $c_j \in R^{1\times n}$ is the j_{th} row of the matrix C, $C^j \in \mathbb{R}^{(m-1)\times n}$ is obtained from the matrix C by deleting j_{th} row c_j, y_j is the j_{th} component of y and $y^j \in \mathbb{R}^{m-1}$ is obtained from the vector y by deleting j_{th} component y_j. Based on this description, m UIO-based residual generator can be constructed as:

$$\left\{ \begin{array}{lll} \dot{z}^j(t) & = & F^j z^j(t) + T^j Bu(t) + K^j y^j(t) \\ r^j(t) & = & (I - C^j H^j) y^j(t) - C^j z^j(t) \end{array} \right. \quad \text{for} \quad j = 1, 2, \cdots, m \tag{3.18}$$

where the parameter matrices must satisfy the following equations:

$$\left.\begin{array}{rcl} H^jC^jE & = & E \\ T^j & = & I - H^jC^j \\ F^j & = & T^jA - K_1^jC^j \quad \text{to be stabilized} \\ K_2^j & = & F^jH^j \\ K^j & = & K_1^j + K_2^j \end{array}\right\} \text{for} \quad j = 1, 2, \cdots, m \tag{3.19}$$

It is clear that each residual generator is driven by all inputs and all but one outputs. When all actuators are fault-free and a fault occurs in the j_{th} sensor, the residual will satisfy the following isolation logic:

$$\left\{\begin{array}{lcl} \|r^j(t)\| & < & T_{SFI}^j \\ \|r^k(t)\| & \geq & T_{SFI}^k \qquad \text{for} \quad k = 1, \cdots, j-1, j+1, \cdots, m \end{array}\right. \tag{3.20}$$

where T_{SFI}^j $(j = 1, \cdots, m)$ are isolation thresholds. A robust and UIO-based sensor fault isolation scheme is shown in Fig.3.2.

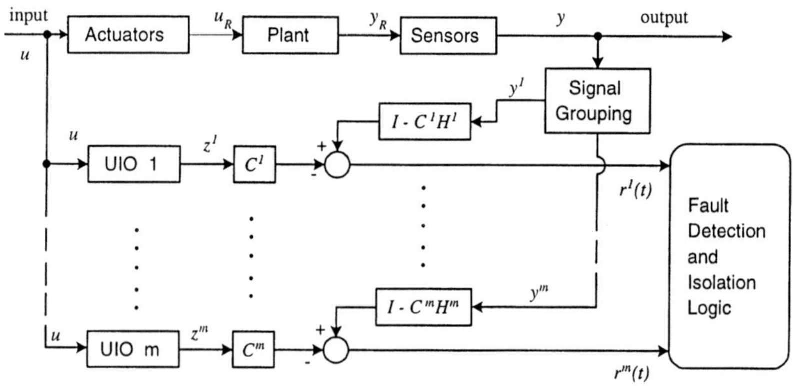

Figure 3.2. A robust sensor fault isolation scheme

3.3.2.2 Robust actuator fault isolation schemes. To design robust actuator fault isolation schemes, all sensors are assumed to be fault-free and the system equation can be described as:

$$\left\{\begin{array}{rcl} \dot{x}(t) & = & Ax(t) + B^iu^i(t) + B^if_a^i(t) + b_i(u_i(t) + f_{ai}(t)) + Ed(t) \\ & = & Ax(t) + B^iu^i(t) + B^if_a^i(t) + E^id^i(t) \\ y(t) & = & Cx(t) \qquad\qquad \text{for} \quad i = 1, 2, \cdots, r \end{array}\right. \tag{3.21}$$

where $b_i \in \mathbb{R}^n$ is the i_{th} column of the matrix B, $B^i \in \mathbb{R}^{n\times(r-1)}$ is obtained from the matrix B by deleting the i_{th} column b_i, u_i is the i_{th} component of u, $u^i \in \mathbb{R}^{r-1}$ is obtained from the vector u by deleting the i_{th} component u_i, and

$$E^i = [E \quad b_i] \qquad d^i(t) = \left[\begin{array}{c} d(t) \\ u_i(t) + f_{ai}(t) \end{array}\right] \qquad \text{for} \quad i = 1, 2, \cdots, r$$

Based on the above system description, r UIO-based residual generators can be constructed as:

$$\begin{cases} \dot{z}^i(t) &= F^i z^i(t) + T^i B^i u^i(t) + K^i y(t) \\ r^i(t) &= (I - CH^i)y(t) - Cz^i(t) \end{cases} \quad \text{for} \quad i = 1, 2, \cdots, r \tag{3.22}$$

The parameter matrices must be satisfy the following equations:

$$\left.\begin{array}{rcll} H^i C E^i &=& E^i & \\ T^i &=& I - H^i C & \\ F^i &=& T^i A - K_1^i C & \text{to be stabilized} \\ K_2^i &=& F^i H^i & \\ K^i &=& K_1^i + K_2^i & \end{array}\right\} \text{for} \quad i = 1, 2, \cdots, r \tag{3.23}$$

One can seen that each residual generator is driven by all outputs and all but one inputs. When all sensors are fault-free and a fault occurs in the i_{th} actuator, the residual will satisfy the following isolation logic:

$$\begin{cases} \|r^i(t)\| &< T_{AFI}^i \\ \|r^k(t)\| &\geq T_{AFI}^k \quad \text{for} \quad k = 1, \cdots, i-1, i+1, \cdots, r \end{cases} \tag{3.24}$$

where T_{AFI}^i $(i = 1, \cdots, r)$ are isolation thresholds. A robust and UIO-based actuator fault isolation scheme is shown in Fig.3.3.

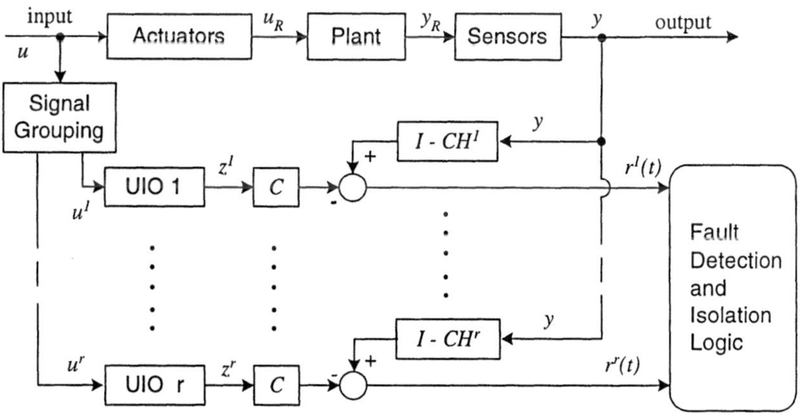

Figure 3.3. A robust actuator fault isolation scheme

<u>**Remarks:**</u> The isolation schemes presented in this section can only isolate a single fault in either a sensor or an actuator, at the same time. This is based on the fact that the probability for two or more faults to occur at the same time is very small in a real situation. If simultaneous faults need to be isolated, the fault isolation scheme should be modified based on a regrouping of faults. Each residual will be designed to be sensitive to one group of faults and insensitive to another group of faults. Frank and Wünnenberg (1989) have

studied this problem. The way of grouping faults is dependent on the system and the faults to be isolated. The isolation of sensor faults is normally possible, however it is impossible to isolate two actuator faults which have the same distribution direction. To isolate such actuator faults, other fault information such as fault frequency distributions should be utilized (Bogh, 1995). FDI schemes are related to particular systems, a *general* scheme cannot expected to suit any system without any modification.

3.3.3 A practical example (Robust actuator fault detection and isolation for a chemical reactor)

Watanabe and Himmelblau (1982) studied the sensor fault detection problem for a well-stirred chemical reactor with heat exchanger. This system is used here to demonstrate the robust actuator fault detection and isolation scheme developed in Section 3.3.2.

3.3.3.1 System representation. The state, input and output vectors for the considered chemical reactor are:

$$x(t) = \begin{bmatrix} x_1(t) \\ x_2(t) \\ x_3(t) \\ x_4(t) \end{bmatrix} = \begin{bmatrix} C_o(t) \\ T_o(t) \\ T_w(t) \\ T_m(t) \end{bmatrix} \qquad u(t) = \begin{bmatrix} u_1(t) \\ u_2(t) \\ u_3(t) \end{bmatrix} = \begin{bmatrix} 3.6C_i(t) \\ 3.6T_i(t) \\ 36T_{wi}(t) \end{bmatrix}$$

$$y(t) = \begin{bmatrix} y_1(t) \\ y_2(t) \\ y_3(t) \end{bmatrix} = \begin{bmatrix} C_o(t) \\ T_o(t) \\ T_w(t) \end{bmatrix}$$

where:

C_o	$\longmapsto$	concentration of the chemical product
T_o	$\longmapsto$	temperature of the product
T_w	$\longmapsto$	temperature of jacket water of heat exchanger
T_m	$\longmapsto$	coolant temperature
C_i	$\longmapsto$	inlet concentration of reactant
T_i	$\longmapsto$	inlet temperature
T_{wi}	$\longmapsto$	coolant water inlet temperature

According to Watanabe and Himmelblau (1982), the system is modeled as:

$$\begin{cases} \dot{x}(t) &= Ax(t) + Bu(t) + Ed(t) \\ y(t) &= Cx(t) \end{cases}$$

where the term $Ed(t)$ is used to represent the nonlinearity in the system, and

$$d(t) = 3.012 \times 10^{12} exp\{-\frac{1.2515 \times 10^7}{T_0}\} = 3.012 \times 10^{12} exp\{-\frac{1.2515 \times 10^7}{x_2(t)}\}$$

$$A = \begin{bmatrix} -3.6 & 0.0 & 0.0 & 0.0 \\ 0.0 & -3.6702 & 0.0 & 0.0702 \\ 0.0 & 0.0 & -36.2588 & 0.2588 \\ 0.0 & 0.6344 & 0.7781 & -1.4125 \end{bmatrix} \qquad E = \begin{bmatrix} 1.0 \\ 20.758 \\ 0.0 \\ 0.0 \end{bmatrix}$$

$$B = \begin{bmatrix} 1 & 0 & 0 \\ 0 & 1 & 0 \\ 0 & 0 & 1 \\ 0 & 0 & 0 \end{bmatrix} \qquad C = \begin{bmatrix} 1 & 0 & 0 & 0 \\ 0 & 1 & 0 & 0 \\ 0 & 0 & 1 & 0 \end{bmatrix}$$

Note that the system matrices are not exactly the same as give by Watanabe and Himmelblau (1982), this is because the time scale has been changed to hours for the sake of convenience.

3.3.3.2 UIOs design and residuals generation. Both control inputs $u_1(t)$ ($C_i(t)$) and $u_2(t)$ ($T_i(t)$) are related to the inlet chemical substance, and any fault in $u_1(t)$ or $u_2(t)$ will cause a similar consequence. Hence it is not necessary to isolate faults between $u_1(t)$ and $u_2(t)$. Two UIOs are designed here, the first UIO is driven by $u_1(t)$ and $u_2(t)$ and the second UIO is driven by $u_3(t)$. These two UIOs are robust to the non-linear factor in $d(t)$.

UIO 1: The dynamic equation for the first UIO is:

$$\dot{z}^1(t) = F^1 z(t) + K^1 y(t) + T^1 [b_1 \quad b_2] \begin{bmatrix} u_1(t) \\ u_2(t) \end{bmatrix}$$

where b_1 and b_2 are the first two columns of B, and the parameter matrices for this UIO are:

$$H^1 = \begin{bmatrix} 21.758 & -1.0 & 0.0 \\ 0.0 & 1.0 & 0.0 \\ 0.0 & 0.0 & 1.0 \\ -2075.8 & 100.0 & 0.0 \end{bmatrix}; \; T^1 = \begin{bmatrix} -20.758 & 1.0 & 0.0 & 0.0 \\ 0.0 & 0.0 & 0.0 & 0.0 \\ 0.0 & 0.0 & 0.0 & 0.0 \\ 2075.8 & -100.0 & 0.0 & 1.0 \end{bmatrix}$$

$$F^1 = \begin{bmatrix} -10 & 0.0 & 0.0 & 0.0702 \\ 0 & -\lambda_1 & 0.0 & 0.0 \\ 0 & 0.0 & -\lambda_2 & 0.0 \\ 0 & 0.0 & 0.0 & -8.4325 \end{bmatrix}$$

$$K^1 = \begin{bmatrix} -278.5724 & 13.3496 & 0.0 \\ 0.0 & 0.0 & 0.0 \\ 0.0 & 0.0 & 0.0 \\ 10031.304 & -475.5956 & 0.7781 \end{bmatrix}$$

The sub-observer for z_2^1 and z_3^1 (element of vector z^1) has no inputs of y, u_1 and u_2, and has no coupling with z_1^1 and z_4^1, hence z_2^1 and z_3^1 will stay at zero if the initial values of z_2^1 and z_3^1 are zero and the observer matrix F^1 is designed to be stable. The full-order UIO can be reduced to:

$$\begin{bmatrix} \dot{z}_1^1 \\ \dot{z}_4^1 \end{bmatrix} = \begin{bmatrix} -10.0 & 0.0702 \\ 0.0 & -8.4325 \end{bmatrix} \begin{bmatrix} z_1^1 \\ z_4^1 \end{bmatrix} + \begin{bmatrix} -20.758 & 1.0 \\ 2075.8 & -100.0 \end{bmatrix} \begin{bmatrix} u_1 \\ u_2 \end{bmatrix} + \begin{bmatrix} -278.57236 & 13.3498 & 0.0 \\ 10031.3035 & -475.5956 & 0.7781 \end{bmatrix} y$$

The state estimation is:

$$\hat{x}^1 = \begin{bmatrix} z_1^1 + 21.758y_1 - y_2 \\ y_2 \\ y_3 \\ z_4^1 - 2075.8y_1 + 100y_2 \end{bmatrix}$$

The residual is generated by:

$$r^1(t) = y_1(t) - \hat{y}_1(t) = y_1(t) - \hat{x}_1(t) = y_2(t) - z_1^1(t) - 20.758y_1(t)$$

UIO 2: The dynamic equation for the second UIO is:

$$\dot{z}^2(t) = F^2 z(t) + K^2 y(t) + T^2 b_3 u_3(t)$$

where b_3 is the third column of B, and the parameter matrices for this UIO are:

$$H^2 = \begin{bmatrix} 1.0 & 0.0 & 0.0 \\ 0.0 & 1.0 & 0.0 \\ 0.0 & 0.0 & 0.0 \\ 0.0 & 0.0 & 40.0 \end{bmatrix}; \; T^2 = \begin{bmatrix} 0.0 & 0.0 & 0.0 & 0.0 \\ 0.0 & 0.0 & 0.0 & 0.0 \\ 0.0 & 0.0 & 1.0 & 0.0 \\ 0.0 & 0.0 & -40.0 & 0.0 \end{bmatrix}$$

$$F^2 = \begin{bmatrix} -\lambda_1 & 0.0 & 0.0 & 0.0 \\ 0.0 & -\lambda_2 & 0.0 & 0.0 \\ 0.0 & 0.0 & -10.0 & 0.2588 \\ 0.0 & 0.0 & 0.0 & -11.7645 \end{bmatrix}$$

$$K^2 = \begin{bmatrix} 0.0 & 0.0 & 0.0 \\ 0.0 & 0.0 & 0.0 \\ 0.0 & 0.0 & -15.9068 \\ 0.0 & 0.6344 & 980.5501 \end{bmatrix}$$

Similar to the first UIO, the UIO 2 can also be reduced as:

$$\begin{bmatrix} \dot{z}_3^2 \\ \dot{z}_4^2 \end{bmatrix} = \begin{bmatrix} -10.0 & 0.2588 \\ 0.0 & -11.7645 \end{bmatrix} \begin{bmatrix} z_3^2 \\ z_4^2 \end{bmatrix} + \begin{bmatrix} 0.0 & 0.0 & -15.9068 \\ 0.0 & 0.6344 & 980.5501 \end{bmatrix} y + \begin{bmatrix} 1.0 \\ -40.0 \end{bmatrix} u_3(t)$$

$$\hat{x}^2 = \begin{bmatrix} y_1 & y_2 & z_3^2 & z_4^2 + 40y_3 \end{bmatrix}^T$$

The residual is generated by:

$$r^2(t) = y_3(t) - \hat{y}_3(t) = y_3(t) - \hat{x}_3(t) = y_3(t) - z_3^2(t)$$

Simulation: The above UIOs is applied to the non-linear chemical reaction process to detect and isolate faulty actuators. The system input and the initial state vectors are:

$$u = \begin{bmatrix} 34.632 \\ 1641.6 \\ 29980 \end{bmatrix} \qquad x(0) = \begin{bmatrix} 0.3412 \\ 525.7 \\ 472.2 \\ 496.2 \end{bmatrix}$$

The initial values for UIOs are:

$$z_1^1(0) = 518.6174; \; z_4^1(0) = -51365.5370; \; z_3^2(0) = 472.2; \; z_4^2(0) = -18391.8$$

The sampling interval is set as 0.05 hour, and the simulation is carried out for $t = 10$ hours. Various types of faults are introduced to the system at $t = 4$ hours. The list of the simulated faults is:

(a) A fault occurs in the inlet reactant when $t > 4$ hour, the fault signal in the first input is $20\% u_1(t)$.

(b) A fault occurs in the inlet reactant when $t > 4$ hour, the fault signal in the second input is $20\% \sin(2(t-4))u_2(t)$.

(c) A fault occurs in the coolant circular when $t > 4$ hour, the fault signal in the third input is $-2\% u_3(t)$.

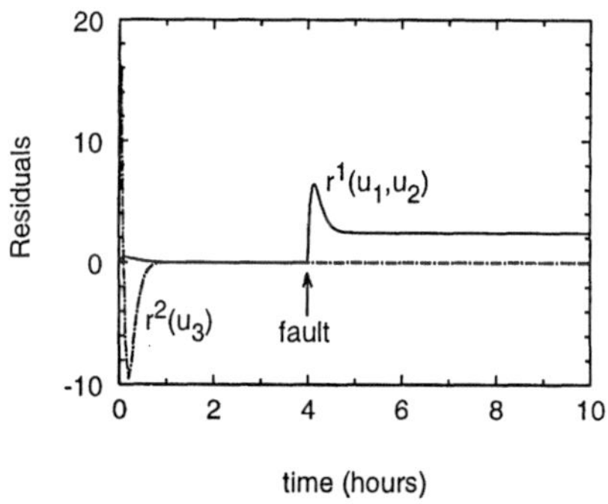

Figure 3.4. UIO residuals when a fault occurs in $u_1(t)$ (without parameter variations)

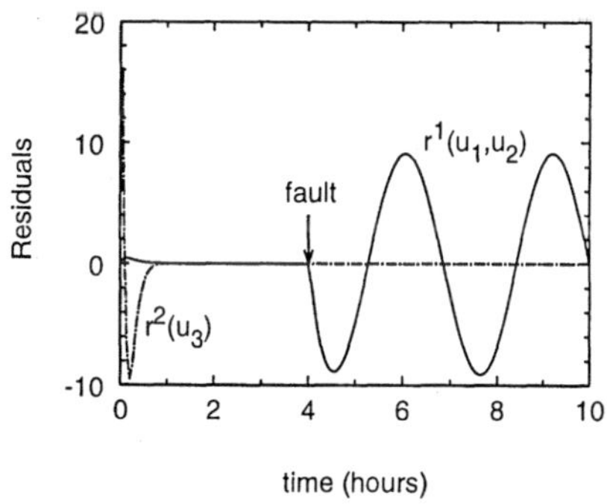

Figure 3.5. UIO residuals when a fault occurs in $u_2(t)$ (without parameter variations)

The simulation results are shown in Figs.3.4–3.6, from which one can seen that the residual is almost zero throughout the 10 hours simulation run for fault-free residuals. The residuals of the respective UIO increase in magnitude considerably, when actuator faults occur at $t = 4$ hours. The faults can be easily isolated using the information provided by residuals.

Robustness analysis: From the above analysis and simulation, we know that the fault detection and isolation scheme is robust to nonlinearity in $d(t)$. The

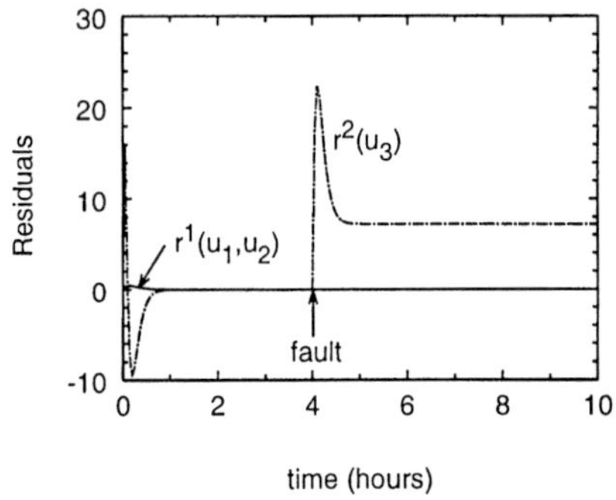

Figure 3.6. UIO residuals when a fault occurs in $u_3(t)$ (without parameter variations)

robustness with respect to parameter variations is analyzed below. The system with parameter variations is described as:

$$\dot{x}(t) = Ax(t) + Bu(t) + Ed(t) + \sum_{i=1}^{4} I_i w_i(x(t), \Delta A)$$

where: I_i is the i_{th} column of identity matrix, w_i represent the variations in i_{th} row elements of A. This equation can be rewritten as:

$$\dot{x}(t) = Ax(t) + Bu(t) + Ed(t) + Ew_1 + I_2(w_2 - 20.758w_1) + I_3w_3 + I_4w_4$$

Parameter variations in the form of Ew_1 and I_3w_3 will not affect the first UIO, because $T^1E = 0$ and $T^1I_3 = 0$. Similarly, the parameter variations in the form of Ew_1 and $I_2(w_2 - 20.758w_1)$ will not affect the second UIO, because $T^2E = 0$ and $T^2I_2 = 0$. In all cases, the sensitivities to process parameter variations have been decreased. The robustness of UIOs to process parameter variations can be assessed by the simulation in which the matrix A is changed to:

$$A = \begin{bmatrix} -4.14 & 0.0 & 0.0 & 0.0 \\ 0.0 & -4.22073 & 0.0 & 0.08073 \\ 0.0 & 0.0 & -36.4401 & 0.2601 \\ 0.0 & 0.9516 & 1.1672 & -2.1188 \end{bmatrix}$$

The residuals for three types of faults are shown in Figs.3.7–3.9, from which one can conclude that the robust FDI scheme can reliably detect and isolate faulty actuators even in the presence of process parameter mismatch.

Remarks: Robust actuator fault detection and isolation based on UIOs has been demonstrated in a chemical reactor example. The UIO is a time-invariant linear filter but can also be applied to a class of non-linear time-variant systems if the non-linear function is separated from the linear function and can be treated as an unknown input term. The robust FDI based on UIOs has also a certain degree of robustness against parameter variations. The application results of this example was reported in Chen and Zhang (1991).

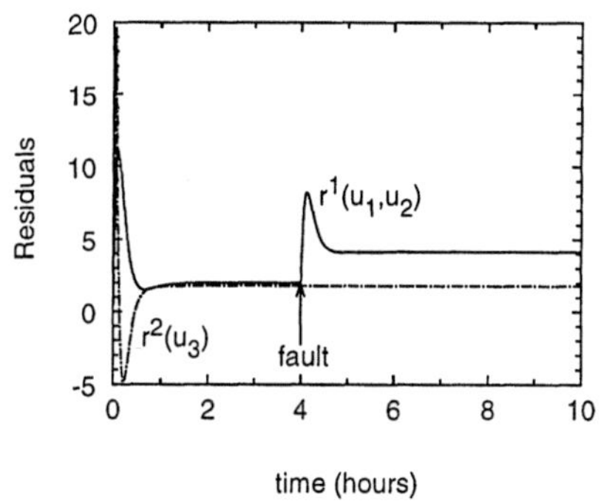

Figure 3.7. UIO residuals when a fault occurs in $u_1(t)$ (with parameter variations)

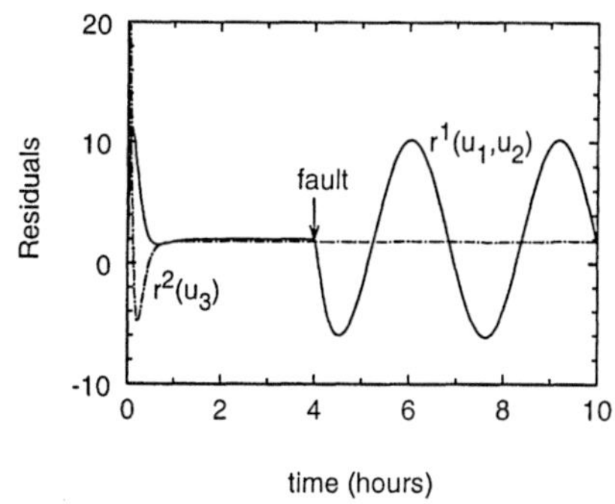

Figure 3.8. UIO residuals when a fault occurs in $u_2(t)$ (with parameter variations)

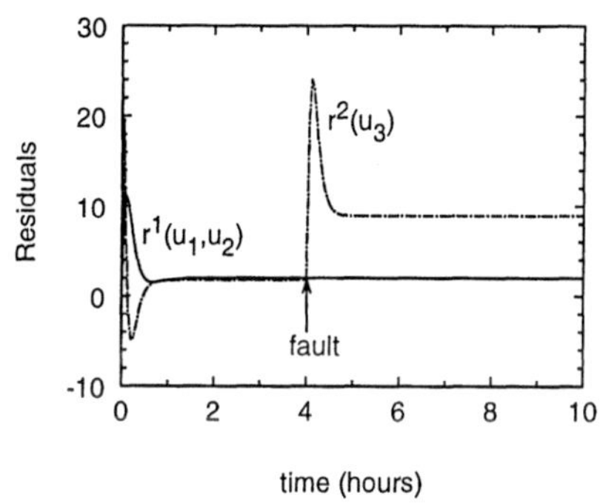

Figure 3.9. UIO residuals when a fault occurs in $u_3(t)$ (with parameter variations)

3.4 Robust Fault Detection Filters and Robust Directional Residuals

Fault detection filters (Beard, 1971) are a particular class of the full-order Luenberger observer with a specially designed feedback gain matrix such that the output estimation error (residual vector) has uni-directional characteristics associated with some known fault directions. To be specific, the residual vector of a fault detection filter is fixed along with a predetermined direction for an actuator fault or lies in a specific plane for a sensor fault. Since the important information required for isolation is contained in the direction of the residual rather than in its time function, the use of a Beard fault detection filter (BFDF) does not require the knowledge of fault modes. The fault isolation task can be facilitated by comparing the residual direction with pre-defined fault signature directions (or planes), and only one (or the minimum number of) observers required for fault isolation due to directional characteristics of the residual. This is the main and most appealing advantage of fault detection filters. However, *the main drawback* of the BFDF is that the robustness problem has not been considered. This section describes a method to design a robust fault detection filter which is based on the combination of UIO and BFDF theories. The method presented in this Section is based on the results presented in Chen, Patton and Zhang (1996). The main principle is that the remaining design freedom, after disturbance de-coupling conditions have been satisfied, can be used to make the residual vector have directional characteristics. A realistic simulation example of isolating faulty sensors in a jet engine system is presented. This is a non-linear system and the linearization error can cause false isolation if the robustness issue is not considered. A way of representing linearization errors as an unknown input term is presented and its distribution is estimated using a least-squares procedure. The simulation results shows that faults are correctly isolated using the technique developed.

3.4.1 Basic principles of fault detection filters

The BFDF was first developed by Beard (1971) using a matrix algebra approach and later reformed by Jones (1973) in a vector space notation. The theory of BFDFs has been extended by many researchers, for example, Massoumnia (1986b) used a geometric interpretation, White and Speyer (1987) improved the design procedure using a spectral approach which is suitable for the isolation of multiple faults. The a closed-form expression of BFDFs using eigenstructure assignment was also proposed (Park and Rizzoni, 1993; Park and Rizzoni, 1994b; Park and Rizzoni, 1994a; Park, Halevi and Rizzoni, 1994; Park, Rizzoni and Ribbens, 1994). Recent developments can be found in Douglas and Speyer (1996) and Chung and Speyer (1998).

In order to describe the BFDF theory, let us consider a system without disturbances in the state space format as:

$$\left\{ \begin{array}{rcl} \dot{x}(t) & = & Ax(t) + Bu(t) + b_i f_{ai}(t) \\ y(t) & = & Cx(t) + I_j f_{sj}(t) \end{array} \right. \tag{3.25}$$

The term $b_i f_{ai}(t)$ $(i = 1, 2, \cdots, r)$ denotes that a fault has occurred in the i_{th} actuator, $b_i \in \mathbb{R}^n$ is the i_{th} column of the input matrix B and is defined as the fault event vector of the i_{th} actuator fault, and $f_{ai}(t)$ is an unknown scalar time-varying function which represents the evolution of the fault. The term $I_j f_{sj}(t)$ $(j = 1, 2, \cdots, m)$ denotes that a fault occurs in the j_{th} sensor, $I_j \in \mathbb{R}^m$ is a unit vector corresponding to a fault with the j_{th} sensor. Note that component faults appear in the system equation in the same way as the actuator fault and hence are not discussed further here.

A BFDF is just a full-order observer and its structure and the residual can be described as:

$$\left\{ \begin{array}{rcl} \dot{\hat{x}}(t) & = & A\hat{x}(t) + Bu(t) + K(y(t) - C\hat{x}(t)) \\ r(t) & = & y(t) - C\hat{x}(t) \end{array} \right. \tag{3.26}$$

where $r \in \mathbb{R}^m$ is the residual vector, $\hat{x} \in \mathbb{R}^n$ is the state estimation, and $K \in \mathbb{R}^{m \times n}$ is the observer gain matrix which has to be specially designed to make the residual have restricted uni-directional properties in the presence of a particular fault. If the state estimation error is defined as: $e(t) = x(t) - \hat{x}(t)$, the residual and $e(t)$ will be governed by the following error system, when a fault occurs in the i_{th} actuator:

$$\begin{cases} \dot{e}(t) &= (A - KC)e(t) + b_i f_{ai}(t) \\ r(t) &= Ce(t) \end{cases} \tag{3.27}$$

When a fault occurs in the j_{th} sensor, the error system will be:

$$\begin{cases} \dot{e}(t) &= (A - KC)e(t) - k_j f_{sj}(t) \\ r(t) &= Ce(t) + I_j f_{sj}(t) \end{cases} \tag{3.28}$$

where k_j is the j_{th} column of the detection filter gain matrix.

The task of BFDF design is to make $Ce(t)$ have a *fixed direction* in the output space responding to either $b_i f_{ai}(t)$ or $k_j f_{sj}(t)$. Both actuator and sensor fault situations can be considered in the following general error system equation:

$$\begin{cases} \dot{e}(t) &= (A - KC)e(t) + l_i \xi_i(t) \\ r(t) &= Ce(t) \end{cases} \tag{3.29}$$

where $l_i \in \mathbb{R}^n$ is called the *fault event direction.* The definition of the isolability of a fault with known direction l_i is given by Beard (1971) as stated below:

Definition 3.2 (Isolability of a fault with a given direction) *The fault associated with l_i in the system described by Eq.(3.29) is isolable if there exists a filter gain matrix K such that:*

(a) *$r(t)$ maintains a fixed direction in the output space, and*

(b) *$(A - KC)$ can be stabilized.*

Condition (a) which guarantees that the residual has uni-directional characteristics, is equivalent to ensuring that the rank of the controllability matrix of (A, l_i) pair is one, i.e:

$$rank[l_i \;\; (A - KC)l_i \;\; \cdots \;\; (A - KC)^{n-1}l_i] = 1$$

Condition (b) ensures the convergence of the filter. In the original definition of Beard (1971), condition (b) requires arbitrarily assignment of eigenvalues of $(A - KC)$. This condition has been modified as the stability requirement is sufficient if the residual response time does not need to specified. This definition was referred to as "fault detectability" by Beard (1971) and others (Jones, 1973; Massoumnia, 1986b; White and Speyer, 1987). In the author's view, the term "isolability" is more appropriate, because the directional property of the residuals is especially desirable for fault isolation purposes, although it can also be used for fault detection. Hence, the BFDF is designed to satisfy the fault isolability.

Here the abbreviation BFDF is reserved for a filter (an observer) with residual having uni-directional properties. If a fault associated with the direction

b_i is isolable, the residual of the BFDF will be fixed in the direction parallel to Cb_i, when a fault occurs in the i_{th} actuator. Similarly, the residual will lie somewhere in the plane defined by Ck_j and I_j, when a fault occurs in the j_{th} sensor.

To isolate faults associated with p isolable fault event directions l_i ($i = 1, \cdots, p$), the following output separability condition (Beard, 1971) must be satisfied.

Definition 3.3 (Output Separability of Faults) *The faults associated with p fault event directions l_i ($i = 1, 2, \cdots, p$) are separable in the residual space if the vectors $Cl_1, Cl_2, \cdots, Cl_p$ are linearly independent.*

Output separability is necessary for a group of faults to be isolated in the residual space according to their signature directions. The directions Cl_i ($i = 1, 2, \cdots, p$) are then known as the *fault signature directions* in the residual space.

Definition 3.4 (Mutual Isolability) *The faults which associated with the fault event vectors l_i (i = 1, 2, $\cdots$, p) are mutually isolable if there exists a filter gain matrix K satisfying the isolability conditions of Definition 3.2 for all l_i ($i = 1, 2, \cdots, p$), i.e.*

$$rank[l_i \ \ (A-KC)l_i \ \ \cdots \ \ (A-KC)^{n-1}l_i] = 1 \qquad for\ all \ \ i = 1, 2, \cdots, p$$

A group of mutually isolable faults can be isolated using the residual generated by a single BFDF by comparing the residual direction with the fault signature directions, when there are no simultaneous faults. The condition for mutual isolability can be found in the well known literature (Beard, 1971; Jones, 1973; White and Speyer, 1987; Park and Rizzoni, 1994b; Douglas and Speyer, 1996; Chung and Speyer, 1998). If a group of faults is not mutually isolable, it can be divided into a number of subgroups and each subgroup is mutually isolable. For such cases, a few BFDFs are required to fulfil the fault isolation task. In any case, only a minimum number of filters are required for fault isolation. This is the most important and appealing advantage of the BFDF approaches.

In conclusion, the task of designing a fault detection filter is to make the residual have a uni-directional property by choosing the gain matrix K. Design techniques can be found in the classical literature on fault detection filters (Beard, 1971; Jones, 1973; White and Speyer, 1987; Park and Rizzoni, 1994b; Douglas and Speyer, 1996; Chung and Speyer, 1998).

3.4.2 Disturbance de-coupled fault detection filters and robust fault isolation

It can be seen that uncertain factors associated with a dynamical system such as disturbances and modeling errors have not be considered in the design of BFDFs. This is the main disadvantage of BFDFs, because uncertain factors are unavoidable in real systems and any FDI scheme has to be made robust against disturbances and modeling errors. Now, consider a system with disturbance term $Ed(t)$ and possible sensor and actuator faults described as:

$$\left\{ \begin{array}{lcl} \dot{x}(t) & = & Ax(t) + Bu(t) + Ed(t) + b_i f_{ai}(t) \\ y(t) & = & Cx(t) + I_j f_{sj}(t) \end{array} \right. \tag{3.30}$$

If a standard BFDF described by Eq.(3.26) is applied to such a system, the state estimation error and residual will be:

$$\begin{cases} \dot{e}(t) &= (A - KC)e(t) + Ed(t) + b_i f_{ai}(t) - k_j f_{sj}(t) \\ r(t) &= Ce(t) + I_j f_{sj}(t) \end{cases} \tag{3.31}$$

It is clear from Eq.(3.31) that all faults and disturbances affect the residual. It is not easy to discriminate between faults and disturbances if this residual is used to detect and isolate faults. Hence, it is necessary to de-couple disturbance effects from the residual for reliable diagnosis.

It has been shown that the disturbances can be de-coupled from the state estimation error using an unknown input observer (see also Section 3.3.1). This inspires us to generate the residual using the unknown input observer described in Eq.(3.2). The residual is thus defined as:

$$r(t) = y(t) - C\hat{x}(t) = (I - CH)y(t) - Cz(t) \tag{3.32}$$

When this UIO-based residual generator is applied to the system described by the model Eq.(3.31), the residual and the state estimation error ($e(t)$) will be:

$$\begin{cases} \dot{e}(t) &= (A_1 - K_1 C)e(t) + T b_i f_{ai}(t) \\ r(t) &= Ce(t) \end{cases} \tag{3.33}$$

when a fault occurs in the i_{th} actuator.

Similarly,

$$\begin{cases} \dot{e}(t) &= (A_1 - K_1 C)e(t) - k_{1j} f_{sj}(t) - h_j \dot{f}_{sj}(t) \\ r(t) &= Ce(t) + I_j f_{sj}(t) \end{cases} \tag{3.34}$$

when a fault occurs in the j_{th} sensor. Where k_{1j} is the j_{th} column of the matrix K_1 and h_j is the j_{th} column of the matrix H. From Eq.(3.33) & (3.34), it can be seen that the disturbance effects have been de-coupled from the residual. This robust (in the disturbance de-coupling sense) residual can be used to detect faults according to a simple threshold logic:

$$\begin{cases} \|r(t)\| < Threshold & \text{for} \quad \text{fault-free case} \\ \|r(t)\| \geq Threshold & \text{for} \quad \text{faulty cases} \end{cases} \tag{3.35}$$

As pointed out in the introduction, fault isolation can be facilitated using uni-directional residual vectors. So, one has to make the residual generated by a UIO, have the directional properties in order to achieve robust fault isolation. From the design of UIOs, it is known that the matrix K_1 can be designed arbitrarily after the robust (in the sense of disturbance de-coupling) conditions have been satisfied. This design freedom can be exploited to make the residual have the uni-directional property.

Comparing the error system Eq. (3.33) with Eq.(3.27), it can be seen that the actuator fault is expressed in the same way for a UIO or a standard BFDF. Hence, the theory for the design of a BFDF (Beard, 1971; Jones, 1973; White and Speyer, 1987; Park and Rizzoni, 1994b; Douglas and Speyer, 1996; Liu and Si, 1997; Chung and Speyer, 1998) can be directly used to design the matrix K_1, if the vector b_i is replaced by Tb_i and the matrix A is replace by A_1.

Comparing the error system Eq.(3.34) with Eq.(3.28), it can be seen that the sensor fault is also expressed in a similar way for *both* the BFDF and UIO, except an extra term $h_j \dot{f}_{sj}(t)$ occurs in the error equation of the UIO. Fortunately, this term can be treated in the same way as an actuator fault. Hence, the theory of BFDF can be adopted for the design of K_1 in the sensor isolation problem. However, it must be pointed out that the residual will lie in a subspace spanned by vectors I_j, Ck_{1j} and Ch_j when the residual uni-directional property has been satisfied. For constant sensor faults, the term $h_j \dot{f}_{sj}(t)$ will disappear from the error system and the residual will lie in the plane spanned by the vectors I_j and Ck_{1j}, this is same as the BFDF.

It is necessary to combine the theory of UIOs with the theory of BFDFs to design a robust (disturbance de-coupled) fault detection filter. The design procedure can be summarized as follows:

- Compute the matrices H and T using Eqs. (3.11) & (3.6), to satisfy disturbance de-coupling conditions.
- Compute A_1 using Eq.(3.12).
- Compute K_1 to satisfy a uni-directional property using the theory of BFDFs.
- Compute the observer gain matrix K using Eqs.(3.8) & (3.4).

The key step is then to design the matrix K_1. Once this matrix is available, the computation of other matrices is very straightforward. The BFDF design procedure can be found in the well known literature (Beard, 1971; Jones, 1973; White and Speyer, 1987; Park and Rizzoni, 1994b; Douglas and Speyer, 1996; Chung and Speyer, 1998) and is not presented in this chapter. To show the basic idea, an ideal situation is discussed now, in which the number of independent measurements is equal to the number of states, i.e. $rank(C) = n$. In this situation, all eigenvalues of the matrix $A_1 - K_1C$ can be assigned to the same value $-\sigma < 0$, i.e.,

$$A_1 - K_1C = -\sigma I$$

This can be achieved by setting K_1 as:

$$K_1 = (A_1 + \sigma I)C^+ \tag{3.36}$$

where C^+ is the pseudo-inverse of C. For this design, the residual will be:

$$\begin{aligned} r(t) &= Ce(t) + I_j f_s(t) \\ &= I_j f_s(t) + Ce^{-\sigma(t-t_0)}e(t_0) \\ &\quad + C\int_{t_0}^{t} e^{-\sigma I(t-\tau)}[Tb_i f_a(\tau) - k_{1j} f_s(t) - h_j \dot{f}_s(t)]d\tau \end{aligned}$$

$$
\begin{aligned}
&= Ce^{-\sigma(t-t_0)}e(t_0) + CTb_i \int_{t_0}^{t} e^{-\sigma(t-\tau)} f_a(\tau)d\tau \\
&\quad + I_j f_s(t) - Ck_{1j} \int_{t_0}^{t} e^{-\sigma(t-\tau)} f_s(\tau)d\tau - Ch_j \int_{t_0}^{t} e^{-\sigma(t-\tau)} \dot{f}_s(\tau)d\tau \\
&= Ce^{-\sigma(t-t_0)}e(t_0) + CTb_i\alpha(t,t_0) \\
&\quad + I_j f_s(t) + Ck_{1j}\beta(t,t_0) + Ch_j\gamma(t,t_0)
\end{aligned}
$$

Clearly, the residual is parallel to CTb_i after the transient has settled down following a fault in the i_{th} actuator. Similarly, the residual will lie in the subspace spanned by vectors I_j, Ck_{1j} and Ch_j, when a fault occurs in the j_{th} sensor.

Due to the residual directional property, the fault can be isolated by comparing the residual direction with the fault signature directions (or subspaces).

Definition 3.5 *The direction of CTb_i is termed a signature direction of the i_{th} actuator fault (Chen, Patton and Zhang, 1996).*

The directional relationship between two vectors CTb_i and $r(t)$ can be quantified by the correlation parameter $CORR_i$:

$$
CORR_i(t) = \frac{|(CTb_i)^T r(t)|}{||CTb_i||_2 ||r(t)||_2} \tag{3.37}
$$

If $CORR_j > CORR_k$, the fault is more likely in the j_{th} rather than in the k_{th} actuator.

Definition 3.6 *The signature subspace of the j_{th} sensor fault is defined as (Chen, Patton and Zhang, 1996):*

$$
R_j = Span\{I_j, Ck_{1j}, Ch_j\} \tag{3.38}
$$

The relationship between the vector $r(t)$ with the subspace R_j can be measured by the relationship between the vector $r(t)$ with its projection $r_j^*(t)$ in the subspace R_j. This is quantified by:

$$
CORR_j(t) = \frac{|(r_j^*)^T r(t)|}{||r_j^*||_2 ||r(t)||_2} \tag{3.39}
$$

where the projection $r_j^*(t)$ of $r(t)$ in R_j is:

$$
r_j^*(t) = \Phi_j(\Phi_j^T\Phi_j)^{-1}\Phi_j^T r(t) \tag{3.40}
$$

where

$$
\Phi_j = [I_j \; Ck_{1j} \; Ch_j]
$$

If $CORR_j > CORR_k$, the fault is more likely in the j_{th} rather than in the k_{th} sensor. The relationship between a residual vector with the signature subspace can also be judged by the *normalized projection distance* which is defined as:

$$
NPD_j = \frac{||r(t) - r_j^*(t)||_2}{||r(t)||_2} \tag{3.41}
$$

when NPD_{j^*} is the smallest one amongst all NPD_j ($j = 1, 2, \cdots, m$), the fault is most likely in the j^*_{th} sensor. The idea of fault isolation by comparing the residual direction with the signature subspace is illustrated in Fig.3.10.

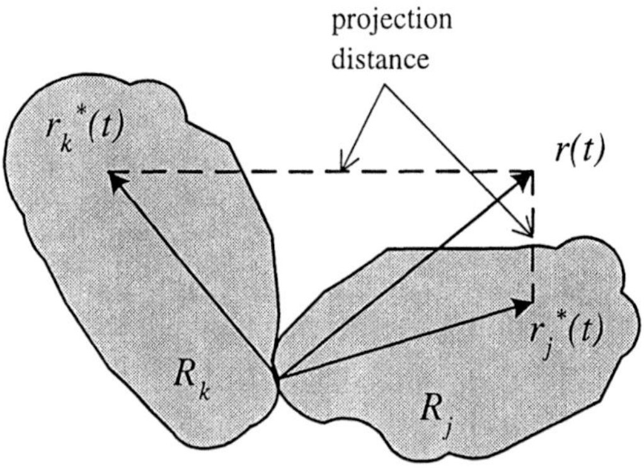

Figure 3.10. Fault isolation based on directional residuals

3.4.3 Robust isolation of faulty sensors in a jet engine system

indexrobust fault isolation!directional residual vector!jet engine example To control a jet engine efficiently and to monitor its health effectively, the sensors have to be perform reliably. However, the sensors in a jet engine work in a very harsh environment and could fail during normal engine operation. This is especially true of the thermocouple (gas temperature) sensors. Hence, the detection of sensor faults in jet engine systems is very important and has become an active research field (Merrill, 1985; Merrill, 1990; Meserole, 1981; Patton and Chen, 1991f). A simplified non-linear dynamic model of a jet engine control system can be described as (Chen, Patton and Zhang, 1996):

$$\begin{cases} \dot{X}_1(t) &= f_1(X_1, X_2, X_3) \\ \dot{X}_2(t) &= f_2(X_1, X_2, X_3) \\ \dot{X}_3(t) &= 10(U - X_3) \end{cases}$$

where:

$$\begin{array}{lcllll} X_1 & = & n_L & \longmapsto & \text{Low pressure rotor speed} \\ X_2 & = & n_H & \longmapsto & \text{High pressure rotor speed} \\ X_3 & = & W_f & \longmapsto & \text{Main burner fuel flow} \\ U & = & W_{fe} & \longmapsto & \text{Fuel flow command} \end{array}$$

The jet engine is a very complicated non-linear dynamic system. Non-linear functions such as $f_1(X_1, X_2, X_3)$ and $f_2(X_1, X_2, X_3)$ cannot written out *analytically*. The system behavior is normally expressed in a non-linear dynamic simulation package (Merrill and Leininger, 1981; Merrill, 1990; Meserole, 1981; Meserole, 1981). This package is capable of simulating the entire operating envelope of the engine, and can also generate linearized models for any operating points. It is useful here to define the following non-dimensional variables:

$$x_1 = \frac{X_1 - X_1^0}{X_1^0} \quad ; \quad x_2 = \frac{X_2 - X_2^0}{X_2^0}$$

$$x_3 = \frac{X_3 - X_3^0}{X_3^0} \quad ; \quad u = \frac{U - U^0}{U^0}$$

where the superscript "0" denotes the values at equilibrium. The system can be linearized around an operating point. If u is small (e.g. 1%), x_1, x_2 and x_3 will be small, i.e. all variables have a small variation around the equilibrium and the following linear model is derived:

$$\begin{cases} \dot{x}(t) &= Ax(t) + bu(t) \\ y(t) &= Cx(t) \end{cases}$$

where the state is $x = [x_1 \; x_2 \; x_3]^T$ and the measurement vector is:

$$y = [x_1 \quad x_2 \quad x_3 \quad p_2 \quad p_4 \quad t_4]^T$$

in which

$$p_2 = \frac{P_2 - P_2^0}{P_2^0} \; ; \; p_4 = \frac{P_4 - P_4^0}{P_4^0} \; ; \; t_4 = \frac{T_4 - T_4^0}{T_4^0}$$

where

$$\begin{array}{lll} P_2 & \longmapsto & \text{High pressure compressor discharge pressure} \\ P_4 & \longmapsto & \text{Turbine discharge pressure} \\ T_4 & \longmapsto & \text{Turbine exit temperature} \end{array}$$

When the equilibrium is set at $N_L = 450(rpm)$, the linear model matrices are:

$$A = \begin{bmatrix} -1.5581 & 0.6925 & 0.3974 \\ 0.2619 & -2.2228 & 0.2238 \\ 0 & 0 & -10 \end{bmatrix} \qquad B = \begin{bmatrix} 0 \\ 0 \\ 10 \end{bmatrix}$$

$$C = \begin{bmatrix} 1 & 0 & 0 \\ 0 & 1 & 0 \\ 0 & 0 & 1 \\ 0.55107 & 0.13320 & 0.30603 \\ 0.55217 & 0.13526 & 0.32912 \\ -0.25693 & -0.23625 & 0.61299 \end{bmatrix}$$

A BFDF described by Eq.(3.26) is designed to isolate sensor faults. If all eigenvalues of the filter are set to -3, the gain matrix can be determined as $K = (3I + A)C^+$ because $rank(C) = 3$. The fault isolation scheme is applied to the non-linear simulation model. A reliable diagnostic scheme should perform well for a wide range of operating conditions, and hence the input is set at $u = 20\%$ in the simulation. The sensor fault is simulated as 2% offset around the normal measurement. In the simulation, we only consider the fault in sensor Nos.1, 2 and 3, i.e. the low pressure rotor speed sensor, the high pressure rotor speed sensor and the main burner fuel flow sensor. After the transient has settled down, the normalized projection distances for different faulty situations are shown in Table 3.2.

Table 3.2. Fault isolation using Beard fault detection filter

Faulty sensor	*No.1*	*No.2*	*No.3*
NPD_1	**0.37090**	0.77783	0.66389
NPD_2	0.93117	0.95527	0.42455
NPD_3	0.96529	**0.71161**	**0.31559**

From Table 3.2, it can be seen that the fault in sensor No.1 (or No.3) can be correctly isolated as the corresponding normalized projection distance NPD_1 (or NPD_3) is the smallest. However, the fault in the sensor No.2 will be mis-reported as a fault in sensor No.3 as NPD_3 is the smallest amongst all normalized projection distances. Moreover, the smallest NPD is not significantly different from other NPDs, and this could make isolation difficult when there is noise in the system.

The example in Table 3.2 illustrates the importance of robustness in fault isolation. The false isolation problem is possibly caused by the linearized errors, as the fault isolation scheme is based on the linear model and this scheme is applied to the original non-linear system. In the model linearization, only the first order terms in the Taylor expansion have been considered. To model a system more accurately, one can consider to the inclusion of the second order terms in the system dynamic equation as follows:

$$\dot{x}(t) = Ax(t) + Bu(t) + Ed(x(t))$$

where the matrices A and B are the same as for the linear model. The term $Ed(x)$ represents modeling errors and the vector $d(x)$ consists of the second order terms of $x(t)$ as:

$$d(x) = [x_1^2 \quad x_2^2 \quad x_3^2 \quad x_1x_2 \quad x_1x_3 \quad x_2x_3]^T$$

The distribution matrix E can be obtained using an identification procedure based on the least-squares method. Given a series of values $u^{(1)}$, $u^{(2)}$, $\cdots$, $u^{(N)}$ for input u, we can obtain the corresponding steady responses $x^{(1)}$, $x^{(2)}$, $\cdots$, $x^{(N)}$ and $d^{(1)}$, $d^{(2)}$, $\cdots$, $d^{(N)}$, which are related by the following steady state equations:

$$\begin{cases} Ax^{(1)} + Bu^{(1)} + Ed^{(1)} & = & 0 \\ Ax^{(2)} + Bu^{(2)} + Ed^{(2)} & = & 0 \\ \cdots & & \cdot \quad \cdot \\ Ax^{(N)} + Bu^{(N)} + Ed^{(N)} & = & 0 \end{cases}$$

If N is greater than the dimension of $d(x)$, the least-squares estimate of the matrix E is given as:

$$E^* = (\Gamma^+ \Psi)^T$$

where Γ^+ is the pseudo-inverse of Γ and

$$\Gamma = \begin{bmatrix} (d^{(1)})^T \\ (d^{(2)})^T \\ \vdots \\ (d^{(N)})^T \end{bmatrix} \qquad \Psi = -\begin{bmatrix} (Ax^{(1)} + Bu^{(1)})^T \\ (Ax^{(2)} + Bu^{(2)})^T \\ \cdots \\ (Ax^{(N)} + Bu^{(N)})^T \end{bmatrix}$$

From the simulation, the following estimate is obtained:

$$E^* = \begin{bmatrix} 1.3293 & 3.4440 & 0.1375 & -5.1304 & -1.7826 & -1.8719 \\ 5.6812 & -0.5281 & -0.3385 & -1.6193 & 0.5229 & 0 \\ 0 & 0 & 0 & 0 & 0 & 0 \end{bmatrix}$$

E^* is *not* a full column rank matrix ($rank(E^*) = 2$) and should be decomposed as $E^* = E_1 E_2$. Here E_1 is a full column matrix and will be used in the robust fault detection filter design.

$$E_1 = \begin{bmatrix} 6.2006 & 2.8639 \\ 4.1048 & -4.3262 \\ 0 & 0 \end{bmatrix}$$

All eigenvalues of the robust fault detection filter are set to -3. Using the design procedure presented in this chapter, with E replaced by E_1, the parameter matrices of the robust fault detection filter are as follows:

$$H = \begin{bmatrix} 0.6117 & -0.1170 & 0 & 0.3215 & 0.3220 & -0.1295 \\ -0.1170 & 0.9382 & 0 & 0.0605 & 0.0623 & -0.1916 \\ 0 & 0 & 0 & 0 & 0 & 0 \end{bmatrix}$$

$$T = \begin{bmatrix} 0 & 0 & -0.1251 \\ 0 & 0 & 0.0783 \\ 0 & 0 & 1.0000 \end{bmatrix}$$

$$K = \begin{bmatrix} -0.0708 & 0.0443 & 0.5658 & 0.1400 & 0.1531 & 0.3540 \\ 0.0443 & -0.0277 & -0.3540 & -0.0876 & -0.0958 & -0.2215 \\ 0.5658 & -0.3540 & -4.5229 & -1.1193 & -1.2239 & -2.8297 \end{bmatrix}$$

This robust fault detection filter is also applied to the non-linear simulation model to isolate faults in sensor Nos.1, 2 and 3. To compare the isolation

Table 3.3. Fault isolation using robust fault detection filter

Faulty sensor	*No.1*	*No.2*	*No.3*
NPD_1	**0.00621**	0.86727	0.90677
NPD_2	0.88625	**0.00213**	0.56602
NPD_3	0.89433	0.02092	**0.00159**

performance with the BFDF, the system and fault simulation have been set as exactly the same. The normalized projection distances for different faulty situations are shown in Table 3.3.

From Table 3.3, one can seen that NPD_i $(i = 1, 2, 3)$ is the smallest one amongst all normalized projection distances when a fault occurs in the i_{th} sensor. Moreover, the smallest NPD is significantly different from other NPDs. This simulation shows that the fault can be correctly isolated using a robust fault detection filter, even in the presence of modeling errors.

Remarks: This section has studied the design of a robust fault detection filter, and its application in the sensor fault isolation problem for a jet engine control system. The jet engine is a highly non-linear system, and hence the linearization error causes unreliable isolation if the robustness issues are not considered at the design. To cope with this problem, this section has developed a second order model to account for the linearization errors. Based on this model, a robust fault detection filter is designed and applied to the non-linear jet engine simulation model and the results show the effectiveness of the robust fault isolation strategy developed in the paper. The technique can be applied to the robust fault isolation for a wide range of systems with uncertain factors.

3.5 Filtering and Robust Fault Diagnosis of Uncertain Stochastic Systems

The problem of detecting and isolating faults in systems with both modeling uncertainty (including unknown disturbances and modeling errors) and noise has not attracted enough research attention, although most systems actually suffer from both modeling uncertainty and noise. This is partly due to a lack of techniques for designing disturbance de-coupling (unknown input de-coupling) optimal (minimum estimation error variance) observers for systems with both unknown disturbances and noise. Recently, some progress has been made in the design of optimal filters for stochastic systems with unknown disturbances. Darouach, Zasadzinski and Keller (1992) proposed an approach for the design of unknown input de-coupled optimal observers by transforming a standard system with unknown inputs into a singular system without unknown inputs, however they only considered time-invariant systems. Chang and Hsu (1993b) also made a contribution in the design of unknown input de-coupled optimal observers for time-invariant systems. Hou and Müller (1993) studied the unknown-input de-coupled filtering for descriptor (singular) systems with unknown inputs. In their study, two transformations were used to remove the unknown inputs. The first transformation transforms the descriptor system with unknown inputs into a descriptor system without unknown inputs, the second step is to transform the singular system into an ordinary system. The filtering algorithm in their approach is very complicated due to the involvement of two transformations. Moreover, the transformation could introduce extra restrictions and result in loss of design freedom.

This section introduces the authors' studies on optimal filtering and robust fault diagnosis for stochastic systems with unknown disturbances (or unknown and inaccessible inputs) (Chen and Patton, 1994b; Chen and Patton, 1996). This section uses an optimal full order observer with a simple structure, with which, the disturbance de-coupling is easily satisfied. This avoids some of the unnecessary and complex computation involved in some unknown input observer design methods. This section proves that the remaining design freedom, after disturbance de-coupling, can be utilized to ensure that the state estimation has the required minimal variance when noise (with known statistics) acts upon the system. This forms a solution for the optimal observer problem when the system has both unknown disturbance and noise. This section also presents the existence condition and the design procedure for the optimal observer. The existence condition for disturbance de-coupling can be easily verified. Unlike other studies (Darouach et al., 1992; Chang and Hsu, 1993b; Keller et al., 1996), this section focuses on time-varying systems. Comparing the algorithm given by Hou and Müller (1993), the filtering algorithm presented in this section is simpler and more straightforward. It should be also pointed out that the optimal observers presented in (Darouach et al., 1992; Chang and Hsu, 1993b; Hou and Müller, 1993) have not as yet been applied to robust fault diagnosis.

The optimal observer proposed in this section is applied to the robust fault diagnosis problem. The optimal output estimation can easily be produced using the principle of disturbance de-coupling state estimation. To detect and isolate faults, the output estimation error is used as a residual which is robust against unknown disturbances and has minimal variance. A hypothesis-testing procedure is then applied to examine the likelihood of residuals, and to indicate whether or not a fault has occurred in the system. A simplified flight control system is used to illustrate the method presented in the section. It has been shown that the state estimation obtained by the developed method is an improvement over the estimation obtained using a standard Kalman filter, when modeling errors occur. This is, of course, an advancement which is not confined to FDI problems. The simulation results also show that the method developed is able to detect faults in the presence of both modeling errors and noise.

3.5.1 Optimal observers for systems with unknown disturbances and noise

Consider the following discrete-time mathematical description of the system:

$$\begin{cases} x_{k+1} &= A_k x_k + B_k u_k + E_k d_k + \zeta_k \\ y_k &= C_k x_k + \eta_k \end{cases} \tag{3.42}$$

where $x_k \in \mathbb{R}^n$ is the state vector, $y_k \in \mathbb{R}^m$ is the output vector, $u_k \in \mathbb{R}^r$ is the known input vector and $d_k \in \mathbb{R}^q$ is the disturbance (or unknown input) vector, ζ_k and η_k are independent zero mean white noise sequences with covariance matrices Q_k and R_k. A_k, B_k, C_k and E_k are known matrices with appropriate dimensions.

The term $E_k d_k$ can be used to describe a number of different kinds of modeling uncertainties, e.g., interconnecting terms in the large scale systems, nonlinear terms in system dynamics (Frank and Wünnenberg, 1989; Chen and Zhang, 1991; Patton and Chen, 1993b), and also linearization and model reduction errors and parameter variations. A detailed study can be found in Chapter 5. It should be pointed out, however, that there are some problems which need to be studied further in the representation of modeling errors as disturbances. One problem is that the distribution matrix could be time varying and this is considered here as the study focuses on stochastic time-varying systems.

In order to estimate the state of the stochastic system with unknown disturbances described by Eq.(3.42), an optimal observer with the following structure is proposed:

$$\begin{cases} z_{k+1} &= F_{k+1} z_k + T_{k+1} B_k u_k + K_{k+1} y_k \\ \hat{x}_{k+1} &= z_{k+1} + H_{k+1} y_{k+1} \end{cases} \tag{3.43}$$

where the matrices F_{k+1}, T_{k+1}, K_{k+1} and H_{k+1} are to be designed to achieve disturbance de-coupling minimum variance estimation. The block diagram to illustrate this optimal observer is shown in Fig.3.11.

When the proposed observer is applied to a stochastic system with unknown disturbances, the state estimation error ($e_k = x_k - \hat{x}_k$) is as follows:

$$\begin{aligned} e_{k+1} &= x_{k+1} - (z_{k+1} - H_{k+1} y_{k+1}) \\ &= (I - H_{k+1} C_{k+1}) x_{k+1} - z_{k+1} - H_{k+1} \eta_{k+1} \\ &= (I - H_{k+1} C_{k+1}) x_{k+1} - H_{k+1} \eta_{k+1} \\ &\quad -[F_{k+1} z_k + T_{k+1} B_k u_k + (K_{k+1}^1 + K_{k+1}^2) y_k] \\ &= (I - H_{k+1} C_{k+1}) x_{k+1} - H_{k+1} \eta_{k+1} - T_{k+1} B_k u_k \\ &\quad -F_{k+1}(x_k - e_k - H_k y_k) - K_{k+1}^1 (C_k x_k + \eta_k) - K_{k+1}^2 y_k \\ &= F_{k+1} e_k - K_{k+1}^1 \eta_k - H_{k+1} \eta_{k+1} \\ &\quad +(I - H_{k+1} C_{k+1}) \zeta_k - [F_{k+1} - (I - H_{k+1} C_{k+1}) A_k + K_{k+1}^1 C_k] x_k \\ &\quad +(I - H_{k+1} C_{k+1}) E_k d_k - [K_{k+1}^2 - F_{k+1} H_k] y_k \\ &\quad -[T_{k+1} - (I - H_{k+1} C_{k+1})] B_k u_k \end{aligned} \tag{3.44}$$

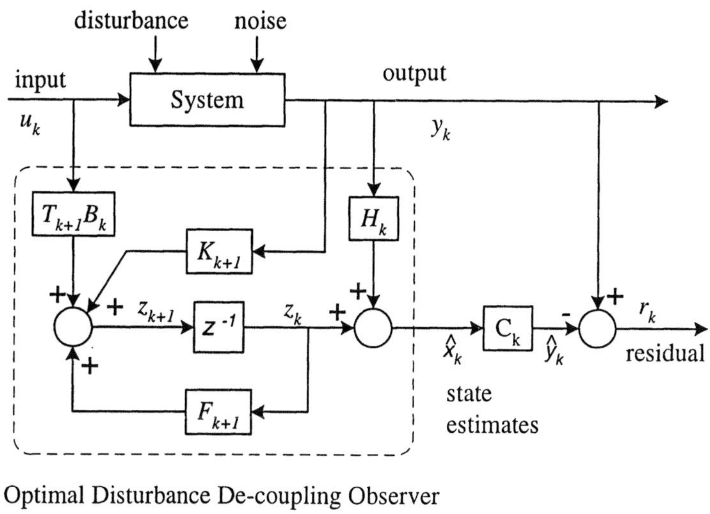

Figure 3.11. Optimal disturbance de-coupling observer and residual

where

$$K_{k+1} = K^1_{k+1} + K^2_{k+1} \tag{3.45}$$

If one can make the following relations hold true:

$$E_k = H_{k+1}C_{k+1}E_k \tag{3.46}$$
$$T_{k+1} = I - H_{k+1}C_{k+1} \tag{3.47}$$
$$F_{k+1} = A_k - H_{k+1}C_{k+1}A_k - K^1_{k+1}C_k \tag{3.48}$$
$$K^2_{k+1} = F_{k+1}H_k \tag{3.49}$$

the estimation error will be:

$$e_{k+1} = F_{k+1}e_k - K^1_{k+1}\eta_k - H_{k+1}\eta_{k+1} + T_{k+1}\zeta_k \tag{3.50}$$

Loosely speaking, if the matrix F_{k+1} is stable, $\mathcal{E}\{e_k\} \to 0$ and $\mathcal{E}\{\hat{x}_k\} \to \mathcal{E}\{x_k\}$ (where $\mathcal{E}\{.\}$ denotes the expectation or mean operator). That is to say, the state estimation will approach the real state asymptotically, in the mean sense. From Eq.(3.50), it can be seen that the unknown disturbance vector has been de-coupled once Eqs. (3.46)–(3.49) hold true. To design the disturbance de-coupled observer, one needs to chose the matrix H_{k+1} to satisfy Eq.(3.46) and to chose the matrix K^1_{k+1} to stabilize the matrix F_{k+1}. Once H_{k+1} and K^1_{k+1} have been chosen, other matrices can be determined using Eqs.(3.47) to (3.49).

Lemma 3.3 *The necessary and sufficient condition for the existence of a solution to Eq. (3.46) is:*

$$rank(C_{k+1}E_k) = rank(E_k) \tag{3.51}$$

The proof is the same as that for Lemma 3.1 (see Section 3.2.1).

Eq. (3.51) is the only condition for achieving disturbance (unknown input) de-coupling. To satisfy this equation, the number of independent rows of the matrix C_{k+1} must not be less than the number of independent columns of the matrix E_k. That is to say, the maximum number of disturbances which can be de-coupled cannot be larger than the number of independent measurements. When condition (3.51) holds true, the general solution for Eq. (3.46) can be constructed as:

$$H_{k+1} = H^0_{k+1} + H^1_{k+1}H^2_{k+1} \tag{3.52}$$
$$H^0_{k+1} = E_k(C_{k+1}E_k)^+ \tag{3.53}$$
$$H^2_{k+1} = I_m - (C_{k+1}E_k)(C_{k+1}E_k)^+ \tag{3.54}$$

and $H^1_{k+1} \in \mathbb{R}^{n\times m}$ can be arbitrarily chosen. To simplify the observer design, the matrix H^1_{k+1} can be set zero for most cases, i.e.,

$$H_{k+1} = E_k(C_{k+1}E_k)^+ \tag{3.55}$$

The stability (or convergence) of the observer is dependent on the matrix F_{k+1}, once the matrix H_{k+1} is obtained, the system dynamic matrix can be determined by:

$$F_{k+1} = A^1_{k+1} - K^1_{k+1}C_k \tag{3.56}$$

where:

$$A^1_{k+1} = A_k - H_{k+1}C_{k+1}A_k \tag{3.57}$$

The matrix K^1_{k+1} should be designed to stabilize the observer. On considering the simplest case, i.e., when the system is time-invariant, the matrix F can easily be stabilized using pole placement if the matrix pair $\{C, A_1\}$ is observable. For time-varying systems the stability is more difficult to verify, however divergence should not be a problem if the eigenvalues of the each matrix F_{k+1} have been assigned to the left side of complex plane via the gain matrix K^1_{k+1}.

It is clearly of interest to know how good the estimate $\hat{x}_k$ is. The variance of this estimation can be measured using the error covariance matrix P_k defined as:

$$P_k = \mathcal{E}\{[x_k - \hat{x}_k][[x_k - \hat{x}_k]^T\} \tag{3.58}$$

From the Eq.(3.50), it is easy to seen that the update of the covariance matrix is:

$$\begin{aligned} P_{k+1} = {} & (A^1_{k+1} - K^1_{k+1}C_k)P_k(A^1_{k+1} - K^1_{k+1}C_k)^T \\ & + K^1_{k+1}R_k(K^1_{k+1})^T + T_{k+1}Q_kT^T_{k+1} + H_{k+1}R_{k+1}H^T_{k+1} \end{aligned} \tag{3.59}$$

The best (optimal) state estimation should have minimal variance. From Eq.(3.59), it can be seen that the covariance matrix of the estimation error is controlled by the matrix K^1_{k+1}. The following theorem is now used to give the design of the matrix K^1_{k+1}, for achieving the minimum variance estimation.

Theorem 3.2 *To make the state estimation error e_{k+1} have the minimum variance, the matrix K^1_{k+1} should be determined by:*

$$K^1_{k+1} = A^1_{k+1} P_k C_k^T [C_k P_k C_k^T + R_k]^{-1} \tag{3.60}$$

Proof: For brevity, some subscripts are omitted in the following proof.

$$\begin{aligned} P_{k+1} &= A^1 P_k (A^1)^T + T Q_k T^T + H R_{k+1} H^T \\ &\quad -K^1 C P_k (A^1)^T - A^1 P_k C^T (K^1)^T + K^1 [C P_k C^T + R_k](K^1)^T \end{aligned}$$

As R_k is a positive definite matrix, $CP_kC^T + R_k$ is also positive definite and there exists an invertible matrix S, such that:

$$SS^T = CP_kC^T + R_k$$

Let $D = A^1 P_k C^T [S^T]^{-1}$, the covariance matrix is:

$$\begin{aligned} P_{k+1} &= A^1 P_k (A^1)^T + H R_{k+1} H^T - DD^T \\ &\quad + [K^1 S - D][K^1 S - D]^T + T Q_k T^T \end{aligned}$$

To minimize $var\{e_{k+1}\} = trace\{P_{k+1}\}$, one should make $K^1 S - D = 0$, this leads to Eq.(3.60) and we have that:

$$P_{k+1} = A^1_{k+1} P'_{k+1} (A^1_{k+1})^T + T_{k+1} Q_k T^T_{k+1} + H_{k+1} R_{k+1} H^T_{k+1} \tag{3.61}$$

where

$$P'_{k+1} = P_k - K^1_{k+1} C_k P_k (A^1_{k+1})^T \tag{3.62}$$

◊ QED

From the above derivation and theorem, the computational procedure for the optimal filtering algorithm can be listed in Table 3.4.

It is important to note that the optimal filtering algorithm proposed in this section is equivalent to a standard Kalman filter for systems *without* unknown disturbances, by setting the matrices $H_{k+1} = 0$ and $T_{k+1} = I$ when there is no disturbance, i.e. $E = 0$. From Eq.(3.52), it can be seen that the solution for the matrix H_{k+1} is not unique as the matrix H^1_{k+1} can be set arbitrarily. By choosing this free matrix H^1_{k+1}, the variance of the estimation error may be decreased slightly further, however this will result in a very complicated algorithm. Hence, it is more practical to fix the solution for H_{k+1} using Eq.(3.55).

Table 3.4. Optimal disturbance de-coupling observer design procedure

---

1° Set initial values: $P_0 = P(0)$, $z_0 = x_0 - C_0E_0(C_0E_0)^+y_0$, $H_0 = 0$ and k=0.

2° Compute H_{k+1} using Eq. (3.55).

3° Compute K^1_{k+1} and P'_{k+1} using Eqs. (3.60) and (3.62).

4° Compute T_{k+1}, F_{k+1}, K^2_{k+1} and K_{k+1} using Eqs. (3.47), (3.48), (3.49) and (3.45).

5° Compute the state estimate $\hat{x}_{k+1}$ and z_{k+1} using Eq. (3.43).

6° Compute P_{k+1} using Eqs. (3.61) & (3.62).

7° Set $k = k + 1$ go to step 2°.

3.5.2 *Robust residual generation and fault detection*

In order to diagnose faults, a fault indicating signal, i.e. residual, can be generated using the output estimation as follows:

$$r_k = y_k - \hat{y}_k = (I - C_kH_k)y_k - C_kz_k \tag{3.63}$$

The system with possible actuator and sensor faults can be described as:

$$\left\{\begin{array}{rcl} x_{k+1} & = & A_kx_k + B_ku_k + E_kd_k + \zeta_k + B_kf^a_k \\ y_k & = & C_kx_k + \eta_k + f^s_k \end{array}\right. \tag{3.64}$$

where $f^a_k \in \mathbb{R}^r$ is the actuator fault vector and $f^s_k \in \mathbb{R}^m$ is the sensor fault vector. For this system, the state estimation error and the residual are governed by the following equations:

$$\left\{\begin{array}{rcl} e_k & = & F_ke_{k-1} + K^1_k\eta_{k-1} - H_k\eta_k + T_k\zeta_{k-1} + K^1_kf^s_{k-1} \\ & & -H_kf^s_k + T_kB_{k-1}f^a_{k-1} \\ r_k & = & C_ke_k + \eta_k + f^s_k \end{array}\right. \tag{3.65}$$

It can be seen that the unknown disturbance term E_kd_k does not affect the residual, i.e. the residual is robust against unknown disturbances. As the state estimation error e_k has minimum variance, the residual is also optimal with respect to noise (with assumed statistics). For the residual, the two hypotheses to be tested can be identified as H_0, the normal mode, and the faulty mode H_1. Under the normal (no fault) condition, the statistics of the residual are:

$$H_0 : \left\{\begin{array}{l} \mathcal{E}\{r_k\} = 0 \\ covariance\{r_k\} = W_k = C_kP_kC_k^T + R_k \end{array}\right. \tag{3.66}$$

When a fault occurs in the system (H_1), the statistics of the residual will be different from the normal mode. The task of fault detection is to distinguish between two hypotheses H_1 and H_0. Any of the well-known hypothesis-testing methods, e.g. Generalized Likelihood Ratio (GLR) testing and Sequential Probability Ratio Testing (SPRT) (Willsky, 1976; Basseville, 1988) can be used to examine the residual and, subsequently to diagnose faults. If one assumes that the noise sequences ζ_k and η_k are Gaussian white, the residual will also have the Gaussian distribution. To construct a detection decision function (the test statistic) λ_k:

$$\lambda_k = r_k^T W_k^{-1} r_k \tag{3.67}$$

which is χ^2 distributed with m degrees of freedom (m is the dimension of r_k). The test for fault detection is then:

$$\begin{cases} \lambda_k \geq T_D & \text{fault} \\ \lambda_k < T_D & \text{no fault} \end{cases} \tag{3.68}$$

where the threshold T_D is determined from the χ^2 distribution table and:

$$Probability\{\lambda_k \leq T_D \mid H_0\} = P_f \tag{3.69}$$

where P_f is the probability of false alarm which is given by the designer.

The detection function λ_k is constructed using only a single sample of the residual. To increase the reliability of statistical testing, a residual sequence over a time window can be used. It is easy to verify that the covariance of $\{r_k, r_{k-1}\}$ is non-zero, i.e. the resulting residual sequence is not white, although both noise signals ζ_k and η_k are white. This will increase the difficulty and complexity in testing the residual sequence, however this penalty is worth paying in order to ensure that the unknown disturbance has been de-coupled from the residual. This is especially true when the unknown disturbance has a more dominant effect on the residual than the noise does.

3.5.3 An illustrative example

The linearized discrete-time model of a simplified longitudinal flight control system is as follows:

$$\begin{cases} x_{k+1} &= A_k x_k + B_k u_k + \zeta_k + E_k d_k \\ y_k &= C_k x_k + \eta_k \end{cases}$$

where the state variables are: pitch angle δ_z, pitch rate ω_z and normal velocity η_y, the control input is elevator control signal. The system parameter matrices are:

$$A_k = \begin{bmatrix} 0.9944 & -0.1203 & -0.4302 \\ 0.0017 & 0.9902 & -0.0747 \\ 0 & 0.8187 & 0 \end{bmatrix} \qquad B_k = \begin{bmatrix} 0.4252 \\ -0.0082 \\ 0.1813 \end{bmatrix}$$

$$C_k = I_{3\times 3} \qquad x = [\eta_y \ \omega_z \ \delta_z]^T$$

The covariance matrices for input and output noise sequences are: $Q_k = diag\{0.1^2,\ 0.1^2,\ 0.01^2\}$ and $R_k = 0.1^2 I_{3\times3}$. The term $E_k d_k$ is used here to represent the parameter perturbation in matrices A_k and B_k:

$$\begin{aligned} E_k d_k &= \Delta A_k x_k + \Delta B_k u_k \\ &= E\{\begin{bmatrix} \Delta a_{11} & \Delta a_{12} & \Delta a_{13} \\ \Delta a_{21} & \Delta a_{22} & \Delta a_{23} \end{bmatrix} x_k + \begin{bmatrix} \Delta b_1 \\ \Delta b_2 \end{bmatrix} u_k\} \end{aligned}$$

with

$$E = \begin{bmatrix} 1 & 0 \\ 0 & 1 \\ 0 & 0 \end{bmatrix}$$

where, Δa_{ij} and Δb_i $(i = 1, 2; j = 1, 2, 3)$ are perturbations in aerodynamic and control coefficients. They are unknown and can be time-varying. The perturbations can affect the estimation accuracy. In this section, their effects on the system have been modeled as unknown disturbances and can be de-coupled from the state estimation using the method given in Section 3.5.1.

The simulation is used to assess the usefulness of the optimal observer for estimating states. In the simulation, the input and initial conditions are set as $u_k = 10$, $x_0 = 0$ and $P_0 = 0.1^2 I_{3\times3}$. The aerodynamic coefficients are perturbed by $\pm 50\%$, i.e. $\Delta a_{ij} = -0.5 a_{ij}$ and $\Delta b_j = 0.5 b_j$. Fig.3.12–Fig.3.14 shows the absolute values of the state estimation errors.

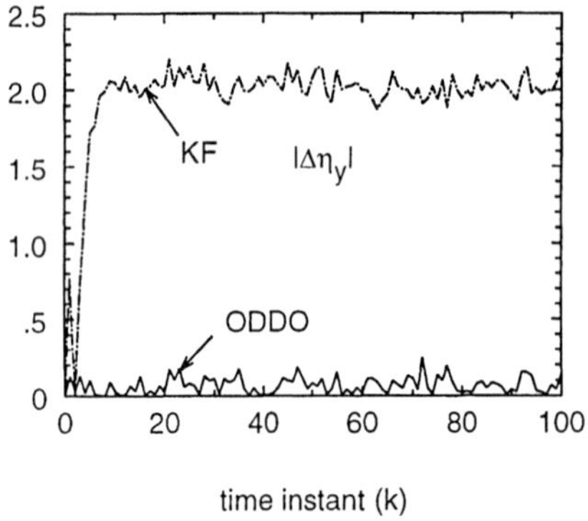

Figure 3.12. The state estimation error absolute values for η_y (ODDO: Optimal Disturbance De-coupling Observer; KF: Kalman Filter)

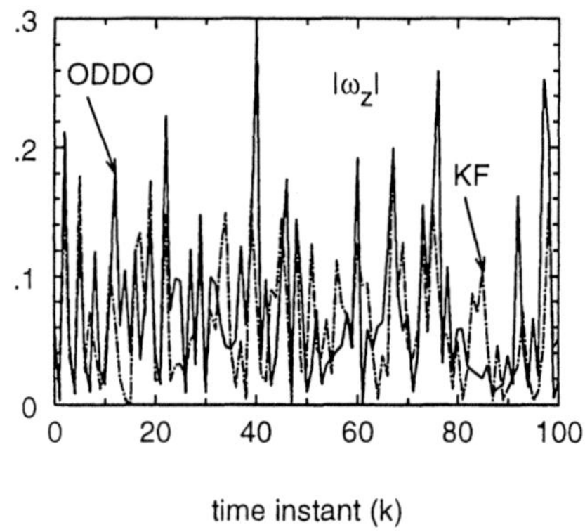

Figure 3.13. The state estimation error absolute values for ω_z (ODDO: Optimal Disturbance De-coupling Observer; KF: Kalman Filter)

The estimation errors achieved by the traditional Kalman filter (not disturbance de-coupled) are also shown in the Fig.3.12–Fig.3.14. It can be seen that the method developed in this section can give better state estimation, even when the system parameters have large perturbations. A number of situations when aerodynamic coefficients have time-varying (e.g. sinusoid function) perturbations (the results are not shown in this section) have also been simulated.

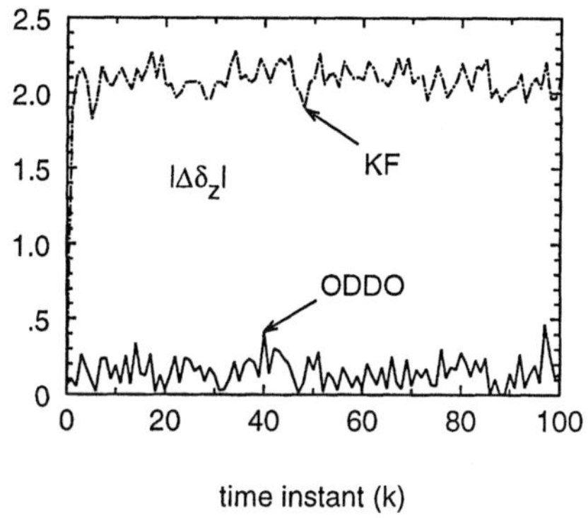

Figure 3.14. The state estimation error absolute values for δ_z (ODDO: Optimal Disturbance De-coupling Observer; KF: Kalman Filter)

For such cases, the estimation error using the Kalman filter is always divergent even if the perturbation magnitude is very small. However, the disturbance de-coupling method given in this section can give satisfactory estimation. This is expected, since the perturbation effects on the estimation error have been de-coupled.

Fig.3.15 shows the detection function λ_k when an incipient (small and slow) fault occurs in the sensor for δ_z. Fig.3.16 shows the fault detection function λ_k when a step fault occurs in the actuator. It can be seen that the faults are detected very reliably by setting a threshold (T_D) on the fault detection function.

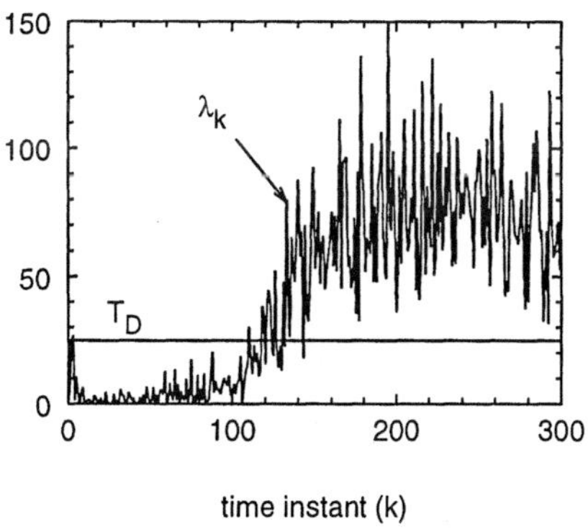

Figure 3.15. The fault detection function when a fault occurs in the sensor for δ_z

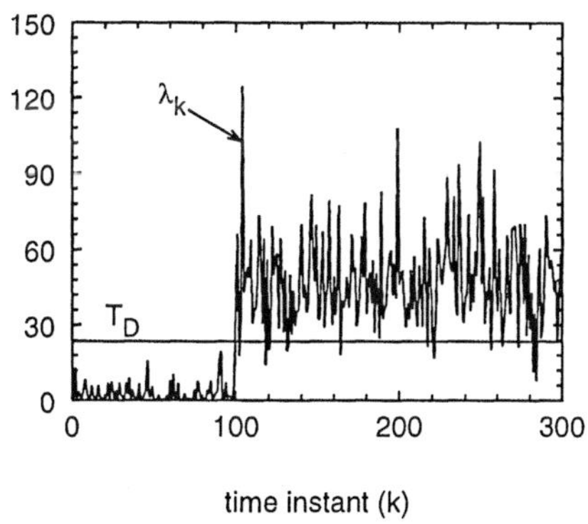

Figure 3.16. The fault detection function when a fault occurs in the actuator

<u>**Remarks:**</u> This section has proposed a systematic approach to designing optimal disturbance de-coupled observers for systems with both unknown disturbance and noise. This optimal observer is used to estimate the system state and to generate residuals for detecting faults in stochastic uncertain systems.

It is the first time such consideration has been addressed and solved in a fault diagnosis design. The method has been applied to detecting sensor and actuator faults in a simplified flight control system and the simulation results show the effectiveness of the method. Considering the extreme difficulty in enhancing the fault diagnosis performance under modeling uncertainty and noise, any improvement in the robustness of residual design is very welcome. The scope of applications of this work extends to a wide range of stochastic uncertain systems and is not confined to the fault diagnosis problem domain.

3.6 Summary

The purpose of this chapter has been the introduction of UIO-based robust residual generation methods. A full-order UIO structure has been described in this chapter. The existence conditions and design procedures for such UIOs have also been introduced and soundly proved. When compare with other techniques in designing UIOs, the existence conditions presented in this chapter are very easy to verify. The design procedure described in this chapter is very straightforward, because it can be implemented using the pole placement routine (*PLACE*) in Control Systems Toolbox for MATLAB, together with a few simple matrix manipulation routines which are also available in MATLAB. The robust FDI schemes based upon UIOs have also been discussed in this chapter. A chemical reactor has been used to illustrate the robust actuator fault detection and isolation schemes.

The main advantage of full-order UIOs over other commonly used reduced-order UIOs is that there is more design freedom available after the unknown input de-coupling conditions have been satisfied. This chapter has exploited the remaining freedom to achieve other performance requirements for FDI, and has introduced a method to design a robust fault detection filter which can generate disturbance de-coupled directional residuals for fault isolation. This is achieved via a combination of the UIO and the BFDF principles. The effectiveness of robust fault detection filters in robust fault isolation has been demonstrated by a highly non-linear jet engine system example. The remaining freedom has been also used in this chapter to produce the minimum variance state estimations and residuals for stochastic systems with unknown disturbances. The optimal disturbance de-coupled observer described in this chapter can be used for the optimal filtering problem for a wide range of uncertain stochastic systems.

Robust FDI based on UIOs have been studied for many years. However, the number of reported applications is very limited. The main argument is that the unknown input distribution matrix, required for designing UIOs, is actually unknown for most practical systems. The chapter has demonstrated, by means of a number of examples, how UIO-based robust FDI methods can be used in practical systems in which the unknown input distribution matrix is not directly known. The success of such application studies could give some guide-lines for real industrial applications.

4 ROBUST RESIDUAL GENERATION BY THE ASSIGNMENT OF OBSERVER EIGENSTRUCTURE

4.1 Introduction

In Chapter 3, various approaches for generating robust residual via unknown input observers have been introduced. The underlying principle of these approaches is to make the state estimation error be independent of disturbances (or unknown inputs). The residual is defined as the (weighted) output estimation error which is a linear transformation of the state estimation error. The residual generated by UIOs is also independent of disturbances, if the disturbance term does not appear in the output equation or the disturbance term in the output equation has been nulled. In model-based FDI, the state estimation is not necessarily needed, because the required information is the diagnostic signal – residual. Hence, it is not necessary to de-couple the state estimation error from disturbances in model-based FDI. A direct approach to design disturbance de-coupled residuals is then required. In this approach, the residual itself is de-coupled from disturbances, however the state estimation error may not be. It can be expected that existing conditions for such a direct approach could be relaxed compared with those required for UIOs.

The most important direct approach to design robust (in the disturbance de-coupling sense) residual generators is the use of eigenstructure assignment in which some left eigenvectors of the observer are assigned to be orthogonal to the disturbance distribution directions. In this way, the residual can be made robust against disturbances. This approach was initially proposed by Patton et al. (1986) and has been studied and developed extensively in Prof. Patton's group (Patton and Willcox, 1987; Patton, 1988; Patton and Kangethe, 1989; Patton and Chen, 1991h; Patton and Chen, 1991c; Patton and Chen, 1991e; Patton and Chen, 1991b; Dalton et al., 1996; Duan, Patton, Chen and Chen, 1997). A mathematically sound treatment and extensions were given by Patton and Chen (Patton and Chen, 1991g; Patton, Chen, Millar and Kiupel, 1991; Patton and Chen, 1992b). The approach has been successfully applied to robust FDI of flight control systems (Patton and Willcox, 1987; Patton and Kangethe, 1989; Shen et al., 1998), jet engine systems (Patton and Chen, 1990; Patton and Chen, 1991f; Patton and Chen, 1991b; Patton and Chen, 1991a; Patton and Chen, 1992d; Patton, Chen and Zhang, 1992; Patton, Zhang and Chen, 1992), nuclear reactors (Patton, Chen and Millar, 1991; Patton, Chen and Millar, 1992), industrial actuators (Jorgensen et al., 1994), pumping systems (Dalton et al., 1996) and autonomous underwater vehicle (Healey, 1998). Note that Daley and Wang (Daley and Wang, 1991; Daley and Wang, 1992; Wang, Kropholler and Daley, 1993) have also presented a different approach to generate robust residuals via the assignment of observer left eigenvectors. Magni *et al* (Magni and Mouyon, 1991; Magni and Mouyon, 1992; Magni, Mouyon and Arsan, 1993; Arsan, Mouyon and Magni, 1994) have also proposed another approach in the robust residual generation by assigning eigenvectors for so-called "one-dimensional" (or elementary) observers. The problem was revisited recently by Shen et al. (1998).

This chapter gives a detailed treatment of the eigenstructure assignment approach for robust residual generation. The principle and existence conditions are presented in a number of theorems, and the design procedure is also given. The remaining design freedom after disturbance de-coupling has been satisfied is used to optimize other performance indices such as fault sensitivity. When the left eigenvectors of the observer are not assignable, the design problem of approximate assignment is discussed.

One of the recent developments in the eigenstructure assignment method for designing robust residual generators is the assignment of some right eigenvectors parallel to the disturbance distribution directions. This method was proposed by Patton and Chen (1991g; 1992b) and extended by Chen (1995) with a complete and sound mathematical treatment. Note that the observer design is a dual of the control design problem. The assignment of right eigenvectors in an observer design is equivalent to the assignment of left eigenvectors in a controller design. There were only a few studies on this problem (Zhang, Slater and Allemang, 1990; Choi, Lee, Kim and Kang, 1995; Choi, 1998). The right eigenvector assignment method developed by Chen (1995) is introduced in this chapter.

The chapter is mainly based on the use of continuous-time system models, although the techniques developed can be directly applied to discrete-time system models. The dead-beat design has unique characteristics in the discrete time domain. To take advantage of the dead-beat design, this chapter also introduces the robust residual generation problem in the discrete time domain. It can be seen that the dead-beat design makes the principle and design procedure very simple. The dead-beat design also gives a direct correspondence between the observer-based and parity relation approaches in residual generation and this phenomenon has been discussed by Patton and Chen (1991h; 1991c; 1991f; 1991e). Three numerical examples are used in this chapter to demonstrate the eigenstructure assignment approach in robust residual generation, and real applications are given in Chapter 5.

4.2 Residual Generation and Responses

In a similar way to Chapter 3, it is also assumed in this chapter that the system is disturbed by an additive unknown input term as follows:

$$\begin{cases} \dot{x}(t) & = & Ax(t) + Bu(t) + R_1 f(t) + Ed(t) \\ y(t) & = & Cx(t) + Du(t) + R_2 f(t) \end{cases} \tag{4.1}$$

where $x(t) \in \mathbb{R}^n$ is the state vector, $y(t) \in \mathbb{R}^m$ is the output vector, $u(t) \in \mathbb{R}^r$ is the known input vector and $d(t) \in \mathbb{R}^q$ is the unknown input (or disturbance) vector, $f(t) \in \mathbb{R}^g$ represents the fault vector which is considered as an unknown time function. A, B, C, D and E are known matrices with appropriate dimensions. The matrices R_1 and R_2 are fault distribution matrices which are known when the designer has been told which faults should be diagnosed. Similar to Chapter 3, the matrix E is assumed to be full column rank.

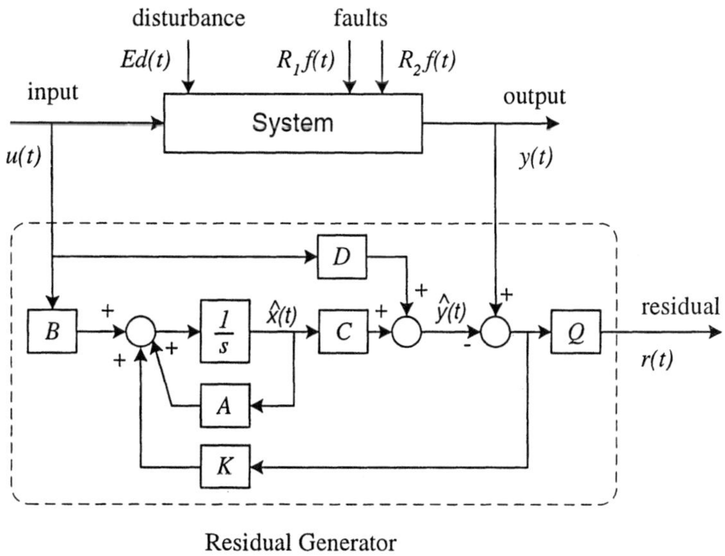

Figure 4.1. Robust observer-based residual generation

The residual generator based on a full-order observer illustrated in Fig. 4.1, is described as:

$$\begin{cases} \dot{\hat{x}}(t) &= (A - KC)\hat{x}(t) + (B - KD)u(t) + Ky(t) \\ \hat{y}(t) &= C\hat{x}(t) + Du(t) \\ r(t) &= Q[y(t) - \hat{y}(t)] \end{cases} \tag{4.2}$$

where $r \in \mathbb{R}^p$ is residual vector, $\hat{x}$ and $\hat{y}$ are state and output estimations. The matrix $Q \in \mathbb{R}^{p \times m}$ is the residual weighting factor. Note that, the residual is a linear transformation of the output estimation error. Hence, the residual dimension p cannot be larger than the output dimension m. This is because the linearly dependent extra residual components do not provide additional useful information in FDI.

When the residual generator represented by Eq.(4.2) is applied to the system described by Eq.(4.1), the state estimation error $(e(t) = x(t) - \hat{x}(t))$, and the residual are governed by the following equations:

$$\begin{cases} \dot{e}(t) &= (A - KC)e(t) + Ed(t) + R_1 f(t) - KR_2 f(t) \\ r(t) &= He(t) + QR_2 f(t) \end{cases} \tag{4.3}$$

where $H = QC$. The Laplace transformed residual response to faults and disturbances is thus:

$$\begin{aligned} r(s) &= QR_2 f(s) + H(sI - A + KC)^{-1}(R_1 - KR_2)f(s) \\ &\quad + H(sI - A + KC)^{-1}Ed(s) \end{aligned} \tag{4.4}$$

One can seen that the residual $r(t)$ and the state estimation error are not zero, even if no faults occur in the system. Indeed, it can be difficult to distinguish the effects of faults from the effects of disturbances acting on the system. The effects of disturbances obscure the performance of FDI and act as a source of false and missed alarms. Therefore, in order to minimize the false and missed alarm rates, one should design the residual generator such that the residual itself becomes de-coupled with respect to disturbances. Chapter 3 has discussed the UIO-based approaches in which the state estimation error $e(t)$ and hence the residual are de-coupled from disturbances. This chapter focuses on the technique which de-couples $r(t)$ from $d(t)$ directly. It is clearly not important whether or not $e(t)$ is de-coupled from $d(t)$ as $e(t)$ itself is not required in robust FDI.

4.3 General Principle for Disturbance De-coupling Design

In order to make the residual $r(t)$ be independent of disturbances, it is necessary to null the entries in the transfer function matrix between the residual and the disturbance. That means:

$$G_{rd}(s) = QC(sI - A + KC)^{-1}Ed(s) = 0 \tag{4.5}$$

This is a special case of the *output-zeroing* problem which is well known in multivariable control theory (Karcanias and Kouvaritakis, 1979). Once E is known, the remaining problem is to find the matrices K and Q to satisfy Eq.(4.5), in addition to choosing the suitable eigenvalues to optimize the FDI performance.

4.3.1 *Disturbance de-coupling design via invariant subspaces*

The solvability condition for matrices Q and K in Eq.(4.5) can be determined in the context of the invariant subspace theory (Morse, 1973; Antsaklis, 1980). The transfer matrix can be expanded[1] as follows:

$$\begin{aligned} H(sI-A_c)^{-1}E &= H[a_1(s)I_n + a_2(s)A_c + \cdots + a_n(s)A_c^{n-1}]E \\ &= [a_1(s)I_p \; a_2(s)I_p \; \cdots \; a_n(s)I_p] \begin{bmatrix} H \\ HA_c \\ \vdots \\ HA_c^{n-1} \end{bmatrix} E \\ &= H[E \; A_cE \; \cdots A_c^{n-1}E] \begin{bmatrix} a_1(s)I_q \\ a_2(s)I_q \\ \vdots \\ a_n(s)I_q \end{bmatrix} \end{aligned} \tag{4.6}$$

where $A_c = A - KC$ and $a_1(s), \cdots, a_n(s)$ are functions of s. From the above relation, it is easy to see that Eq.(4.5) can be solved by satisfying one of the following conditions:

(a) If the $\{H \, , \, A_c\}$ - invariant subspace lies in the left zero space of E, Eq. (4.5) holds true.

(b) If the $\{A_c \, , \, E\}$ - invariant subspace contained in the right zero space of H, Eq.(4.5) holds true.

The above two conditions give general guide-lines for designing disturbance de-coupling residuals (Patton and Willcox, 1987), however it is not easy to achieve these conditions without further assistance of design tools such as eigenstructure assignment.

4.3.2 *Disturbance de-coupling design via eigenstructure assignment*

In multivariable systems, there is extra design freedom available beyond eigenvalue assignment (Moore, 1976) and which can be used to assign eigenvectors to achieve the required system performances. In the residual generator design problem, the design freedom is used to assign the observer eigenstructure (eigenvalues and eigenvectors) to achieve disturbance de-coupling property. To

[1] This expansion can be proved by using the Taylor expansion of $\frac{1}{1-x}$ and the matrix Cayley-Hamilton theorem.

study this technique, two Lemmas which relate to the properties of the system observer eigenstructure, should be introduced.

Lemma 4.1 *A give left eigenvector l_i^T which is corresponding to eigenvalue λ_i of A_c is always orthogonal to the right eigenvector v_j corresponding to the remaining (n-1) eigenvalue λ_j of A_c where $\lambda_i \neq \lambda_j$ (Patton and Kangethe, 1989).*

Proof: For the left eigenvector l_i^T of A_c, we have:

$$l_i^T A_c = \lambda_i l_i^T \quad \text{for} \quad i = 1, 2, \cdots, n$$

Post-multiplying both side of the above equation by v_j $(j \neq i)$:

$$l_i^T A_c v_j = \lambda_i l_i^T v_j \quad \text{for} \quad i = 1, 2, \cdots, n; \;\; j \neq i$$

As the vector v_j is right eigenvector of A_c, we have $A_c v_j = \lambda_j v_j$, and the above equation can be rewritten as:

$$\lambda_j l_i^T v_j = \lambda_i l_i^T v_j \quad \text{for} \quad i = 1, 2, \cdots, n; \;\; j \neq i$$

Hence, if $\lambda_i \neq \lambda_j$, the only solution to the above equation is the trivial solution and it thus follows that:

$$l_i^T v_j = 0 \quad \text{for} \quad i \neq j \tag{4.7}$$

i.e. the left and right eigenvectors corresponding to mutually distinct eigenvalues are orthogonal.

◊ **QED**

Lemma 4.2 *Any transfer function matrix can be expanded in term of eigenstructure:*

$$(sI - A_c)^{-1} = \frac{v_1 l_1^T}{s - \lambda_1} + \frac{v_2 l_2^T}{s - \lambda_2} + \cdots + \frac{v_n l_n^T}{s - \lambda_n} \tag{4.8}$$

where v_i and l_i^T are right and left eigenvectors of A_c respectively, corresponding to the eigenvalue λ_i.

Note that this Lemma is only valid for cases when all eigenvectors of the observer are distinct, however this requirement does not impose any restriction on the observer design for FDI.

Proof: Define the left eigenvector and right eigenvector matrices as:

$$L = \begin{bmatrix} l_1^T \\ l_2^T \\ \vdots \\ l_n^T \end{bmatrix} \qquad V = [v_1, v_2, \cdots, v_n]$$

According to Lemma 4.1, we have the following relation:

$$LV = \begin{bmatrix} l_1^T v_1 & 0 & \cdots & 0 \\ 0 & l_2^T v_2 & \cdots & 0 \\ \vdots & \vdots & \ddots & \vdots \\ 0 & 0 & \cdots & l_n^T v_n \end{bmatrix}$$

If vectors l_i and v_i $(i = 1, 2, \cdots, n)$ are properly scaled, the above equation become:

$$LV = I_n$$

This means that:

$$L = V^{-1}$$

It is well-known that, the matrix A_c can be decomposed as:

$$A_c = V \Lambda V^{-1}$$

where $\Lambda = diag\{\lambda_1, \lambda_2, \cdots \lambda_n\}$. From the above equation, we have:

$$e^{A_c t} = V e^{\Lambda t} V^{-1} = \sum_{i=1}^{n} e^{\lambda_i t} v_i l_i^T$$

this leads to:

$$(sI - A_c)^{-1} = Laplace\left\{e^{A_c t}\right\} = Laplace\left\{\sum_{i=1}^{n} e^{\lambda_i t} v_i l_i^T\right\} = \sum_{i=1}^{n} \frac{v_i l_i^T}{s - \lambda_i}$$

◊ **QED**

Based on Lemma 4.2, Eq.(4.5) can be rewritten as:

$$G_{rd}(s) = \sum_{i=1}^{n} \frac{H v_i l_i^T E}{s - \lambda_i} \tag{4.9}$$

Thus, it can be noted that the disturbance de-coupling is possible *if and only if*:

$$R_i = H v_i l_i^T E = 0 \quad \text{for} \quad i = 1, 2, \cdots, n \tag{4.10}$$

This implies that:

$$\sum_{i=1}^{n} R_i = H \left(\sum_{i=1}^{n} v_i l_i^T \right) E = HVLE = HE = QCE = 0 \tag{4.11}$$

Hence, one of the necessary conditions for designing disturbance de-coupled residuals is given by the above equation and restated in the following theorem:

Theorem 4.1 *A necessary condition for achieving disturbance de-coupling design is:*

$$QCE = HE = 0 \tag{4.12}$$

If $CE = 0$, any residual weighting matrix can satisfy this necessary condition. However, this is not always the case. Loosely speaking, the column number of E cannot be larger than the independent row number of C to satisfy the above necessary condition, i.e. the number of independent disturbances can be de-coupled cannot larger than the number of independent measurements. If this necessary condition cannot satisfied, an approximate de-coupling procedure should be used, this is to approximate the matrix E by a lower rank matrix. This problem is studied in Chapter 5.

A general solution for Eq.(4.12) is given by:

$$Q = Q_1[I_m - CE(CE)^+] \tag{4.13}$$

where $Q_1 \in \mathbb{R}^{p \times m}$ is an arbitrary design matrix and $(CE)^+$ is the pseudo-inverse of CE and is given by the following equation if $rank(CE) = q$

$$(CE)^+ = [(CE)^T(CE)]^{-1}(CE)^T \tag{4.14}$$

The maximum independent row number of the matrix Q satisfying Eq.(4.12) is $m - rank(CE)$. As the linearly dependent rows do not provide any useful information, hence the row of the residual weighting matrix Q is normally chosen as:

$$p = m - rank(CE) \leq m \tag{4.15}$$

4.4 Disturbance De-coupling by Assigning Left Eigenvectors

The first method which was initially proposed and developed by Patton *et al* (Patton et al., 1986; Patton and Kangethe, 1989; Patton and Chen, 1991g) for disturbance de-coupling design via eigenstructure assignment is to assign left observer eigenvectors orthogonal to all columns of E. This method is summarized by the following theorem:

Theorem 4.2

The sufficient conditions for satisfying the disturbance de-coupling requirement Eq.(4.5) are:

(1) $QCE = 0$.

(2) *All rows of the matrix $H = QC$ are left eigenvectors of $(A - KC)$ corresponding to any eigenvalues.*

Proof: According to condition (2), the matrix H is constructed as:

$$H = \begin{bmatrix} l_1^T \\ \vdots \\ l_p^T \end{bmatrix}$$

where l_i^T $(i = 1, 2, \cdots, n)$ are the left eigenvectors of $A - KC$. Using the relation given in Lemma 4.1, we have:

$$Hv_i = 0 \quad \text{for} \quad i = p + 1, \cdots, n$$

where v_i $(i = 1, 2, \cdots, n)$ are the right eigenvectors of $A - KC$. According to condition (1) in Theorem 4.2, we have:

$$l_i^T E = 0 \quad \text{for} \quad i = 1, \cdots, p$$

From Lemma 4.2, the transfer matrix from the disturbance to the residual is expressed as:

$$G_{rd}(s) = \sum_{i=1}^{n} \frac{(Hv_i) l_i^T E}{s - \lambda_i} = \sum_{i=1}^{p} \frac{Hv_i (l_i^T E)}{s - \lambda_i} = 0$$

◇ **QED**

The main principle utilized in this proof can be illustrated graphically by Fig.4.2. This diagram shows the orthogonal relationships of eigenvectors and matrices H and E. According to condition (2) in Theorem 4.2, the rows of the matrix H are orthogonal to the lower partition of the right eigenvectors and hence the lower partition is nulled. Similarly, the top partition part is also nulled due to condition (1).

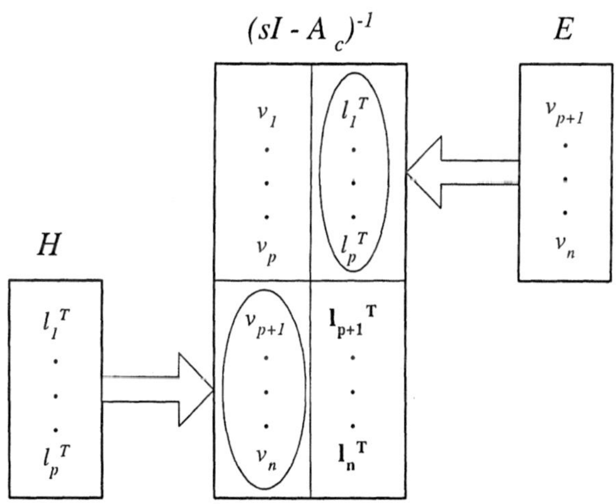

Figure 4.2. Disturbance de-coupling via eigenvector assignment

The procedure for the design of the disturbance de-coupling residual generator via left eigenvector assignment is thus as follows:

(a) Compute the residual weighting matrix Q so that $QCE = 0$.

(b) Determine the eigenstructure of the observer: The eigenvalues of the observer are chosen according to the desired dynamic property of residuals. The rows of QC must be the p left eigenvectors of the observer. The remaining $(n - p)$ left eigenvectors will be chosen so that one can ensure a design with good conditioning.

(c) Compute the gain matrix K using suitable eigenstructure assignment technique.

The observer feedback eigenstructure assignment problem can be handled by means of a transformation of the dual control form. On assignment of the right eigenvectors to the dual control problem, these eigenvectors become the left eigenvectors of the observer system (Andry, Chung and Shapiro, 1984; Sobel and Banda, 1989; Burrows and Patton, 1992; Liu and Patton, 1998). The assignment of the right eigenvectors for the control problem is a well-developed technique (Moore, 1976; Fahmy and O'Reilly, 1982; Andry, Shapiro and Chung, 1983; Kautsky, Nichols and Van Dooren, 1985; Roppenecker, 1986; Mudge and Patton, 1988; Owens, 1988; Owens and O'Reilly, 1989; White, 1991; Burrows, Patton and Szymanski, 1989; Burrows and Patton, 1991; Sobel, Shapiro and Andry, 1994). A comprehensive treatment and the design toolbox can be found in the recent book by Liu and Patton (1998).

The assignability condition is that, for each eigenvalue λ_i, the corresponding left eigenvector l_i^T must belong to the row subspace spanned by $[C(\lambda_i I - A)^{-1}]$. That is to say the vector l_i should lie in the column subspace spanned by $[(\lambda_i I - A^T)^{-1}C^T]$.

If l_i lies in the subspace $span\{[(\lambda_i I - A^T)^{-1}C^T]\}$, a vector w_i exists which satisfies the following equation:

$$l_i = P(\lambda_i)w_i \quad \text{for} \quad i = 1, \cdots, p \tag{4.16}$$

where:

$$P(\lambda_i) = (\lambda_i I - A^T)^{-1}C^T \quad \text{for} \quad i = 1, \cdots, p \tag{4.17}$$

One can inspect if l_i is in the subspace $span\{P(\lambda_i)\}$ by comparing l_i with its projection in this subspace, denoted by:

$$l_i^* = P(\lambda_i)w_i^* \quad \text{for} \quad i = 1, \cdots, p \tag{4.18}$$

where:

$$w_i^* = [P(\lambda_i)^T P(\lambda_i)]^{-1} P(\lambda_i)^T l_i \quad \text{for} \quad i = 1, \cdots, p \tag{4.19}$$

if $l_i = l_i^*$, l_i is in $span\{P(\lambda_i)\}$ and is assignable. Otherwise, an approximate procedure must be taken, i.e. to replace l_i by its projection l_i^*. In observer-based residual generator design, there are no other restrictions on the choice of eigenvalues apart from stability. Hence, one can chose stable eigenvalues to minimize the distance between a required eigenvector with its projection in the assignable subspace. The approximate disturbance de-coupling can be achieved by minimizing the following performance index:

$$\begin{aligned} J_1 &= \sum_{i=1}^{p} \|l_i - l_i^*\|_2 \\ &= \sum_{i=1}^{p} \|\{I_n - P(\lambda_i)[P(\lambda_i)^T P(\lambda_i)]^{-1} P(\lambda_i)^T\} l_i\|_2 \end{aligned} \tag{4.20}$$

where l_i^T $(i = 1, \cdots, p)$ are the required left eigenvectors to be assigned for the disturbance de-coupling design. It is possible the J_1 can be made zero by properly chosen eigenvalues λ_i $(i = 1, \cdots, p)$.

Because $(l_i^*)^T$ $(i = 1, \cdots, p)$ are left eigenvectors of $A - KC$ corresponding to eigenvalues λ_i, we have:

$$(l_i^*)^T (A - KC) = \lambda_i (l_i^*)^T \quad \text{for} \quad i = 1, \cdots, p \tag{4.21}$$

i.e.

$$l_i^* = (\lambda_i I - A)^{-1} C^T K^T l_i^* \quad \text{for} \quad i = 1, \cdots, p \tag{4.22}$$

Comparing Eq.(4.22) with Eq.(4.18), we have:

$$w_i^* = K^T l_i^* \quad \text{for} \quad i = 1, \cdots, p \tag{4.23}$$

For a disturbance de-coupling design, only p left eigenvectors are specified, the remaining $n - p$ eigenvectors can be chosen freely from the assignable subspace, i.e.

$$l_i = (\lambda_i I - A)^{-1} C^T w_i \quad \text{for} \quad i = p + 1, \cdots, n \tag{4.24}$$

where

$$w_i = K^T l_i \quad \text{for} \quad i = p + 1, \cdots, n \tag{4.25}$$

Hence, the observer feedback gain matrix is computed by:

$$K = [W L^{-1}]^T = [WV]^T = V^T W^T \tag{4.26}$$

where

$$W = [w_1^* \cdots w_p^*; \;\; w_{p+1} \cdots w_n] \in \mathbb{R}^{m \times n}$$

$$L = [l_1^* \cdots l_p^*; \;\; l_{p+1} \cdots l_n] \in \mathbb{R}^{n \times n}$$

and $V = L^{-1}$ is the right eigenvector matrix. Note that the first p eigenvalues corresponding to the required eigenvectors l_i^T $(i = 1, \cdots, p)$ must be real because all these eigenvectors are real-valued. The remaining $n - p$ eigenvalues and corresponding eigenvectors can be real as well as complex-conjugate.

Disturbance de-coupling does not place any restriction on the choice of eigenvectors l_i^T $(i = p+1, \cdots, n)$ and corresponding eigenvalues λ_i $(i = p+1, \cdots, n)$. Hence, these free parameters can be used to maximize the fault effect on the residual. Consider the transfer function between residuals and faults as:

$$\begin{aligned} G_{rf}(s) &= QR_2 + H(sI - A + KC)^{-1}(R_1 - KR_2) \\ &= QR_2 + H \sum_{i=1}^{n} \frac{v_i l_i^T}{s - \lambda_i} (R_1 - KR_2) \\ &= QR_2 + H \sum_{i=1}^{p} \frac{v_i l_i^T}{s - \lambda_i} (R_1 - KR_2) \end{aligned} \tag{4.27}$$

As pointed out in Section 2.7, the most important factor in fault detectability is the steady-state gain matrix $G_{rf}(0)$, hence a performance index to be maximized for increasing fault detectability, is defined as:

$$J_2(\Lambda, \overline{W}) = ||QR_2 + H\sum_{i=1}^{p} \frac{v_i l_i^T}{-\lambda_i}(R_1 - V^T W^T R_2)||_F \tag{4.28}$$

where $||\cdot||_F$ denotes the Frobenius norm, the following two matrix are designing parameters.

$$\Lambda = [\lambda_1, \ \cdots, \ \lambda_n]; \ \overline{W} = [w_{p+1}, \ \cdots, \ w_n]$$

To maximize the fault effect and, subsequently fault detectability, the performance index $J_2(\Lambda, \overline{W})$ should be maximized. The optimization problem can be solved by any suitable numerical search method. The genetic algorithm is a generic optimization technique because it has minimum degree of problem dependence, and hence can be used to solve this problem. In Chapter 6, the use of genetic algorithms is discussed in details.

The maximization of $J_2(\Lambda, \overline{W})$ is a constrained optimization problem because all elements of Λ must be in the left hand side of the complex plan. To remove this constraint, the eigenvalues λ_i are assumed in a pre-defined wide region $[L_i, U_i]$ and introduce a simple transformation (Burrows and Patton, 1991):

$$\lambda_i = L_i + (U_i - L_i)sin^2(z_i) \tag{4.29}$$

where $z_i \in \mathbb{R}$ $(i = 1, \cdots, n)$ can be freely chosen. Now, the performance index J_2 is a function of the parameters $Z = [z_1, \ \cdots, \ z_n]$ and $\overline{W}$.

If the required left eigenvectors are assignable, the performance index J_1 is zero and only the index J_2 needs to be maximized. If the assignability conditions cannot be satisfied, one alternative is to assign all the columns of E as right eigenvectors and this is studied in Section 4.6. Another alternative way is to use approximate de-coupling, i.e. to minimize J_1. The best FDI performance can be achieved by maximizing J_2 and minimizing J_1, this is a multi-objective optimization problem and can be solved by minimizing a single mixed objective. The objectives can be mixed-up in one of the following ways:

$$J(Z, W) = \frac{J_1(Z, W)}{J_2(Z, W)} \tag{4.30}$$

$$J(Z, W) = \alpha_1 \sum_{i=1}^{p} ||l_i - l_i^*||_2 + \frac{\alpha_2}{||QR_2 + H\sum_{i=1}^{p} \frac{v_i l_i^T}{-\lambda_i}(R_1 - V^T W^T R_2)||_F} \tag{4.31}$$

The multi-objective optimization problem can also be solved via the method of inequalities which is discussed in Chapter 6.

4.5 Robust Fault Detection Observer Design Using Parametric Eigenstructure Assignment Approach

There are many approaches for assigning the left observer eigenvectors (right eigenvector assignment in controller design). The design algorithms and associated MATLAB toolbox of eigenstructure assignment can be found in the book by Liu and Patton (1998). In this section, a parametric approach for eigenstructure assignment is presented (Duan et al., 1997). By using this approach, solutions of the weighted matrix and the eigenvectors of the observer are sought simultaneously, and therefore, the degrees of design freedom is fully provided. An example is worked out with the proposed algorithm and a general solution with eight design parameters is obtained, which well demonstrates the effect of the proposed approach.

Lemma 4.3 *Let $\{C, A\}$ be observable, then, for any group of distinct self-conjugate complex numbers λ_i, $i = 1, 2, \cdots, n$, there always exist a matrix $K \in \mathbb{R}^{n\times m}$ and a non-singular matrix $L \in \mathbb{C}^{n\times n}$ such that*

$$A - KC = L^{-1} diag\{\lambda_1, \lambda_2, \cdots, \lambda_n\} L \tag{4.32}$$

Moreover, all such matrices are parameterized by

$$K = L^{-1}W \tag{4.33}$$

and

$$L = \begin{bmatrix} l_1^T \\ l_2^T \\ \vdots \\ l_n^T \end{bmatrix}, \quad W = \begin{bmatrix} w_1^T \\ w_2^T \\ \vdots \\ w_n^T \end{bmatrix} \tag{4.34}$$

with

$$l_i = N(\lambda_i) g_i\,,\ w_i = M(\lambda_i) g_i\,,\ i = 1, 2, \cdots, n \tag{4.35}$$

where $M(s)$ and $N(s)$ is a pair of right coprime polynomial matrices of dimensions $n \times m$ and $m \times n$, respectively, and satisfy the following right-coprime factorization

$$(A^T - sI)^{-1} C^T = N(s) M^{-1}(s) \tag{4.36}$$

and $g_i \in \mathbb{C}^m$, $i = 1, 2, \cdots, n$, is a group of parameter vectors satisfying the following two constraints:

C1: $det\{[N(\lambda_1)g_1\, N(\lambda_2)g_2 \cdots N(\lambda_n)g_n]\} \neq 0$

C2: $g_i = \bar{g}_j$ *if and only if* $\lambda_i = \bar{\lambda}_j$

Remarks:

(1) Existence of the pair of polynomial matrices $M(s)$ and $N(s)$ satisfying the right-coprime factorization (4.36) is ensured by the observability of the matrix pair $\{C,\ A\}$. For solution of the right-coprime factorization (4.36), several methods have been given in Duan (1993).

(2) Existence of parameters $g_i \in \mathbb{C}^m$, $i = 1, 2, \cdots, n$, satisfying constraints **C1** and **C2** are guaranteed by general pole assignment result and the distinctness of the closed-loop eigenvalues λ_i, $i = 1, 2, \cdots, n$. Indeed, in the case of

$m > 1$, parameters $g_i \in \mathbb{C}^m$, $i = 1, 2, \cdots, n$, satisfying these two constraints are not unique, and hence can be properly chosen to meet some further specifications of the robust detection problem.

(3) It is only because of simplicity and convenience that we quote a simplified version of the eigenstructure assignment result in Duan (1993). Whilst as a matter of fact, the distinctness of the closed-loop eigenvalues λ_i, $i = 1, 2, \cdots, n$, can be removed, and each λ_i may possess arbitrary algebraic and geometric multiplicities.

4.5.1 Robust FDI design using parametric eigenstructure assignment

Let $A_c = A - KC$ have distinct eigenvalues, then it follows from Lemma 4.3 that the robust fault detection condition (4.5) holds if

i) $q_i^T CE = 0$, $i = 1, 2, \cdots, p$

ii) all $q_i^T C$, $i = 1, 2, \cdots, p$, are left eigenvectors of matrix A_c corresponding to p number eigenvalues of A_c.

Noting that all the left eigenvectors $l_i^T \in \mathbb{C}^n$, $i = 1, 2, \cdots, n$, of matrix A_c given in Lemma 4.3 have the same structure, so to guarantee the second condition above, without loss of generality, we may require

$$g_i^T N^T(\lambda_i) = q_i^T C, \quad i = 1, 2, \cdots, p \tag{4.37}$$

In view of the fact that $N(s)$ has full rank for all $s \in \mathbb{C}$, we see that the above group of equations have solutions with respect to g_i, $i = 1, 2, \cdots, p$, if and only if

$$rank\left\{\left[N(\lambda_i) \;\; C^T q_i\right]\right\} = m \quad i = 1, 2, \cdots, p \tag{4.38}$$

When the conditions in (4.38) hold, it is clear that solutions to the group of linear equations in (4.37) are

$$g_i = \left[N^T(\lambda_i) N(\lambda_i)\right]^{-1} N^T(\lambda_i) C^T q_i, \quad i = 1, 2, \cdots, p \tag{4.39}$$

Equations $q_i^T CE = 0$, $i = 1, 2, \cdots, p$ state that q_i^T, $i = 1, 2, \cdots, p$, are all belong to the left kernel space of (CE), that is, $q_i^T \in LKer(CE)$. Let ξ_i^T, $i = 1, 2, \cdots, \mu$, be a group of basis of $LKer(CE)$, then Equations $q_i^T CE = 0$, $i = 1, 2, \cdots, p$ are equivalent to

$$q_i = \alpha_{i1}\xi_1 + \alpha_{i2}\xi_2 + \cdots + \alpha_{i\mu}\xi_\mu, \quad i = 1, 2, \cdots, p \tag{4.40}$$

where α_{ij}, $i = 1, 2, \cdots, p$, $j = 1, 2, \cdots, \mu$, is a group of real numbers. Substituting (4.40) into (4.38) and (4.39) we now have

$$rank\left\{\left[N(\lambda_i) \;\; C^T \sum_{j=1}^{\mu} \alpha_{ij}\xi_j\right]\right\} = m, \quad i = 1, 2, \cdots, p \tag{4.41}$$

and

$$g_i = \left[N^T(\lambda_i)N(\lambda_i)\right]^{-1} N^T(\lambda_i)C^T \sum_{j=1}^{\mu} \alpha_{ij}\xi_j, \quad i = 1, 2, \cdots, p \tag{4.42}$$

It follows from the above reasoning we have the following theorem for solution to our robust fault detection problem.

Theorem 4.3 *Assume that E is a full column rank matrix, C is a full row rank matrix, $\{C, A\}$ is an observable pair, μ is the dimension of $LKer(CE)$, and ξ_i, $i = 1, 2, \cdots, \mu$, a basis of $LKer(CE)$. Then the robust fault detection problem has a solution if there exist (a) a group of complex numbers λ_i, $i = 1, 2, \cdots, n$; (b) a group of real scalars α_{ij}, $i = 1, 2, \cdots, p$, $j = 1, 2, \cdots, \mu$, and (c) a group of vectors $g_i \in \mathbb{C}^m$, $i = 1, 2, \cdots, n$, such that the following four requirements are met:*

i) *λ_i, $i = 1, 2, \cdots, n$, are distinct complex numbers with $Re(\lambda_i) < 0$, $i = 1, 2, \cdots, n$, and both the two groups λ_i, $i = 1, 2, \cdots, p$, and λ_j, $j = p + 1, p + 2, \cdots, n$, are self-conjugate;*

ii) *for any pair of $i, j \in \{p + 1, p + 2, \cdots, n\}$, $g_i = \bar{g}_j$ holds if and only if $\lambda_i = \bar{\lambda}_j$;*

iii) *the rank conditions in (4.41) hold;*

iv) *the matrix L given by (4.34), (4.35) and (4.41) is non-singular.*

When these conditions are met, the solution to this problem is given by (4.34), (4.35) and (4.40).

Based on the above theorem, an algorithm for robust fault detection can be given in Table 4.1.

As soon as the weighting matrix Q and the observer gain matrix K are obtained using this algorithm, the residual generate observer can be formed and implemented.

4.5.2 *An illustrative example*

Example 4-1: Consider the system with the parameters (Hou and Müller, 1994b):

$$A = \begin{bmatrix} 0 & 3 & 4 \\ 1 & 2 & 3 \\ 0 & 2 & 5 \end{bmatrix} \quad C = \begin{bmatrix} 0 & 1 & 0 \\ 0 & 0 & 1 \end{bmatrix}, \quad E = \begin{bmatrix} 1 & 0 \\ 0 & 1 \\ 0 & 0 \end{bmatrix}$$

It is clear that $\{C, A\}$ is an observable pair, E is full column rank and C is full row rank. Now let us apply Robust Fault Detection Design Algorithm to this system.

Table 4.1. Robust fault detection design algorithm

Step 1: Find the pair of right coprime polynomial matrices $M(s)$ and $N(s)$ satisfying the right-coprime factorization (4.36).

Step 2: Determine the dimension μ of $LKer(CE)$. If $\mu = 0$, the problem does not have a solution, terminate the algorithm. Otherwise, calculate a basis ξ_i, $i = 1, 2, \cdots, \mu$, for $LKer(CE)$.

Step 3: Form the general expressions of the parameter vectors g_i, $i = 1, 2, \cdots, p$, through formula (4.42).

Step 4: Form the general expression for the left eigenvector matrix L according to formula (4.34) & (4.34).

Step 5: Select a group of parameters λ_i's, α_{ij}'s and g_i, $i = p + 1, p + 2, \cdots, n$, satisfying the four requirements in Theorem (4.3).

Step 6: Calculate q_i^T, $i = 1, 2, \cdots, p$ according to (4.40) and g_i, $i = 1, 2, \cdots, p$, according to (4.42) based on the parameters λ_i's and α_{ij}'s obtained in Step 5.

Step 7: Calculate matrices L and W based on the parameters obtained in Step 5 and those g_i, $i = 1, 2, \cdots, p$, obtained in Step 6.

Step 8: Calculate the observer gain matrix K according to (4.33) based on the matrices L and W given in Step 7.

Step 1: Using a method given in Duan (1993), we can obtain

$$N(s) = \begin{bmatrix} 1 & 0 \\ s & 0 \\ 0 & 1 \end{bmatrix}, \quad M(s) = \begin{bmatrix} -3 + s(s-2) & -2 \\ -4 - 3s & s - 5 \end{bmatrix}$$

Step 2: Note that

$$CE = \begin{bmatrix} 0 & 1 \\ 0 & 0 \end{bmatrix}$$

we have $p = 1$, and a basis for $LKer(CE)$ may be taken as

$$\xi^T = \xi_1^T = \begin{bmatrix} 0 & 1 \end{bmatrix}$$

Step 3: Denote α_{11} simply by α, then though some simple deductions we have

$$g_1 = \left[N^T(\lambda_1)N(\lambda_1)\right]^{-1} N^T(\lambda_1)C^T\alpha\xi = \begin{bmatrix} 0 \\ \alpha \end{bmatrix}$$

Step 4: Denote

$$g_i = \begin{bmatrix} x_{i1} \\ x_{i2} \end{bmatrix}, \quad i = 2, 3$$

the L matrix can be obtained from Eqs.(4.34)&(4.35) as:

$$L = \begin{bmatrix} 0 & 0 & \alpha \\ x_{21} & \lambda_2 x_{21} & x_{22} \\ x_{31} & \lambda_3 x_{31} & x_{32} \end{bmatrix}$$

Step 5: Notice

$$\begin{bmatrix} N(\lambda_1) & C^T \alpha \xi \end{bmatrix} = \begin{bmatrix} 1 & 0 & 0 \\ \lambda_1 & 0 & 0 \\ 0 & 1 & \alpha \end{bmatrix}$$

we see that the rank condition (4.41) holds for all $\lambda_1 \in \mathbb{C}$ and $\alpha \in \mathbb{R}$. From $\det(L) = \alpha(\lambda_3 - \lambda_2)x_{21}x_{31}$ and notice the distinctness requirement on the observer eigenvalues λ_i, $i = 1, 2, 3$, we know that the non-singularity of the matrix L is equivalent to: $\alpha \neq 0$, $x_{21} \neq 0$, $x_{31} \neq 0$. For a specific solution, we may simply choose a group of distinct, self-conjugate observer eigenvalues λ_i, $i = 1, 2, 3$ with negative real parts, and select a group of parameters α and x_{ij}, $i = 2, 3$; $j = 1, 2$, satisfying the non-singularity requirements. However, in order to show the effect of our parametric approach, we will not specify the parameters at this stage so as to obtain a general solution.

Step 6: It follows from (4.40) that the weighting matrix is

$$Q = q_1^T = \begin{bmatrix} 0 & \alpha \end{bmatrix}, \quad \alpha \neq 0$$

and follows from Step 3 and Step 5 we have

$$g_1 = \begin{bmatrix} 0 \\ \alpha \end{bmatrix}, \quad \alpha \neq 0$$

Step 7: Again from (4.34) and (4.35) we have

$$W = \begin{bmatrix} -2\alpha & (\lambda_1 - 5)\alpha \\ -(3 - \lambda_2(\lambda_2 - 2))x_{21} - 2x_{22} & -(3\lambda_2 + 4)x_{21} - (\lambda_2 - 5)x_{22} \\ -(3 - \lambda_3(\lambda_3 - 2))x_{31} - 2x_{32} & -(3\lambda_3 + 4)x_{31} - (\lambda_3 - 5)x_{32} \end{bmatrix}$$

Step 8: The observer gain can be obtained through (4.33) as we have

$$L^{-1} = \frac{1}{\det(L)} \begin{bmatrix} \lambda_2 x_{21} x_{32} - \lambda_3 x_{22} x_{31} & \alpha \lambda_3 x_{31} & -\alpha \lambda_2 x_{21} \\ x_{22} x_{31} - x_{21} x_{32} & -\alpha x_{31} & \alpha x_{21} \\ \det(L) & 0 & 0 \end{bmatrix}$$

$$K = \begin{bmatrix} k_{11} & k_{12} \\ k_{21} & k_{22} \\ k_{31} & k_{32} \end{bmatrix} = \frac{-1}{\alpha(\lambda_3 - \lambda_2)x_{21}x_{31}} \begin{bmatrix} k'_{11} & k'_{12} \\ k'_{21} & k'_{22} \\ k'_{31} & k'_{32} \end{bmatrix}$$

with

$$\left\{\begin{array}{lcl} k'_{11} & = & \alpha(\lambda_2-\lambda_3)(\lambda_2\lambda_3+3)x_{21}x_{31} \\ k'_{21} & = & \alpha(\lambda_3-\lambda_2)(\lambda_2+\lambda_3-2)x_{21}x_{31} \\ k'_{31} & = & -2\alpha^2(\lambda_3-\lambda_2)x_{21}x_{31} \\ k'_{12} & = & \alpha\lambda_2(\lambda_1+\lambda_3-10)x_{21}x_{32}-k''_{12} \\ k'_{22} & = & \alpha(\lambda_1+\lambda_2-10)x_{22}x_{31}-\alpha(\lambda_1+\lambda_3-10)x_{21}x_{32}+k''_{22} \\ k'_{23} & = & \alpha^2(\lambda_1-5)(\lambda_3-\lambda_2)x_{21}x_{31} \end{array}\right.$$

where

$$\left\{\begin{array}{lcl} k''_{12} & = & \alpha\lambda_3(\lambda_1+\lambda_2-10)x_{22}x_{31}+4\alpha(\lambda_2-\lambda_3)x_{21}x_{31} \\ k''_{22} & = & 3\alpha(\lambda_2-\lambda_3)x_{21}x_{31} \end{array}\right.$$

The solutions to the robust fault detection problem are given by Q and K with parameters α and x_{ij}, $i=2,3$; $j=1,2$, satisfying non-singularity conditions. The observer eigenvalues λ_i, $i=1,2,3$, can be chosen to fit into either of two cases: (i) λ_i, $i=1,2,3$, are distinct real negative numbers; or (ii) λ_1 is real negative, and $\lambda_2=\bar{\lambda}_3$, $Re(\lambda_2)<0$, $Im(\lambda_2)>0$. For the second case, one can chose:

$$\lambda_2=\bar{\lambda}_3=a+bj,\quad g_2=\bar{g}_3=\begin{bmatrix} x_1+y_1j \\ x_2+y_2j \end{bmatrix}$$

where a, b and x_i, y_i, $i=1,2$, are real numbers. With these notations, the following relations are clearly hold:

$$x_{21}=x_1+y_1j,\ \ x_{22}=x_2+y_2j,\ \ x_{31}=x_1-y_1j,\ \ x_{32}=x_2-y_2j$$

Note that neither $\alpha\lambda_2(\lambda_1+\lambda_3-10)x_{21}$ or $\alpha\lambda_3(\lambda_1+\lambda_2-10)x_{31}$ is zero, it can be easily seen that k_{12} and k_{22} are in fact completely arbitrary since both x_{22} and x_{32} are arbitrary.

By specifying the parameters in the general solutions Q and K, the solutions can be determined. For the case of distinct real observer eigenvalues, by choosing $\alpha=x_{21}=x_{31}=1$, $x_{22}=x_{32}=0$, $\lambda_1=-1$, $\lambda_2=-2$, $\lambda_3=-3$, we obtain

$$L=\begin{bmatrix} 0 & 0 & 1 \\ 1 & -2 & 0 \\ 1 & -3 & 0 \end{bmatrix},\quad W=\begin{bmatrix} -2 & -6 \\ 5 & 2 \\ 12 & 5 \end{bmatrix}$$

and for this group of parameters the observer gain is

$$K=-L^{-1}W=\begin{bmatrix} 9 & 4 \\ 7 & 3 \\ 2 & 6 \end{bmatrix}$$

and the weighting matrix is

$$Q=[0\quad 1]$$

For the case of distinct complex observer eigenvalues, by choosing $\lambda_1=-5$, $\lambda_2=\bar{\lambda}_3=-1+j$ and $\alpha=x_{21}=x_{31}=1$, $x_{22}=x_{32}=0$, we obtain the following special observer gain

$$K = \begin{bmatrix} 5 & 4 \\ 4 & 3 \\ 2 & 10 \end{bmatrix}$$

and the weighting matrix is the same as that in the first special solution.

The design approach presented in this section is parametric in the sense that it fully utilizes the design freedom and gives general parameterizations of solutions. These parameters, of course, can be used to achieve some other specifications of the system.

4.6 Disturbance De-coupling by Assigning Right Eigenvectors

If the left eigenvector assignability conditions are not satisfied, an alternative approach can be used is to assign the columns of the matrix E as right eigenvectors of the observer dynamics. This approach is given by the following theorem:

Theorem 4.4 *The sufficient conditions for satisfying the disturbance de-coupling requirement Eq.(4.5) are:*

(1) $QCE = 0$.

(2) *All columns of the matrix E are right eigenvectors of $(A - KC)$ corresponding to any eigenvalues.*

Patton and Kangethe (1989) pointed out the possibility of assigning columns of the matrix E as right eigenvectors for disturbance de-coupling design, however they did not describe an algorithm for achieving this. This approach only became implementable when Chen and Patton (Patton and Chen, 1991g; Patton and Chen, 1992b; Chen, 1995) proposed a new algorithm for assigning observer right eigenvectors. The assignment of the right observer eigenvectors (left eigenvector of dual controller) is a relatively new problem, only considered by few investigators (Zhang et al., 1990; Choi et al., 1995; Choi, 1998). The assignment method proposed by Chen and Patton is thus presented and extended in this section.

Theorem 4.5 *A vector v_i can be assigned as a right eigenvector of $(A - KC)$ corresponding to λ_i only if one of the following necessary conditions is satisfied:*

(1) *v_i is not the right eigenvector of A corresponding to λ_i and $Cv_i \neq 0$.*

or

(2) *v_i is the right eigenvector of A corresponding to λ_i and $Cv_i = 0$*

Proof: For the right eigenvector v_i of $(A - KC)$, we have

$$(A - KC)v_i = \lambda_i v_i$$

This leads to: $$KCv_i = (A - \lambda_i I)v_i$$

The assignment of v_i as the right eigenvector of $A - KC$ is to find the matrix K to satisfy this equation. This equation has solutions only if either condition in Theorem 4.5 holds true. ◊ **QED**

For the cases when a number of right eigenvectors must be assigned, the gain matrix K must satisfy a set of equations. If one wants to assign all columns e_i $(i = 1, 2, \cdots, q)$ of E as the right eigenvectors of $(A - KC)$ corresponding to eigenvalues λ_i, the following equation must be satisfied.

$$KCe_i = (A - \lambda_i I)e_i \qquad \text{for} \quad i = 1, 2, \cdots, q \tag{4.43}$$

Therefore

$$KCE = A_\lambda \tag{4.44}$$

where $$A_\lambda = [(A - \lambda_1 I)e_1 \ (A - \lambda_2 I)e_2 \ \cdots \ (A - \lambda_q I)e_q] \tag{4.45}$$

Now, the right eigenvector assignment problem is to solve the Eq.(4.44) whilst ensuring that the observer is stable.

Lemma 4.4 *The necessary and sufficient condition for a solution of Eq.(4.44) to exist is:*

$$rank(CE) = rank(\begin{bmatrix} A_\lambda \\ CE \end{bmatrix}) \tag{4.46}$$

Subject to this condition, the general form of the solution to Eq.(4.44) is:

$$K = A_\lambda (CE)^+ + K_1[I_m - CE(CE)^+] \tag{4.47}$$

where $K_1 \in \mathbb{R}^{n \times m}$ is an arbitrary design matrix and $(CE)^+$ is the pseudo-inverse of CE. When $rank(CE) = q$, $(CE)^+$ is given by:

$$(CE)^+ = [(CE)^T(CE)]^{-1}(CE)^T \tag{4.48}$$

Proof: Eq.(4.44) has solutions *iff* any row of the matrix A_λ is a linear combination of rows of the matrix (CE). Hence Eq.(4.46) is the necessary and sufficient condition for a solution of Eq.(4.44) to exist. It can be easily verified that the matrix K given by Eq.(4.47) is a solution of Eq.(4.44).

◊ **QED**

Remarks: A matrix equation can be decomposed into a number of linear equations, and hence Eq.(4.44) can be decomposed into nq equations with nm parameters to be determined. When $m > q$, the solutions for these equations normally exist. Some detailed discussion about the solution of matrix equations can be found in Basilevsky (1983, Chapter 6).

When the all q $(\leq m)$ eigenvalues λ_i $(i = 1, 2, \cdots, q)$ are set as the same, i.e.,

$$\lambda_1 = \lambda_2 = \cdots = \lambda_q = \lambda$$

the necessary and sufficient condition for solving Eq.(4.44) is simpler and can be given by the following Lemma.

Lemma 4.5 *If q eigenvalues to be assigned to the corresponding q columns of E are same and this eigenvalue is not an eigenvalue of A, the necessary and sufficient condition for solution of Eq.(4.44) to exist is:*

$$rank(CE) = rank(E) \tag{4.49}$$

Note that the condition given in this Lemma is the same as that given in Lemma 3.1 and the method of proof used in Lemma 3.1 can be used to prove this Lemma. Therefore, this proof is not presented here. The similarity between Lemma 4.5 and Lemma 3.1 demonstrates the correspondence between unknown input observers with eigenstructure assignment in robust residual generation.

Theorem 4.6 *The necessary and sufficient conditions to assign all columns of E as right eigenvectors of $(A - KC)$ with a stabilizing feedback gain K are:*

(i) $rank(CE) = rank(\begin{bmatrix} A_\lambda \\ CE \end{bmatrix})$.

(ii) $(C_1,\ A_1)$ *is a detectable pair, where:*

$$\begin{aligned} A_1 &= A - A_\lambda (CE)^+ C \\ C_1 &= [I_m - CE(CE)^+]C \end{aligned}$$

Proof: All columns of E are *right* eigenvectors of $(A - KC)$ *iff* Eq.(4.44) holds true. Eq.(4.44) has solutions *iff* the condition (i) is true. For a general solution given by Eq.(4.47), the system dynamics will be:

$$A - KC = A - A_\lambda (CE)^+ C - K_1 [I_m - CE(CE)^+]C = A_1 - K_1 C_1$$

Hence, the observer can be stabilized *iff* the condition (ii) holds.

◊ **QED**

Now, the right eigenvector assignment problem is to find a matrix K_1 which assigns eigenvalues of the observer dynamic matrix $(A - KC) = (A_1 - K_1 C_1)$ in the left hand side of complex plane. This is only possible when $(C_1,\ A_1)$ is a *detectable pair.* The problems of assessing the detectability and assigning eigenvalues of a detectable pair have been studied in Section 3.2. As q eigenvalues λ_i $(i = 1, 2, \cdots, q)$ have been assigned as the eigenvalues of $(A - KC) = (A_1 - K_1 C_1)$ in the assignment of right eigenvectors, the maximum number of eigenvalues of $(A - KC) = (A_1 - K_1 C_1)$ that can be moved by changing the design matrix K_1 is $n - q$. This is proved via the following Lemma.

Lemma 4.6 *The eigenvalues λ_i $(i = 1, 2, \cdots, q)$, which used in the assignment of right eigenvectors e_i $(i = 1, 2, \cdots, q)$ for $(A - KC)$, are unobservable modes of the pair $(C_1,\ A_1)$.*

Proof: As the vector e_i $(i = 1, 2, \cdots, q)$ are right eigenvectors of $(A - KC) = (A_1 - K_1 C_1)$ corresponding to eigenvalues λ_i $(i = 1, 2, \cdots, q)$, we have:

$$\{A - A_\lambda (CE)^+ C - K_1[I_m - CE(CE)^+]C\}e_i = \lambda_i e_i \quad \text{for} \quad i = 1, 2, \cdots, q$$

This equation holds true for any arbitrary matrix K_1, if we set $K_1 = 0$, we have:

$$\{\lambda_i I - [A - A_\lambda (CE)^+ C]\}e_i = \{\lambda_i I - A_1\}e_i = 0 \quad \text{for} \quad i = 1, 2, \cdots, q$$

Therefore,

$$K_1[I_m - CE(CE)^+]Ce_i = K_1 C_1 e_i = 0 \quad \text{for} \quad i = 1, 2, \cdots, q$$

As this relation is valid for any matrix K_1, thus,

$$C_1 e_i = 0 \quad \text{for} \quad i = 1, 2, \cdots, q$$

Hence,

$$\begin{bmatrix} \lambda_i I - A_1 \\ C_1 \end{bmatrix} e_i = 0 \quad \text{for} \quad i = 1, 2, \cdots, q$$

i.e., the eigenvalues λ_i $(i = 1, 2, \cdots, q)$, are unobservable modes of the pair $(C_1,\ A_1)$, and the maximum number of eigenvalues of $(A_1 - K_1 C_1)$ that can be moved by K_1 is $n - q$.

◊ **QED**

The eigenvalues for right eigenvector assignment in disturbance de-coupling design are *not* unique. Moreover, the solution for the matrix K_1 is also not unique, even if the eigenvalues have been fixed, due to the multivariable nature. The design freedom beyond right eigenvector assignment can be utilized to maximize the fault effect on residuals, as discussed in Section 4.6. The matrix K_1 can also be parameterized via eigenstructure in the design. The problem of maximizing fault effects utilizing the remaining design freedom is studied in future research and is not discussed here.

4.7 Dead-Beat Design for Robust Residual Generation

The observer-based residual generation techniques developed for continuous-time system models can also be used for the systems described by discrete-time models. However, some special characteristics such as dead-beat design are only valid for discrete-time domain. The dead-beat design can make the derivation of the disturbance de-coupling principle very simple and gives very prompt residual responses. Consider systems described by discrete-time models:

$$\begin{cases} x(k+1) & = & Ax(k) + Bu(k) + R_1 f(k) + Ed(k) \\ y(k) & = & Cx(k) + Du(k) + R_2 f(k) \end{cases} \tag{4.50}$$

For this system, a discrete observer is used to generate residuals:

$$\begin{cases} \hat{x}(k+1) &= (A-KC)\hat{x}(k) + (B-KD)u(k) + Ky(k) \\ \hat{y}(k) &= C\hat{x}(k) + Du(k) \\ r(k) &= Q[y(k) - \hat{y}(k)] \end{cases} \tag{4.51}$$

The Z-transformed residual response to faults and disturbances is thus:

$$\begin{aligned} r(z) &= QR_2 f(z) + H(zI - A + KC)^{-1}(R_1 - KR_2)f(z) \\ &\quad + H(zI - A + KC)^{-1}Ed(z) \end{aligned} \tag{4.52}$$

The transfer matrix between the residual and the disturbance can be expanded as:

$$H(zI - A_c)^{-1}E = z^{-1}H(I + A_c z^{-1} + A_c^2 z^{-2} \cdots + \cdots)E \tag{4.53}$$

where $A_c = A - KC$ and $H = QC$. It can be seen that this transfer matrix is nulled if the following sufficient conditions are satisfied:

$$HE = 0 \tag{4.54}$$

$$HA_c = 0 \tag{4.55}$$

Choose H and K in such a way that the rows of H are the left eigenvectors of A_c corresponding to zero-valued eigenvalues, Eq.(4.55) then holds true. The Eq.(4.54) means that the left eigenvectors to be assigned are orthogonal to the disturbance directions, and the residual weighting matrix Q will be computed using this equation.

Alternatively, the disturbance de-coupling can also be achieved using the following sufficient conditions:

$$HE = 0 \tag{4.56}$$

$$A_c E = 0 \tag{4.57}$$

Eq.(4.57) holds true when each column of E is assigned as a right eigenvector of A_c corresponding to a zero-valued eigenvalue. Eq.(4.56) will determine the residual weighting matrix Q.

Because of the assignment of zero-valued eigenvalues, the residual will have dead-beat (minimum-time) transient performance and this feature can be exploited to good use in the aim to provide a high sensitivity to soft (incipient) faults.

When the left eigenvector assignment condition in Eq.(4.55) for disturbance de-coupling holds true, the residual response to faults will be:

$$\begin{aligned} r(z) &= QR_2 f(z) + H(zI - A + KC)^{-1}(R_1 - KR_2)f(z) \\ &= QR_2 f(z) + z^{-1}H(R_1 - KR_2)f(z) \end{aligned} \tag{4.58}$$

i.e.

$$r(k) = QR_2 f(k) + H(R_1 - KR_2)f(k-1) \tag{4.59}$$

Hence, the fault signal is transmitted directly into the residual, i.e. the residual response to faults is very fast and this can avoid the detection delay. When a fault occurs in the i_{th} element of fault vector $f(k)$ and other elements of $f(k)$ are zeros, the residual will be:

$$r(k) = [QR_2]_i f_i(k) + [H(R_1 - KR_2)]_i f_i(k-1) \tag{4.60}$$

where $[QR_2]_i$ is the i_{th} column of QR_2 and $[H(R_1 - KR_2)]_i$ is the i_{th} column of $H(R_1 - KR_2)$. This equation shows that the residual vector lies in a fixed subspace, i.e.,

$$r(k) \in \mathcal{S}_i = span\{[QR_2]_i,\ [H(R_1 - KR_2)]_i\} \tag{4.61}$$

This relation shows the robust residual has a directional property which can be used for fault isolation. The fault can be isolated by comparing the residual direction with the fault signature subspace $\mathcal{S}_i$ $(i = 1, \cdots, g)$ as reported by Chen and Patton (Patton and Chen, 1991h; Patton and Chen, 1991c). If the fault function is constant, the residual will be parallel to the vector $[QR_2 + H(R_1 - KR_2)]_i$ and the fault isolation will be easier to achieve. Note that the problem of fault isolation using robust directional residual vectors has been studied in Section 3.4.2.

From the residual generation relations given in Eq.(4.51), the computational form of the residual is:

$$r(z) = [Q - H(zI - A_c)^{-1}K]y(z) - [QD + H(zI - A_c)^{-1}(B - KD)]u(z) \tag{4.62}$$

If the left eigenvector assignment condition in Eq.(4.55) (not the right eigenvector assignment condition) holds true, $H(zI - A_c)^{-1} = z^{-1}H$. Thus the computational form of the residual vector $r(z)$ can be re-written as

$$r(z) = (Q - z^{-1}HK)y(z) - [QD + z^{-1}H(B - KD)]u(z) \tag{4.63}$$

i.e.

$$r(k) = [Q \quad -HK] \begin{bmatrix} y(k) \\ y(k-1) \end{bmatrix} - [QD \quad H(B - KD)] \begin{bmatrix} u(k) \\ u(k-1) \end{bmatrix} \tag{4.64}$$

It can be seen that Eq.(4.64) is a 1_{st} order parity equation (parity relation) (Chow and Willsky, 1984; Lou et al., 1986; Patton and Chen, 1991e) which can be implemented directly to generate residuals for FDI. This residual generation method using the 1_{st} order parity relation is illustrated in Fig.4.3.

It is very interesting to note that the disturbance de-coupling is achieved by the assignment of left observer eigenvectors, however the robust residual generator can be implemented in the form of the parity relation given by Eq.(4.64). That is to say that the observer is not required in robust residual generation, and this has significance for real-time application aspects. The direct link between eigenvector assignment and parity relations was discovered by Patton and Chen (Patton and Chen, 1991h; Patton and Chen, 1991c; Patton and

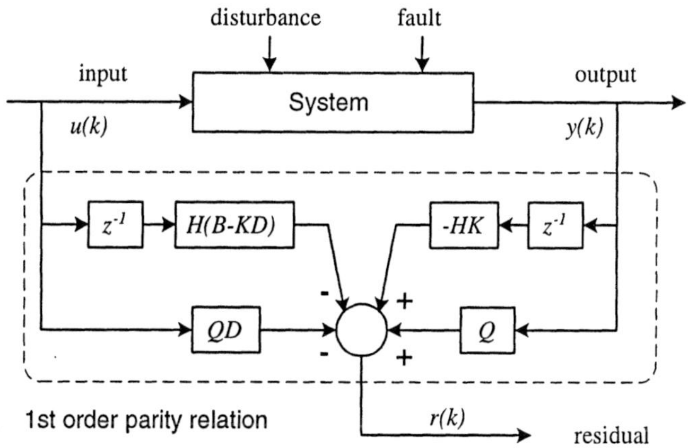

Figure 4.3. Robust residual generation via the 1st-order parity relation

Chen, 1991f; Patton and Chen, 1991e; Patton and Chen, 1992b; Patton, Chen, Millar and Kiupel, 1991).

Note that the link between eigenvector assignment and the parity relation approach cannot be derived for the right eigenvector assignment case (Eq. (4.57)).

4.8 Two Numerical Examples in Eigenstructure Assignment

Example 4-2: Consider the discrete-time system given by,

$$A = \begin{bmatrix} 0.25 & 0 & 0 \\ 0 & 0.5 & 0 \\ 0 & 0 & 0.375 \end{bmatrix} \qquad B = \begin{bmatrix} 0 \\ 1 \\ 1 \end{bmatrix}$$

The disturbance distribution and the measurement matrices are:

$$E = \begin{bmatrix} 1 \\ 1 \\ 0 \end{bmatrix} \qquad C = \begin{bmatrix} 1 & 1 & 0 \\ 0 & 1 & 1 \end{bmatrix}$$

The weighting matrix Q to satisfy $QCE = 0$ can be easily found as:

$$Q = [-1 \quad 2]$$

so that, the desired left eigenvector is:

$$H = QC = [-1 \;\; 1 \;\; 2]$$

corresponding to the eigenvalue 0. This left eigenvector is assignable as H^T belongs to the subspace $span\{-A^TC^T\}$. The remaining two eigenvalues are chosen as $\{0, 0.1\}$. Using the eigenstructure assignment technique (Mudge and

Patton, 1988; Burrows et al., 1989; Burrows and Patton, 1991; Liu and Patton, 1998), the gain matrix is derived as:

$$K = \begin{bmatrix} 0.0165 & -0.3330 \\ 0.4670 & 0.6661 \\ -0.3502 & -0.1246 \end{bmatrix}$$

It can be seen that $H(A - KC) = 0$ and $QCE = 0$, i.e., the de-coupling conditions (4.54) & (4.55) are satisfied and:

$$H(zI - A_c)^{-1}E = 0$$

The z-transform of the residual in response to the sensor fault $f_s(t)$ and actuator fault $f_a(t)$ will be:

$$\begin{aligned} r(z) &= [Q - QC(zI - A_c)^{-1}K]f_s(z) + QC(zI - A_c)^{-1}Qf_a(z) \\ &= [-1 \;\; 2]f_s(z) - [-0.249 \;\; 0.749]z^{-1}f_s(z) - z^{-1}f_a(z) \end{aligned}$$

Clearly, the disturbance term is not present and the residual is only a function of the faults. This means that a robust design has been achieved. According to Eq.(4.64), the computational form of the residual can be:

$$r(z) = [-1 \;\; 2]y(z) - [-0.249 \;\; 0.749]z^{-1}y(z) - z^{-1}u(z)$$

i.e.

$$r(k) = [-1 \;\; 2 \;\; 0.249 \;\; -0.749]\begin{bmatrix} y(k) \\ y(k-1) \end{bmatrix} - u(k)$$

This is a 1_{st} order parity relation.

Example 4-3: Now consider changing the matrix A to

$$A = \begin{bmatrix} 0.3 & 0 & 0 \\ 0 & 0.6 & 0 \\ 0 & 0 & 0.9 \end{bmatrix}$$

In this case, the required left eigenvector of the observer H is not assignable (as H^T does not belong to subspace $span\{-A^TC^T\}$). We must use the alternative approach of assigning right eigenvectors, as given in Section 4.4. The eigenvalues are chosen as $\{0, 0, 0.1\}$. The observer right eigenvector can then be assigned as a single column of E (corresponding to eigenvalue 0), in this case, the resulting gain matrix computed using right eigenvector assignment is:

$$K = \begin{bmatrix} 0.098304 & 0.103392 \\ 0.589304 & -0.596608 \\ -0.8 & 1.6 \end{bmatrix}$$

The z-transform of the corresponding residual response to actuator and sensor faults is:

$$r(z) = \frac{(-1 + 1.2z^{-1} - 0.27z^{-2})(2 - 2.7z^{-1} + 0.81z^{-2})}{1 - 0.1z^{-1}}f_s(z) + \frac{3 - 1.8z^{-1}}{1 - 0.1z^{-1}}f_a(z)$$

The disturbance de-coupling has also been achieved. However, although this residual signal is robust to disturbances, it is a recursive structure and does not directly correspond to a parity relation.

4.9 Conclusion and Discussion

This chapter has discussed the robust (in the sense of disturbance de-coupling) residual generation via observer eigenstructure assignment. The disturbance de-coupling is achieved by the assignment of either left or right observer eigenvectors. Given a design problem, the designer can check the assignability to decide the assignment of left or right eigenvectors. If the number of independent disturbances to be de-coupled is smaller than the number of independent measurements, a disturbance de-coupling solution is very likely achievable via either left or right eigenvector assignment. If the required eigenstructure (left or right) is not perfectly assignable, an approximate approach should be taken. That is to chose assignable eigenvectors close, in a least-squares sense, to the desired eigenvectors. This can be achieved via the left eigenvector assignment. In this situation, the residual is not de-coupled from disturbances but has a low sensitivity to disturbances due to approximate de-coupling.

The chapter discusses mainly the robust residual generation problem. For fault isolation, one way is to design structured residual sets and this can be done using an approach similar to that presented in Section 3.3. For the dead-beat design, when the rows of the matrix $H = QC$ are assigned as left eigenvectors of the observer corresponding to zero-valued eigenvalues, the residual can be generated by a 1_{st} order parity relation, and the resulting residual has directional property which can be used for fault isolation.

5 DETERMINATION OF DISTURBANCE DISTRIBUTION MATRICES FOR ROBUST RESIDUAL GENERATION

5.1 Introduction

It is difficult to develop a highly accurate model of a complex system and hence the interesting question is just what is a reasonable model to enable good performance in FDI. It would be attractive to develop a robust FDI technique which is insensitive to modeling uncertainty, without the use of a very accurate model. However, in order to design a robust FDI scheme, one should have a description (i.e. some information or knowledge) about the system uncertainty, e.g. its distribution matrix or spectral bandwidth, etc. Furthermore, this description should provide assistance for robust FDI design, i.e. it can be handled in a systematic manner.

As pointed out in Chapters 3 & 4, a typical description for the system uncertainty makes use of the concept of "unknown inputs" acting upon a nominal linear model of the system as described by:

$$\left\{ \begin{array}{rcl} \dot{x}(t) & = & Ax(t) + Bu(t) + R_1 f(t) + Ed(t) \\ y(t) & = & Cx(t) + Du(t) + R_2 f(t) \end{array} \right. \tag{5.1}$$

where the disturbance term $Ed(t)$ is used to represent uncertainties acting upon the system, in which the vector $d(t) \in \mathbb{R}^q$ is an unknown "input" or "disturbance" vector. The distribution matrix $E \in \mathbb{R}^{n \times q}$ is assumed known. In robust model-based FDI, this description of the system uncertainty is defined as *structured* uncertainty.

It is clear from Eq.(5.1) that $Ed(t)$ and $R_1 f(t)$ act on the system in the same way, and thus one cannot discriminate between their effects unless the structure of E is known. It is therefore a common practice to assume that E is known, in so called *robust* FDI approaches which are based on the disturbance de-coupling principle (see Section 2.11). Once E is known, the residual can be made to have the disturbance de-coupling (robust) property, i.e. the residual is totally de-coupled from the disturbance (uncertainty). The robust residual can then be used to achieve reliable FDI. The de-coupling design can be achieved using the unknown input observer (see Chapter 3), or alternatively using eigenstructure assignment (see Chapter 4), or frequency domain approaches (Ding and Frank, 1991; Frank, 1991a; Frank and Ding, 1993; Frank and Ding, 1994; Qiu and Gertler, 1993), or orthogonal parity equation approaches (Gertler, Fang and Luo, 1990; Gertler, 1991; Gertler and Kunwer, 1993; Gertler, 1998).

The theories underlying the robust residual generation based on the disturbance de-coupling principle have been well developed, but for real applications the following problems remain unsolved:

- How well can the term $Ed(t)$ characterize the real uncertainty, if there is no knowledge of the uncertainty?
- How can the term $Ed(t)$ and the structure of E be determined, even approximately?

This chapter answers the above questions and provides some simulation examples to test some developed theoretical results. These question must be answered, otherwise the application domain of the disturbance de-coupling approach for robust FDI is very limited. In fact, very few researchers have presented the application results of robust FDI.

As mentioned above, a primary requirement for disturbance de-coupled robust FDI methods is that the disturbance distribution matrix must be known. However, in most practical systems the uncertainty can be expressed in many different ways (e.g. modeling errors) and the distribution matrix E is not known. To apply the disturbance de-coupling robust residual generation techniques to systems with wide ranging uncertainties such as modeling errors and parameter variations, an approximate distribution matrix E is needed to represent the effects of uncertainty. Within the framework of international research on this subject, there have been few attempts to address the problem of determining this distribution matrix. Until recently, this lack of information obstructed the application of disturbance de-coupling for robust FDI in real engineering systems. The work of determining the disturbance distribution matrix has been led by Patton & Chen (Patton and Chen, 1991f; Patton and Chen, 1991b; Patton, Chen and Zhang, 1992; Patton, Zhang and Chen, 1992). They have demonstrated their techniques on a jet engine system (Patton and Chen, 1991f; Patton and Chen, 1991b; Patton and Chen, 1991a; Patton and Chen, 1992b; Patton, Chen and Zhang, 1992; Patton, Zhang and Chen, 1992; Chen, 1995), a nuclear reactor core (Patton, Chen and Millar, 1991; Patton, Chen and Millar, 1992), an industrial actuator (Jorgensen et al., 1994) and a pumping system (Dalton et al., 1996). The technique proposed by Patton and Chen (1991f) was later used by Shields *et al* (Shields, 1994; Yu and Shields, 1994; Yu, Shields and Mahtani, 1994a; Yu et al., 1994b; Yu et al., 1996). The problem has been attracting world-wide attention and other investigators have followed this line of research, e.g. Gertler and Kunwer (1993), Gertler (1994), Keviczky, Bokor, Szigeti and Edelmayer (1993), Saif and Guan (1993), and Edelmayer, Bokor, Szigeti and Keviczky (1997). Note that the determination of the optimal disturbance distribution matrix E is a common problem for all disturbance de-coupling robust residual generation approaches including the orthogonal parity equation approach (Gertler, Fang and Luo, 1990; Gertler, 1991; Gertler and Kunwer, 1993) and frequency domain approach (Ding and Frank, 1991; Frank, 1991a; Frank and Ding, 1993; Frank and Ding, 1994; Qiu and Gertler, 1993).

This chapter presents the research developments surrounding the determination of the disturbance distribution matrix for robust residual generation. A number of approaches for obtaining this matrix (albeit approximate) for real uncertain systems are described. An example of a 17_{th} thermodynamic order simulation model of a jet engine system is used to illustrate some approaches developed.

Clearly, the basis for the model-based FDI technique is the use of mathematical models. The model used should have certain accuracy. In order to make a diagnosis algorithm robust against modeling uncertainty, some knowledge about the modeling uncertainty should be available. Otherwise, what do we need a model for if an algorithm can be made robust enough without *a priori* modeling information? This highlights the need to make some modeling assumptions. To be useful in a robust design, these assumptions should be easily handled in a systematic manner. The disturbance representation of uncertainty can be handled by means of the unknown input observer or the eigenstructure assignment. However, this assumption is not realistic, i.e., the distribution matrix cannot be obtained directly. In practice, we can make some more realistic assumptions about uncertainty, for example, parameters of the system are within a certain bound, etc. However, these assumptions are not normally easy to handle in designing robust FDI algorithms. The aim of this chapter is to introduce some techniques to bridge the gap between theoretical assumption and practical reality. This aim is fulfilled by approximate modeling of uncertainty, in which a disturbance description with an approximate distribution matrix is used to approximately model uncertainty. A number of situations covering a wide range of possibilities for uncertainty are considered in the following sections.

5.2 Direct Determination & Optimization of Disturbance Distribution Matrix

In most situations, the distribution matrix is not readily available. However, there are cases for which some *a priori* knowledge about uncertainty is available and can be used for a direct derivation of the distribution matrix E. This

Section discusses a number of situations in which some realistic assumptions about uncertainty can be used for this direct derivation. Normally, this directly obtained matrix has a high rank (i.e. too many disturbances or unknown inputs) and disturbance de-coupling is not achievable. Hence, a low rank matrix which approximates the distribution is used in the design of optimally robust residual generators. This is an unknown input consolidation procedure, i.e., the unknown inputs with closed directions are combined and hence the number of unknown inputs is reduced. Note that, in some situations, the matrix E can be determined by simple inspection. If the uncertain factors appear in the i_{th} row of matrices A and B, it is most likely the matrix E should contain a column as follows:

$$[\begin{array}{ccccccc} 0 & \cdots & 0 & \underset{i_{th}}{1} & 0 & \cdots & 0 \end{array}]^T$$

This *direct inspection method* for determining the matrix E was used in the example presented in Section 3.5.3 and also showed by Saif and Guan (1993) and Hou and Müller (1994b). This method may not be very effective, however it is simple and can be useful for some systems.

5.2.1 Noise and additive non-linearity

Consider the following dynamic equation of the monitored system:

$$\dot{x}(t) = Ax(t) + Bu(t) + G\mu(t) + Sg(x(t), u(t), t)) \tag{5.2}$$

where $\mu(t)$ is a noise or external disturbance vector. In this equation, the non-linearity is considered as an additive non-linear term $Sg(x(t), u(t), t))$, i.e., the system dynamics can be separated into linear and non-linear parts. This kind of non-linear dynamic structure exists in some non-linear chemical processes (Watanabe and Himmelblau, 1982; Chen and Zhang, 1991) and has been used in Section 3.3.3. For the system described above, the uncertainty can be modeled as an *additive* term $Ed(t)$ and where:

$$Ed(t) = [G \quad S] \left[\begin{array}{c} \mu(t) \\ g(x(t), u(t), t)) \end{array} \right] \tag{5.3}$$

5.2.2 Bilinear systems

The study of bilinear systems has theoretical importance because they are a special class of non-linear systems. Many practical non-linear systems such as ecological systems, nuclear systems, hydraulic systems and heat exchanger systems can be modeled by a bilinear system model(Yu et al., 1994a; Yu et al., 1996):

$$\dot{x}(t) = A_0x(t) + Bu(t) + \sum_{i=1}^{r} A_iu_i(t)x(t) \tag{5.4}$$

where $u_i(t)$ $(i = 1, \cdots, r)$ is the i_{th} component of $u(t)$, and A_i $(i = 0, 1, \cdots, r)$ and B are known matrices. The non-linear term can be treated as the disturbance term with the distribution matrix and the unknown input vector as follows:

$$E = [A_1 \ \ A_2 \ \ \cdots \ \ A_r] \qquad d(t) = \begin{bmatrix} u_1(t)x(t) \\ \vdots \\ u_r(t)x(t) \end{bmatrix} \tag{5.5}$$

A linear disturbance de-coupled residual generator can be designed to generate robust residuals for FDI. This avoids the complexity involved in the design of bilinear observers (Shields, 1994; Yu and Shields, 1994; Yu et al., 1994a; Yu et al., 1994b; Yu et al., 1996; Yu and Shields, 1996; Yu and Shields, 1997; Yang and Saif, 1997).

5.2.3 *Model reduction*

Most systems can have significantly higher order dynamics than their models. Consider, for example, the system described by a higher order model as:

$$\begin{bmatrix} \dot{x}(t) \\ \dot{x}_h(t) \end{bmatrix} = \begin{bmatrix} A_{11} & A_{12} \\ A_{21} & A_{22} \end{bmatrix} \begin{bmatrix} x(t) \\ x_h(t) \end{bmatrix} + \begin{bmatrix} B_1 \\ B_2 \end{bmatrix} u(t) \tag{5.6}$$

where $x(t) \in \mathbb{R}^n$ is a partial state vector corresponding to dominant dynamic part of the system. $x_h(t)$ represents the higher order dynamics in the system, and frequently neglected in practice. For ease of design and implementation in control and fault diagnosis, the following reduced-order model is used to approximate this system:

$$\begin{aligned} \dot{x}(t) &= Ax(t) + Bu(t) + (A_{11} - A)x(t) + (B_1 - B)u(t) + A_{12}x_h(t) \\ &= Ax(t) + Bu(t) + Ed(t) \end{aligned} \tag{5.7}$$

where:

$$Ed(t) = [(A_{11} - A) \ \ (B_1 - B) \ \ A_{12}] \begin{bmatrix} x(t) \\ u(t) \\ x_h(t) \end{bmatrix} \tag{5.8}$$

A typical application of this partitioned state-space structure arises when comparing a *reduced order* model with the full-scale system, for example, in an observer used for FDI. For this case, the nominal model represented by (A, B) is the reduced order model and the remaining modeling errors are considered to be lumped together within an additive term $Ed(t)$. It is assumed that the n reduced order state variables correspond to N state variables of the full-scale system.

5.2.4 *Parameter perturbations*

A system model with time-varying parameter perturbation can be described as:

$$\dot{x}(t) = (A + \Delta A(t))x(t) + (B + \Delta B(t))u(t) \tag{5.9}$$

The parameter perturbations considered in the robust control field are sometimes approximated as:

$$\Delta A(t) \approx \sum_{i=1}^{N} a_i(t) A_i \qquad \Delta B(t) \approx \sum_{i=1}^{N} b_i(t) B_i$$

where A_i and B_i are known constant matrices with proper dimensions, $a_i(t)$ and $b_i(t)$ are unknown scalar time-varying factors. In this case, the modeling error can be approximated by the disturbance term as:

$$E_1 d_1(t) = \Delta A(t)x(t) + \Delta B(t)u(t) = [A_1 \ \cdots \ A_N \ B_1 \ \cdots \ B_N] \begin{bmatrix} a_1(t)x(t) \\ \cdots \\ a_N(t)x(t) \\ b_1(t)u(t) \\ \cdots \\ b_N(t)u(t) \end{bmatrix}$$

In the robust control field, the disturbance matrices ΔA and ΔB are sometimes assumed to satisfy the following so-called matched perturbation condition:

$$[\Delta A \ \ \Delta B] = E\Sigma(t)\,[F_a \ \ F_b]$$

where E, F_a and F_b are known matrices with compatible dimensions, $\Sigma(t)$ is a block diagonal time varying matrix which represents the parameter uncertainty. In this situation, the disturbance distribution matrix is E and the unknown input vector is $d(t) = \Sigma(t)[F_a x(t) + F_b u(t)]$.

Now, consider the situation where the system matrices are functions of the parameter vector $\alpha \in \mathbb{R}^k$:

$$\dot{x}(t) = A(\alpha)x(t) + B(\alpha)u(t) \tag{5.10}$$

If the parameter vector is perturbed around the nominal value $\alpha = \alpha_0$, this equation can be expanded as:

$$\dot{x}(t) = A(\alpha_0)x(t) + B(\alpha_0)u(t) + \sum_{i=1}^{k}\{\frac{\partial A}{\partial \alpha_i}\delta\alpha_i x + \frac{\partial B}{\partial \alpha_i}\delta\alpha_i u\} \tag{5.11}$$

In this case, the distribution matrix and unknown input vector are:

$$E = [\frac{\partial A}{\partial \alpha_1} \mid \frac{\partial B}{\partial \alpha_1} \mid \cdots \mid \frac{\partial A}{\partial \alpha_k} \mid \frac{\partial B}{\partial \alpha_k}] \tag{5.12}$$

$$d(t) = [\delta\alpha_1 x^T \mid \delta\alpha_1 u^T \mid \cdots \mid \delta\alpha_k x^T \mid \delta\alpha_k u^T]^T \tag{5.13}$$

5.2.5 Low rank approximation of distribution matrix

Section 4.3 has shown that one of the necessary conditions to design robust residuals (in the disturbance de-coupling sense) using eigenstructure assignment, is to find a matrix $H \in \mathbb{R}^{p\times n}$ which satisfies the following equation:

$$HE = 0 \tag{5.14}$$

where p is the residual dimension and can be chosen by the designer. To satisfy this equation, the rank of the matrix E must be less than its row number (i.e. the system order n). Chapters 3 & 4 have also shown that the maximum number of independent disturbances ($= rank(E)$) cannot be larger than the maximum independent measurement number m. This discussion highlights the point that the most critical condition for achieving disturbance de-coupling in the residual generation is:

$$rank(E) \leq m \tag{5.15}$$

It has been shown that the distribution matrix can be derived directly from the available uncertainty information. If $rank(E) \leq m$, Eq.(5.14) has solutions and exact de-coupling is possible. However, for most situations, this matrix obtained does not satisfy the rank condition (5.15), and thus approximate de-coupling must be taken. The procedure will be to compute a matrix E^* that is as close as possible to E, and $rank(E^*) = q \leq m$, i.e. to find the solution of following optimization problem:

$$\min \|E - E^*\|_F^2 \qquad \text{subject to}: \; rank(E^*) = q \leq m \tag{5.16}$$

Here $\| \cdot \|_F^2$ denotes the Frobenius norm, defined as the root of the sum of squares of the entries of the associated matrix. The matrix E^* is thus chosen so that the sum of the squared distances between the columns of E and E^* is minimized, subject to the constraint: $rank(E^*) \leq m$.

The problem of approximating a matrix by a low rank matrix was first suggested by Eckart and Young (1936). More recently, Tufts, Kumaresan and Kirsteins (1982) and Lou et al. (1986) demonstrated its use. This optimization problem can be solved via the *Singular value Decomposition* (SVD) (Golub and Van Loan, 1989) of E:

$$E = S\Sigma T^T \tag{5.17}$$

where

$$\Sigma = \left[\begin{array}{c|c} diag\{\sigma_1, \cdots, \sigma_k\} & 0 \\ \hline 0 & 0 \end{array}\right] \tag{5.18}$$

and S and T are orthogonal matrices, k is the rank of the matrix E, and $\sigma_1 \leq \sigma_2 \leq \cdots \leq \sigma_k$ are the singular values of E. According to the theorem given by Eckart and Young (1936) (shown in Appendix E, also see Tufts et al. (1982) and Lou et al. (1986)), a low rank approximation for the matrix E which minimizes $\|E - E^*\|_F^2$ is given by:

$$E^* = S\hat{\Sigma}T^T \tag{5.19}$$

where

$$\hat{\Sigma} = \left[\begin{array}{c|c} diag\{0, \cdots, 0, \sigma_{k-q}, \cdots, \sigma_k\} & 0 \\ \hline 0 & 0 \end{array}\right] \tag{5.20}$$

and q is the rank of the matrix E^* which is not larger than m, in order to satisfy the disturbance de-coupling conditions. To achieve approximate disturbance de-coupling design, the matrix H should be made to satisfy the relation $HE^* = 0$. It is easy to see that an orthogonal solution for the matrix H is:

$$H^* = \begin{bmatrix} s_1^T \\ s_2^T \\ \vdots \\ s_{k-q-1}^T \end{bmatrix} \tag{5.21}$$

where $s_1, \cdots, s_{k-q-1}$ are the first $k - q - 1$ columns of S. Once again, the residual dimension p can be freely chosen by the designer. If $p < k - q - 1$, the residual weighting matrix H can be constructed by choosing any p rows from the optimal matrix H^*. If $p > k - q - 1$, any extra rows of H should be linear combinations of the rows in H^*. This does not provide any independent information, hence p should not be larger than $k - q - 1$. The greater the residual dimension, the more information one can obtain. Hence, an optimal solution is to set $p = k - q - 1$.

An alternative statement of the optimization problem can be given as follows. Assume that:

$$E = [e_1 \; e_2 \; \cdots \; e_{n_1}] \tag{5.22}$$

where e_i is the i_{th} column of the matrix E. An ideal matrix H should make $He_i = 0$ for all $i = 1, 2, \cdots, n_1$. This is not always possible. Hence, it makes sense to choose a matrix H that is "as orthogonal as possible" to all e_i ($i = 1, 2, \cdots, n_1$), i.e. to make each He_i ($i = 1, 2, \cdots, n_1$) *as close to zero as possible.* As orthogonality is a directional property, it is not affected by the magnitude of H. There is no loss of generality in applying an orthogonal constraint to the matrix H, i.e. $HH^T = I$. The optimization criterion can then be defined as:

$$J = \sum_{i=1}^{n_1} \|He_i\|_2^2 \tag{5.23}$$

The optimal solution for H follows by minimizing J, subject to $HH^T = I$. Lou et al. (1986) showed that the choice of H given in (5.21) also minimizes J, yielding the minimum value as

$$J^* = \sum_{i=1}^{q} \sigma_i^2 \tag{5.24}$$

This new statement of the optimization problem provides some very useful insight as J^* can be used as a *robustness measure* which is clearly relative to the rank number q of matrix E^* and the singular values of the matrix E.

It will typically be the case that some components of the unknown input vector d are larger than others. Furthermore, certain components of the unknown input vector have more effect on the residual. To take account of this, the varied attention must be paid to the different components of the disturbance signal in the optimization procedure. For example, if the j_{th} component of the disturbance is significantly larger than the i_{th} component, the term He_j will be more important than the term He_i. Hence, the criterion J must be replaced by:

$$J_1 = \sum_{i=1}^{n_1} \alpha_i ||He_i||_2^2 \tag{5.25}$$

where α_i $(i = 1, 2, \cdots, n_1)$ are positive weighting factors. The relative magnitudes of the α_i correspond to relative magnitudes of components of the disturbance weighting. By rewriting the weighted optimization criterion as:

$$J_1 = \sum_{i=1}^{n_1} ||H(\sqrt{\alpha_i} e_i||_2^2 \tag{5.26}$$

this optimization problem can be solved using the procedure described above, but with e_i replaced by $\sqrt{\alpha_i} e_i$ and with E replaced by E', where

$$E' = [\sqrt{\alpha_1} e_1 \;\; \sqrt{\alpha_2} e_2 \;\; \cdots \;\; \sqrt{\alpha_{n_1}} e_{n_1}]$$

5.2.6 Bounded uncertainty

Now, consider the case when the full-order system model is *not* available. An identification procedure is used to obtain the nominal model $\{A_0, B_0, C_0, D_0\}$ with the estimation error $\{\Delta A, \Delta B, \Delta C, \Delta D\}$. Normally, ΔA and ΔB are unknown but bounded:

$$A_1 \leq \;\; \Delta A \;\; \leq A_2 \tag{5.27}$$

$$B_1 \leq \;\; \Delta B \;\; \leq B_2 \tag{5.28}$$

where A_1, A_2, B_1 and B_2 are known and $\Delta A \leq A_2$ denotes that each element of ΔA is not larger than the corresponding element of A_2. This typifies the case where the uncertainty is bounded. Consider ΔA and ΔB in a finite set of possibilities, say $\{\Delta A_i, \Delta B_i\}$ $(i = 1, 2, \cdots, M)$ within the interval $A_1 \leq \Delta A \leq A_2$ and $B_1 \leq \Delta B \leq B_2$. This might involve choosing representative points, reflecting desired weighting on the likelihood or importance of particular sets of parameters. In this situation, a set of unknown input distribution matrices is obtained:

$$E_i = [\Delta A_i, \; \Delta B_i] \qquad i = 1, 2, \cdots, M \tag{5.29}$$

In order to make the disturbance de-coupling valid for a wide range of model parameter variations, an optimal matrix E^* should be as close as possible to all $E_i (i = 1, 2, \cdots, M)$. The optimization problem is thus defined as:

$$\min_{\{s.t. \; rank(E^*) \leq m\}} ||E^* - [E_1 \;\; E_2 \;\; \cdots \;\; E_M]||_F^2 \tag{5.30}$$

E^* is then used to design disturbance de-coupling robust residual generators. As E^* is close to all E_i, approximate de-coupling is achieved over the whole range of parameter variations.

5.3 Estimation of Disturbance and Disturbance Distribution Matrix

In some cases, there is insufficient available knowledge about the state space model of the system and all we can get is a linearized low order model with matrices (A, B, C, D). In order to account for unavoidable modeling errors, it is assumed that the system is described as:

$$\begin{cases} \dot{x}(t) &= Ax(t) + Bu(t) + d_1(t) \\ y(t) &= Cx(t) + Du(t) \end{cases} \tag{5.31}$$

where $d_1(t)$ is used to represent modeling errors. If the vector $d_1(t)$ can be obtained, one may be able to decompose $d_1(t)$ into $Ed(t)$ with E a structured matrix. It seems reasonable to add $d_1(t)$ to account for all uncertainties in the model. But can we determine $d_1'(t)$ with sufficient accuracy? How should we decompose $d_1(t)$ into $Ed(t)$ with E a structured matrix, to involve the disturbance de-coupling concept? The following sections provide answers to these questions.

5.3.1 *Estimation of disturbance vector using an augmented observer*

The state of an augmented observer can be used to estimate the direction of the disturbance direction E. The first step is to assume that $d_1(t)$ is a slowly time-varying vector, so that the system model can be re-written in augmented form as:

$$\begin{bmatrix} \dot{x}(t) \\ \dot{d}_1(t) \end{bmatrix} = \begin{bmatrix} A & I \\ 0 & 0 \end{bmatrix} \begin{bmatrix} x(t) \\ d_1(t) \end{bmatrix} + \begin{bmatrix} B \\ 0 \end{bmatrix} u(t) \tag{5.32}$$

$$y(t) = [C \quad 0] \begin{bmatrix} x(t) \\ d_1(t) \end{bmatrix} + Du(t) \tag{5.33}$$

If we have the true system input and output data $\{u(t), y_t(t)\}$, an observer based on the model described by Eqs.(5.32) & (5.33) can be used to estimate the disturbance vector $d_1(t)$. Once $\hat{d}_1(t)$ has been obtained, it is possible to obtain *some information* about the distribution matrix E. The problem that could arise is that the augmented system may not be observable. The observability matrix for this system is:

$$W_0 = \begin{bmatrix} C & 0 \\ CA & C \\ CA^2 & CA \\ \vdots & \vdots \\ CA^n & CA^{n-1} \end{bmatrix} = \begin{bmatrix} C & 0 \\ 0 & C \\ 0 & CA \\ \vdots & \vdots \\ 0 & CA^{n-1} \end{bmatrix} \begin{bmatrix} I_n & 0 \\ A & I_n \end{bmatrix}$$

As the second matrix on the right hand side of the above equation is a full rank matrix, it is easy to see that:

$$rank(W_0) = rank(C) + rank(\begin{bmatrix} C \\ CA \\ \vdots \\ CA^{n-1} \end{bmatrix})$$

The system shown in Eqs.(5.32)&(5.33) is observable *if and only if* $rank(W_0) = 2n$. From the above equation, it is clear that this system is observable *if and only if* $rank(C) = n$ and the matrix pair (C, A) is observable. The requirement $rank(C) = n$ limits the use of this technique for estimating the disturbance vector, as it requires that the system has n (state dimension) independent measurements. There is a logical explanation of this requirement. When we want to estimate the modeling uncertainty without any *a priori* knowledge about it, information is needed from additional measurements. For the FDI purpose, there are sometimes a large number of measurements available and the dynamics of the system can be approximated by a relatively low order model. Hence, the condition $rank(C) = n$ is not a strict constraint for some FDI problems.

5.3.2 *Derivation of disturbance distribution matrix*

Section 5.3.1 presented the method for determining $d_1(t)$, but the final goal is to express $d_1(t)$ as:

$$Ed(t) = d_1(t) \tag{5.34}$$

Generally speaking, there are many combinations of E and d, but for the robust FDI methods considered here, we only need to know the structure of E, and $d(t)$ can be chosen arbitrarily. There are two possibilities: one is that E is a vector and $d(t)$ is an arbitrary scalar function; another is that E is a matrix and $d(t)$ is an arbitrary vector function.

Using the augmented observer, one can get the estimation of the disturbance vector $d_1(t)$ as $\{\hat{d}_1(1), \hat{d}_1(2), \hat{d}_1(3), \cdots, \hat{d}_1(M)\}$. If the direction of the vector $\hat{d}_1(i)$ changes slightly for all $i = 1, 2, \cdots, M$, it can be believed that E is a vector and $d(t)$ is an arbitrary scalar function. In this case, the matrix E can be approximated as: $E = \frac{1}{M}\sum_{k=1}^{M} \hat{d}_1(k)$.

It is very likely the case that when $d_1(k)$ cannot be assumed to be a constant direction vector, i.e., the directions of $\hat{d}_1(i)$ are very much different for all $i = 1, 2, \cdots, M$. In this case, it is still possible to express the vector $d_1(k)$ as: $d_1(k) = Ed(k)$, where $E \in \mathbb{R}^{n \times q}$ is a constant matrix, $d(k) \in \mathbb{R}^q, d_1(k) \in \mathbb{R}^n$ and $q \leq n$. In the robust FDI method, E must be row rank deficient in order to have a left annihilating matrix H such that the equation $HE = 0$ holds true. This is one of the conditions for achieving robust FDI. To find the optimal distribution matrix, the following matrix can be constructed:

$$\Omega = [\hat{d}_1(1), \quad \hat{d}_1(2), \quad \cdots \quad , \hat{d}_1(M)] \tag{5.35}$$

The maximum rank of Ω is n, i.e. there are at most n linear independent columns. Hopefully, there are some vectors in Ω which are very close to other vectors (or nearly close to a combination of other vectors) and can be neglected. The q most linearly-independent columns of Ω can then be used to construct E, i.e.

$$E = [\hat{d}_1(i), \quad \hat{d}_1(j), \quad \cdots, \quad \hat{d}_1(k)] \in \mathbb{R}^{n\times q} \tag{5.36}$$

The procedure of the derivation of a low rank approximation to the matrix Ω is discussed now. One way to find the q most linearly-independent columns is to calculate the generalized angles between these vectors, i.e., $\angle(\hat{d}_1(i), \hat{d}_1(j))$ $(i, j = 1, \cdots, M; j \neq i)$. If a vector $\hat{d}_1(i)$ has very small generalized angle with other vectors, then $\hat{d}_1(i)$ can be discarded. The matrix E in (5.36) can be used to satisfy $HE = 0$. When E is constructed in this way it has advantage that all rows of the matrix H is orthogonal to almost every unknown input direction. If the remaining single direction is very near to other directions, then all rows of the matrix H are also *almost orthogonal* to it. It should be expected that almost all unknown inputs along these directions can be eliminated.

The way of obtaining the matrix E explained above involves the calculation of the generalized angles between vectors $\{\hat{d}_1(1), \hat{d}_1(2), \hat{d}_1(3), \cdots, \hat{d}_1(M)\}$ which is a complex and time-consuming procedure.

Another way to obtain a rank q matrix E is first to use an approximation matrix Ω_0 with the same dimension as Ω, such that:

$$\min_{rank(\Omega_0)=q} ||\Omega - \Omega_0||_F^2 \tag{5.37}$$

The solution to this optimization problem is readily obtained using the Singular Value Decomposition of Ω. Suppose that:

$$\Omega = U[diag(\sigma_1, \cdots, \sigma_n), \quad 0]V^T \tag{5.38}$$

where $\sigma_1 \geq \sigma_2 \geq \cdots \geq \sigma_n$ are the singular values of Ω. Ω_0 is then constructed as:

$$\Omega = U[diag(\sigma_1, \cdots, \sigma_q, \quad 0, \cdots, \quad 0), \quad 0]V^T \tag{5.39}$$

where q is determined by the magnitude of σ_i $(i = q+1, \cdots, n)$ such that: $\sigma_n \leq \sigma_{n-1} \leq \cdots \leq \sigma_{q+1} \leq \epsilon$. ϵ is a small number determined by the designer. The error of the approximation can be calculated as: $||\Omega - \Omega_0||_F^2 = \sum_{i=q+1}^n \sigma_i^2$. For a good approximation we should have that: $\sum_{i=1}^q \sigma_i^2 \gg \sum_{i=q+1}^n \sigma_i^2$. The second step is to obtain the required distribution matrix E is to decompose the rank deficient matrix Ω_0 as:

$$\Omega_0 = \Omega_1\Omega_2 \tag{5.40}$$

by the rank decomposition (see Appendix E). where $\Omega_1 \in \mathbb{R}^{n\times q}$ is a full column rank matrix, and $\Omega_2 \in \mathbb{R}^{q\times M}$. From the definition of Ω, one can obtain:

$$\begin{aligned} \Omega &= [\hat{d}_1(1), \quad \hat{d}_1(2), \quad \cdots \quad, \hat{d}_1(M)] \\ &= [Ed(1), \quad Ed(2), \quad \cdots \quad, Ed(M)] \\ &= E[d(1), \quad d(2), \quad \cdots \quad, d(M)] \end{aligned} \tag{5.41}$$

However,

$$\Omega \approx \Omega_0 = \Omega_1 \Omega_2 \tag{5.42}$$

Hence, an optimal approximation for the matrix E is Ω_1.

5.3.3 Estimation of disturbance vector using de-convolution

FDI algorithm design and the determination of the disturbance distribution matrix in the discrete-time domain can be carried in a similar way to that of the continuous-time domain. However, some special properties exist in discrete-time design. The discrete-time model described the following equation is considered here:

$$\left\{ \begin{array}{rcl} x(k+1) & = & Ax(k) + Bu(k) + d_1(k) \\ y(k) & = & Cx(k) + Du(k) \end{array} \right. \tag{5.43}$$

where $d_1(k)$ is used to account for all modeling uncertainties. The matrices $\{A, B, C, D\}$ are known nominal model parameters. $\{u(k)\}$ is the model input which is identical to the system input. $\{y(k)\}$ is the model output which is normally not equal to the true system output $\{y_t(k)\}$ due to modeling uncertainty. The task here is to determine the additional term $d_1(k)$ using the nominal model parameters $\{A, B, C, D\}$ and real system inputs and outputs: $\{u(k), y_t(k)\}$. After an estimate of the vector $d_1(k)$ is obtained, it is possible to decompose it into $Ed(k)$ with E as a structured matrix for disturbance de-coupling FDI design.

From Eq.(5.43), it can be seen that:

$$\begin{array}{rcl} y(k) & = & Cx(k) + Du(k) \\ & = & C[Ax(k-1) + Bu(k-1) + d_1(k-1)] + Du(k) \\ & = & \cdots\cdots\cdots\cdots \\ & = & CA^k x(0) + \sum_{i-1}^{k} CA^{i-1} Bu(k-i) \\ & & + \sum_{i=1}^{k} CA^{i-1} d_1(k-i) + Du(k) \end{array} \tag{5.44}$$

Define $\tilde{y}(k)$ as the modeling output error (i.e. the difference between true system output and model output):

$$\begin{array}{rcl} \tilde{y}(k) & = & y_t(k) - y(k) \\ & = & y_t(k) - CA^k x(0) - \sum_{i=1}^{k} CA^{i-1} Bu(k-i) \\ & & - \sum_{i=1}^{k} CA^{i-1} d_1(k-i) - Du(k) \\ & = & y^*(k) - \sum_{i=1}^{k} CA^{i-1} d_1(k-i) \end{array} \tag{5.45}$$

where

$$y^*(k) = y_t(k) - CA^k x(0) - \sum_{i=1}^{k} CA^{i-1}Bu(k-i) - Du(k) \tag{5.46}$$

If $x(0)$ is known, $y^*(k)$ can be calculated from Eq.(5.46). Therefore, in the following it will be assumed that $y^*(k)$ is known. A good model should represent the system behavior accurately, this means that the output modeling error should be zero, i.e.

$$\tilde{y}(k) \longrightarrow 0 \tag{5.47}$$

This is a starting point for computing the disturbance vector $d_1(k)$. Let $k = 1, \cdots, M$ and $C_i = CA^{i-1}$, from Eqs.(5.45), (5.46) and (5.47), one can get:

$$\left\{ \begin{array}{rcl} C_1 d_1(0) & = & y^*(1) \\ C_1 d_1(1) + C_2 d_1(0) & = & y^*(2) \\ \cdots\cdots & = & \cdots\cdots \\ C_1 d_1(M-1) + \cdots + C_M d_1(0) & = & y^*(M) \end{array} \right. \tag{5.48}$$

When $rank(C) = n$, and $m \geq n$ ($C \in \mathbb{R}^{m \times n}$), the solution for $d_1(k)$ is derived form Eq.(5.48) as:

$$\left\{ \begin{array}{rcl} \hat{d}_1(0) & = & C^+ y^*(1) \\ \hat{d}_1(k) & = & C^+[y^*(k+1) - \sum_{i=0}^{k-1} C_{k+1-i}\hat{d}_1(i)] \end{array} \right. \tag{5.49}$$

where C^+ is an inverse of C for $m = n$ ($C^+ = C^{-1}$), or a pseudo-inverse of C when $m > n$ ($C^+ = (C^T C)^{-1} C^T$).

To determine the disturbance vector $d_1(k)$, the number of independent measurements should not be smaller than the state number. This requirement is the as same as the requirement in the augmented observer approach and may limit its application. When $rank(C) = g < n$, the number of independent equations (gM) is less than the number of unknown variables (nM) in Eq.(5.48), therefore the solution of $d_1(k)$, $k = 1, \cdots, M$ cannot be uniquely determined using the system input and output data. A good approximation is to let $(n-g)$ components of $d_1(k)$ be zero, i.e.

$$d_1(k) = \left[\begin{array}{c} d_2(k) \\ 0 \end{array} \right] \begin{array}{c} n \\ n-g \end{array} \tag{5.50}$$

and then solve for $d_2(k)$. For this purpose, the term $C_i d_1(k)$ in Eq.(5.48) can be decomposed as follows:

$$C_i d_1(k) = [C_i' \quad C_i''] \left[\begin{array}{c} d_2(k) \\ 0 \end{array} \right] = C_i' d_2(k) \tag{5.51}$$

where $C_i' \in \mathbb{R}^{m \times g}$, $d_2(k) \in \mathbb{R}^g$. Using C_i' and $d_2(k)$ to replace C_i and $d_1(k)$, one can obtain the solution of $d_2(k)$:

$$\left\{ \begin{array}{rcl} \hat{d}_2(0) & = & (C')^+ y^*(1) \\ \hat{d}_2(k) & = & (C')^+[y^*(k+1) - \sum_{i=0}^{k-1} C'_{k+1-i}\hat{d}_2(i)] \end{array} \right. \tag{5.52}$$

Substituting (5.52) into (5.50) the solution for $d_1(k)$ $(k = 1, \cdots, M)$ can be obtained. A physical explanation of this approximation is that $(n-g)$ components of $d_1(k)$ cannot be observed by $y(k)$ and they also cannot be determined from $y(k)$.

From Eqs.(5.46), (5.49) & (5.52), it can be seen that the computing and memory requirements for determining $d_1(k)$ are increasing very significantly when the time index k increase. This growing complexity makes the implementation of algorithms very difficult. This estimation approach is not very practical and some modification and simplification measures must be taken.

Now, assume that the disturbance $d_1(k)$ is a constant bias vector, i.e. $d_1(k) = d_1$ for all k. From Eq.(5.48) and the definition of C_i, the following equation can be derived:

$$\left\{ \begin{array}{rcl} Cd_1 & = & y^*(1) \\ Cd_1 + CAd_1 & = & y^*(2) \\ \cdots\cdots & = & \cdots\cdots \\ Cd_1 + \cdots + C_A^{M-1}d_1 & = & y^*(M) \end{array} \right. \tag{5.53}$$

This equation can be rewritten as:

$$\underbrace{\begin{bmatrix} C \\ C + CA \\ \cdots \\ C + CA + \cdots + CA^{M-1} \end{bmatrix}}_{G} d_1 = \underbrace{\begin{bmatrix} y^*(1) \\ y^*(2) \\ \cdots \\ y^*(M) \end{bmatrix}}_{Y} \tag{5.54}$$

where $G \in \mathbb{R}^{mM \times n}, d_1 \in \mathbb{R}^n, Y \in \mathbb{R}^{mM}$. There exists a least-squares solution for d_1 if and only if $rank(G) = n$. The rank of G can then be determined as follows:

$$\begin{bmatrix} C \\ C + CA \\ \cdots \\ C + CA + \cdots + CA^{M-1} \end{bmatrix} = \begin{bmatrix} I_m & 0 & \cdots & 0 \\ I_m & I_m & \cdots & 0 \\ \vdots & \vdots & \ddots & 0 \\ I_m & I_m & \cdots & I_m \end{bmatrix} \begin{bmatrix} C \\ CA \\ \vdots \\ CA^{M-1} \end{bmatrix}$$

Because the first matrix on the right hand side of the above equation is full rank, it is easy to see that: $rank(G) = n$ if and only if $M \geq n$ and the matrix pair (C, A) is observable. Hence, for an observable system, one can estimate the constant disturbance vector via the use of a limited number of computations and low memory if M is not a very large integer number.

$$d_1 = (G^T G)^{-1} G^T Y \tag{5.55}$$

In this section, it has been assumed that the initial state vector $x(0)$ is known *a priori*. However, this is not always true. Hence, some approximation must be made when $x(0)$ is unknown. Consider the system output vector:

$$y(k) = CA^{\mu}x(k-\mu) + \sum_{i=1}^{\mu} CA^{i-1}Bu(k-i) + \sum_{i=1}^{\mu} CA^{i-1}d_1(k-i) \tag{5.56}$$

For a large μ and for $k > \mu$, one has $A^{\mu} \to 0$ and $CA^{\mu}x(k-\mu) \approx 0$. In this case, one can get:

$$\sum_{i=1}^{\mu} CA^{i-1}d_1(k-i) = y^*(k) \tag{5.57}$$

Where:

$$y^*(k) \approx y(k) - \sum_{i=1}^{\mu} CA^{i-1}Bu(k-i) \tag{5.58}$$

Assume that the disturbance vector $d_1(k)$ is a piece-wise constant vector, i.e.

$$d_1(k-1) = d_1(k-2) = \cdots = d_1(k-\mu)$$

Eq.(5.57) now can be re-written as:

$$[C\sum_{i=1}^{\mu} A^{i-1}]d_1(k-1) = y^*(k) \tag{5.59}$$

Once again, a unique solution for the disturbance vector $d_1(k)$ exists if and only if $rank(C) = n$. This requires that the independent output dimension is not smaller than the state dimension.

The de-convolution method presented in this section can be used, in some cases, to estimate the disturbance vector. However, there are certain limitations to this method and some further research is still needed.

5.4 Optimal Distribution Matrix for Multiple (Varied) Operating Points

Real process plants normally work at different operating points. The operating point of the system varies according to different plant conditions. This is especially true for the analysis of non-linear systems because they are normally linearized around a wide range of operating points. In the design of model-based FDI schemes, investigators often use a single model for ease of implementation. The success of the single FDI design depends on its robustness properties. When using a single model in this way, different modeling errors arise corresponding to different operating points and even the structure of these errors or perturbations can be quite different! Using the terminology outlined in this chapter, it can be said that different operating points correspond to different disturbance distribution matrices. One way to achieve good robustness is to make disturbance de-coupling conditions hold true (in an optimal sense) for all disturbance distribution matrices. This can be done by using a single optimal disturbance distribution matrix to approximate all disturbance distribution matrices.

Consider that the system works at a wide range of operating points, corresponding to different unknown input distribution matrices, E_i $(i = 1, 2, \cdots, M)$. It is attractive to be able design a single FDI scheme for a whole range (or a set) of operating points. In order to make the disturbance de-coupling hold for all operating points, the following relations should be satisfied:

$$HE_i = 0, \quad for \quad i = 1, 2, \cdots, M \tag{5.60}$$

or:

$$H[E_1 \;\; E_2 \;\; \cdots \;\; E_M] = HP = 0 \tag{5.61}$$

If $rank(P) \leq m$, Eq.(5.61) has solutions and exact de-coupling at all operating points is achievable. Otherwise, approximate de-coupling must be used. This is equivalent to the solution of Eq.(5.14) and can be solved by defining the following optimization problem:

$$\min \|P - P^*\|_F^2 \quad \text{subject to:} \quad rank(P^*) \leq m \tag{5.62}$$

This problem can be solved using the singular value decomposition of P as described in Section 5.2.5. The matrices H and P^* should ensure that a fixed FDI scheme is effective for different operating points.

5.5 Modeling and FDI for a Jet Engine System

Modern engines and control systems have become very complex to meet ever-increasing performance requirements. The rapid increase in complexity has made it difficult to build sufficiently reliable, low-cost, light-weight hydromechanical controls. If faults occur, the consequences can be extremely serious. For example, the operator can either be presented with incorrect information or he may find it difficult to locate and diagnose a fault quickly enough to take any appropriate corrective action, as described by Merrill *et al* (Merrill, 1985; Merrill, 1990; Merrill, DeLaat and Abdelwahab, 1991). This highlights a great need for simple and yet highly reliable methods for detecting and isolating faults in the jet engine.

Engine sensors work in a harsh environment and fault probabilities are high, thus making the sensors the least reliable components of the system. In order to improve the reliability of the engine sensors, analytical and hardware redundancy schemes have been investigated extensively (Merrill, 1985; Merrill, 1990; Merrill et al., 1991). The low reliability of the engine sensor module requires that augmentation of the analytical structures be used in order to provide the reliability necessary to cope with ever increasing engine complexity. For example, the rapid changes that occur in the digital fuel control system must be reflected effectively in the sensor system for the accurate detection of faults and the discrimination of false alarms. The inclusion of a fault monitoring system as an integral part of the control system provides the digital control with the necessary information about the faulty sensors. The information is used to decide when to activate an accommodation filter, with the function of reconfiguring the control laws in order to compensate for the occurrence of a sensor malfunction and thus maintain the integrity of the control system. This makes the digital control system attain an acceptable level of reliability.

As discussed in Chapter 1, traditional approaches to FDI in the wider application context are based on hardware redundancy methods which use multiple lanes of sensors, computers and software to measure and/or control a particular variable. A typical jet engine has a degree of redundancy in hardware (eg duplex fuel lines, actuators and speed sensors), however some components, for example the temperature-sensing thermocouple pods, are only available in simple configuration. Moreover, triplex or higher indices of redundancy are not at all realistic. Multiple redundancy is harder to achieve due to lack of operating space. Such schemes would also be costly and very complex to maintain. Severe operating conditions also limit the reliability of engine hardware (e.g. sensors) to the extent that it may not be worthwhile using hardware redundancy alone as a means of diagnosing malfunctions.

The model-based FDI (analytical redundancy) is normally implemented in software form in a computer, and hence very flexible and practical. This is certainly the case for jet engine reliability. Hardware redundancy results in more costly, heavier, less practical, and less potentially reliable systems than various analytical redundancy strategies. Because cost, weight, and reliability are important issues in turbine engine control systems design, much research interest has been focused on model-based strategies.

5.5.1 Background on fault diagnosis for jet engine systems

The use of model-based approaches for diagnosing faults in jet engine systems has become a very active research topic for theoretical and practical reasons, for example as reported by Merrill (Merrill, 1985; Merrill, 1990; Merrill et al., 1991) and Duyar *et al* (Duyar, Eldem and Saravanan, 1990; Duyar and Merrill, 1992; Duyar, Eldem, Merrill and Guo, 1994). Much of the work in the USA has been of a contract nature under NASA Lewis and in collaboration with Pratt & Whitney (Fort Lauderdale) and GE Gas Turbine Engines (Cincinnati). The most comprehensive and practically feasible study is the NASA Lewis program first reported by Beattie, La Prad, McGlone, Rock and Ahkter (1981). Beattie et al. (1981) surveyed a wide range of FDI schemes, and selected a Kalman Filter (KF) with a Generalized-Likelihood Ratio Testing (GLR)-based scheme as a candidate for further development. Later the whole scheme was rig tested, as reported by Merrill *et al* (Merrill, DeLaat, Kroszkewicz and Abdelwahab, 1987; Merrill, DeLaat and Bruton, 1988; Merrill, DeLaat, Kroszkewicz and Abdelwahab, 1988; Merrill et al., 1991). This study has shown that the theory of sensor FDI could be used in practical turbofan sensor systems.

A gas turbine engine is a very non-linear system whose dynamics are rather uncertain and difficult to model mathematically. Modeling errors and system dynamic uncertainty present a challenge to FDI designs due to the general requirement for robustness. In this context, robustness means that the global (i.e. over the operating range of the process, in this case a jet engine) capability for discrimination between faults and unmodeled effects must be well maintained. Some work in the USA e.g. by Emami-Naeini et al. (1986), arising from the original NASA contract, addresses the robustness problem for FDI. In this work the authors go to the extent of including integral-action feedback according to the *internal model principle* to compensate for the effect of so-called "standoff" biases commonly encountered in the application of observer-based estimation for FDI. This leads to an improved tracking of the states (and inherent robustness) but limited ability to detect and identify *slow drift* fault types. They showed that a suitable compromise can be met through an appropriate choice of integral action time. Also working in the USA, Duyar *et al* (Duyar et al., 1990; Duyar, Eldem, Merrill and Guo, 1991; Duyar and Merrill, 1992; Duyar et al., 1994) used an alternative approach to solve the robustness problem. They attempted to derive accurate linearized models of jet engine systems via the α–Canonical form parameterization identification method and the non-linear dynamic simulation data. The method has been applied in FDI schemes for the Space Shuttle Main Engine in a project with NASA Lewis and Pratt & Whitney. Under certain conditions, the identified linearized models are suitable for the FDI purpose. Although it can be argued (and, indeed it is the view held here) that the complexity involved in *total identification* of the system is unjustifiably complex for the task in hand.

Other investigators, Goodwin *et al* (Smed, Carlsson, de Sonza and Goodwin, 1988; Villaneuva, Merrington, Ninness and Goodwin, 1991; Ninness and Goodwin, 1991) for example, have used an alternative approach to study the FDI problem for jet engine systems, based on system identification methods. The robustness issue is tackled by considering unmodeled dynamics in the identification procedure. Viswanadham, Taylor and Luce (1987) also studied this subject using a frequency-domain design technique. Piercy (1989) deals with the problem of maximizing the analytical redundancy of an FDI scheme, based on model-based detection filters. His work examines the efficiency of FDI methods and proposes some new ideas of design based on over-measured jet engine systems. However, he did not consider robustness problem. The research led by Prof. Patton is aimed at keeping in step with the very latest developments world-wide in this subject and on the provision of diagnosis schemes which can be applied very easily in real engine systems. This research emphasizes robustness issues using the eigenstructure assignment technique in designing observer-based residual generators. To use robust approaches, the sensitivity to faults in actuators and sensors in fault decision signals (or residuals) is maximized over the appropriate dynamic range of operation. The residual response to uncertain disturbance effects, for example due to modeling errors, is at best nulled or otherwise optimally minimized. The research on jet engine FDI at Prof. Patton's group have been widely published, see: Patton (1989), Patton and Kangethe (1989), Patton and Chen (1990), Patton and Chen (1991a), Patton and Chen (1991b), Patton and Chen (1991f), Patton and Chen (1992a), Patton and Chen (1992d), Patton, Chen and Zhang (1992), Patton, Zhang and Chen (1992), Chen (1995), Patton and Chen (1997) and Patton and Chen (1999).

5.5.2 Jet engine system description

The gas turbine can be described essentially as a heat engine which uses atmospheric air as a working medium to generate propulsive thrust and mechanical power (Patton, 1989; Patton and Kangethe, 1989; Chen, 1995; Patton and Chen, 1999). The central unit of the mechanical arrangement comprises two main rotating parts, the compressor and the turbine, and one or more chambers.

The gas turbine engine provides a continuous operation cycle which characterizes the phases of energy exchange which affect the gas mass as it passes through the generator. The phases can be expressed as a variation of the gas pressure against volume. The compressor has the task of converting the mechanical energy of the turbine into pressure energy of the air mass flowing through it. The combustion chamber allows the formation of the fuel-air mixture, in turn, depend on the flight conditions. The primary function of the turbine is to drive the compressor using energy extracted from the hot, accelerated exhaust gas. Further mechanical energy generated during the gas expansion phase is used to drive various accessories such as the fuel pump, oil pump and the electric generator.

The control system has the function of coordinating the main burner fuel flow and the propelling exhaust nozzle. There are other control variables such as inlet variable flaps and rear compressor variable vanes. Under normal operation the control lever selects a desired fuel flow rate which, in turn determines the engine speed. The fuel flow is proportional to the exhaust nozzle area. The coordination of the fuel flow and the size of the exhaust nozzle area is particularly necessary for the afterburner operation. Also, if the turbine jet has a thrust reversal an additional control lever is used to give instinctive control of engine power during the thrust reversal operation.

The jet engine illustrated in Fig.5.1 has the measurement variables N_L, N_H, T_7, P_6, T_{29}. N denotes a compressor shaft speed, P denotes a pressure, whilst T represents a measured temperature. The system has two control inputs, the main engine fuel flow rate u_1 and the exhaust nozzle area u_2.

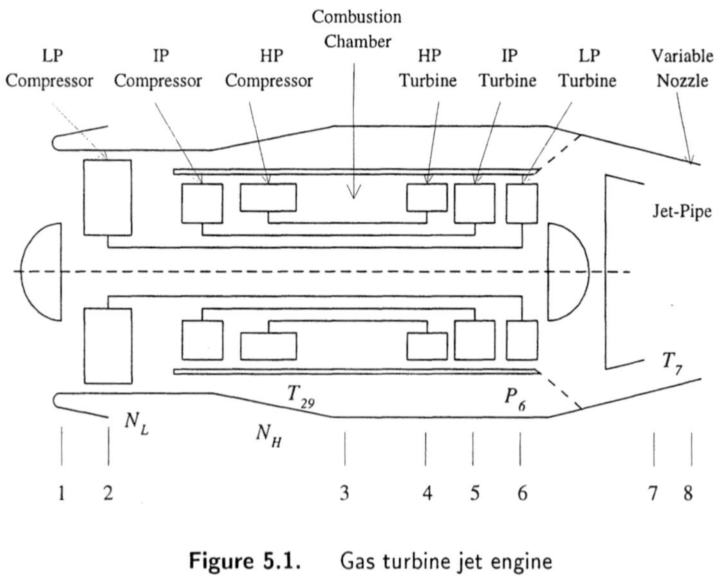

Figure 5.1. Gas turbine jet engine

For the purpose of model-based FDI an accurate representation of the dynamic behavior of the jet engine is required. Modeling of a jet engine is a very difficult problem. One important difficulty lies in the fact that a fully non-linear jet engine system has an iterative structure which means that the equations cannot be written down in differential-algebraic equation form. Fortunately, a non-linear thermodynamic simulation package of the jet engine has been kindly supplied by Lucas Aerospace Ltd. This is a highly non-linear dynamic system which has grossly different steady-state operation over the entire range of spool speeds, flow rates and nozzle areas. Conceptual state space 17_{th} order linearized models at different operating points can be generated using this simulation package. The model has 17 state variables; these include pressure, air and gas mass flow rates, shaft speeds, absolute temperatures and static pressure. The linearized 17_{th} models at different operating points are utilized as a testbed for the evaluation of FDI schemes. Each high order linear model is then further simplified as a low order linear model using balanced model reduction. The linearization error and model reduction are treated as the modeling uncertainty.

In the study presented in this chapter, the nominal operating point is set at 70% of the demanded high spool speed (N_H). For practical reasons and convenience of design, a 5_{th} order model is used to approximate the 17_{th} order model. The model reduction and other errors are represented by the disturbance term $Ed(t)$. The 5th order model matrices are:

$$A = \begin{bmatrix} -78 & 294 & -22 & 21 & -29 \\ 7 & -28 & 2 & -2 & 3 \\ -1325 & 5326 & 526 & 221 & -477 \\ 1081 & -4445 & 377 & -463 & 403 \\ 2152 & -8639 & 781 & -575 & 782 \end{bmatrix}$$

$$B = \begin{bmatrix} -0.0072 & 0.0030 \\ 0.0035 & 0.0003 \\ 1.2185 & -0.0329 \\ 1.3225 & 0.0201 \\ -0.0823 & 0.0244 \end{bmatrix} \qquad C = I_{5\times5} \qquad D = 0_{5\times2}$$

5.5.3 *Application of direct computation and optimization method*

As explained in Section 4.3, one of the necessary steps for the robust residual generation design procedure is to find a matrix H to satisfy Eq.(5.14) (i.e. $HE = 0$) (when the matrix E has been given). The emphasis here is on the derivation of the matrix E which corresponds to uncertainty arising from the application of the lower (5_{th}) order model to the full (17_{th}) order plant. As discussed in Section 5.2.3, this matrix is determined by a comparison of the full-order model and the reduced model. According to Eq.(5.8), the matrix E is obtained as:

$$E = [E_1 \quad E_2 \quad E_3 \quad E_4] \times 10^3$$

where:

$$E_1 = \begin{bmatrix} 0.076 & -0.294 & 0.022 & -0.021 & 0.029 \\ -0.008 & 0.026 & -0.001 & 0.002 & -0.003 \\ 1.309 & -5.024 & 0.305 & -0.333 & 0.478 \\ -1.031 & 4.152 & -0.255 & 0.274 & -0.403 \\ -2.146 & 8.637 & -0.787 & 0.611 & -0.842 \end{bmatrix}$$

$$E_2 = \begin{bmatrix} 0.0 & 0.0 & 0.0 & 0.0 & 0.0 \\ 0.0 & 0.0 & 0.0 & 0.0 & 0.0 \\ 0.0 & 0.0 & 0.093 & 0.005 & 0.003 \\ 0.0 & 0.0 & -0.073 & -0.015 & -0.008 \\ 0.0 & 0.0 & 0.0 & -0.001 & -0.002 \end{bmatrix}$$

$$E_3 = \begin{bmatrix} 0.0 & 0.004 & -0.003 & 0.0 & 0.0 \\ 0.0 & 0.0 & 0.0 & 0.0 & 0.0 \\ 0.003 & 0.0 & 0.0 & 0.004 & 0.0 \\ -0.001 & 0.0 & 0.0 & -0.013 & 0.0 \\ -0.001 & -0.001 & -0.009 & -0.003 & 0.0 \end{bmatrix}$$

$$E_4 = \begin{bmatrix} 0.0 & -0.0013 & 0.0 & 0.0 \\ 0.0 & -0.0002 & 0.0 & 0.0 \\ 0.0 & 0.0269 & 0.0 & 0.0 \\ 0.0 & -0.0804 & 0.0 & 0.0 \\ -0.0169 & 0.0025 & 0.0126 & -0.0091 \end{bmatrix}$$

It can be easily checked that $Rank(E) = 5 = n$, and hence Eq.(5.14) has no solution, the optimization procedure must be employed. The singular values of E are:

$$\sigma_1 = 1, \quad \sigma_2 = 5, \quad \sigma_2 = 60, \quad \sigma_2 = 198, \quad \sigma_2 = 11268$$

The matrices S and T are omitted for brevity. According to the optimization method presented in Section 5.2.5, an optimal q rank approximation for the matrix E is to set $n-q$ smallest singular values as zero. A rank 4 approximation for E is thus given as:

$$E^* = S[diag(0,\ 5,\ 60,\ 198,\ 11268)\ \ 0_{5\times 14}]T^T$$

Based on this matrix, an observer-based robust residual generator can be designed. To simplify the observer design, all eigenvalues are chosen as -100. In this case, the gain matrix $K = -(100I_{5\times 5} + A)$ as C is an identity matrix. The designed robust FDI algorithm is used to detect faulty sensors in the jet engine. The engine data are simulated by the 17_{th} linearized model.

Fig.5.2 shows the output estimation error norm which is very large, and *cannot* be used to detect the fault reliably. This represents the non-robust design situation. Fig.5.3 shows the fault-free residual. Compared with the output estimation error, the residual is very small, i.e., *disturbance de-coupling* is achieved. This robust design can be used to detect incipient faults. In order to evaluate the power of the robust FDI design, a small fault is added to the exhaust gas temperature measurement (T_7); this simulates the effect of an *incipient fault*, here the effect of which is too small to be noticed in the measurements.

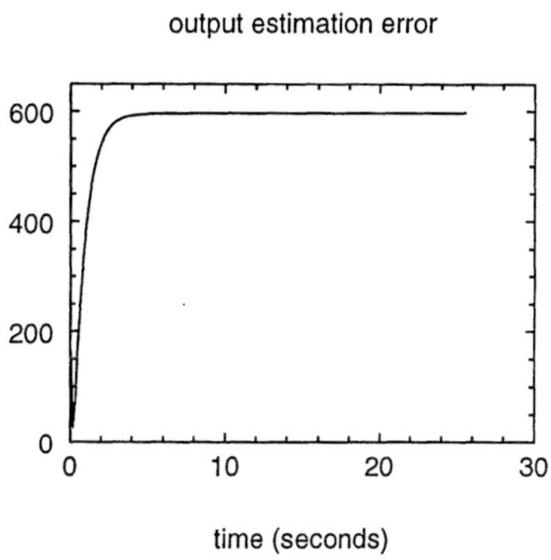

Figure 5.2. Norm of the output estimation error

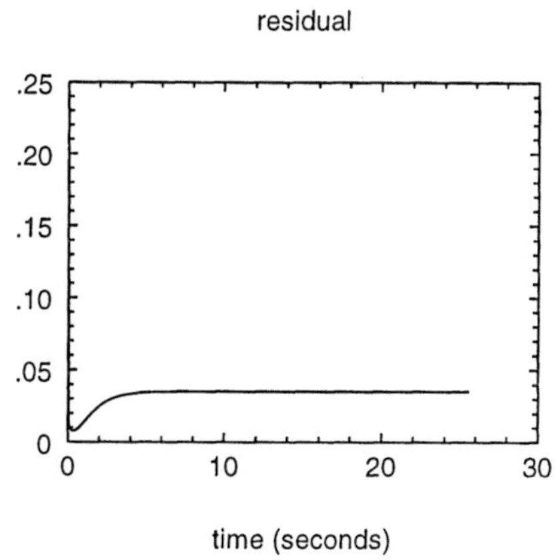

Figure 5.3. Absolute value of the fault-free residual

Fig.5.4 shows the faulty output of the temperature measurement (T_7) and the corresponding residual. The fault is very small compared with the output, and consequently, is not detectable in the measurement. It can be seen that the residual has a very significant increase when a fault has occurred on the system measurement. A threshold can easily be placed on the residual signal to declare the occurrence of faults. Note that the initial peak in the response is not shown in Fig.5.4, this is because FDI is normally carried out after the initial transient has been settled down. To compare the robust design with the non-robust design, the output estimation error which represents a non-robust design is shown in Fig.5.5. The result in this figure cannot easily be used to detect a fault.

The situation when faults occur in the pressure sensor for P_6 is also simulated and the result shown in Fig.5.6 also demonstrates the efficiency of the robust residual in the role of robust FDI.

5.5.4 Application of augmented observer method

The 5_{th} order jet engine linearized model is now discretized for a sampling period of $T = 0.026s$. The model matrices are:

$$A = \begin{bmatrix} -0.981 & 7.532 & -0.598 & 0.486 & -0.698 \\ 0.284 & -0.083 & 0.078 & -0.062 & 0.093 \\ -6.859 & 28.916 & -2.056 & 1.608 & -2.261 \\ 1.224 & -5.661 & 0.402 & -0.319 & 0.414 \\ 13.266 & -53.405 & 4.739 & -3.771 & 5.367 \end{bmatrix}$$

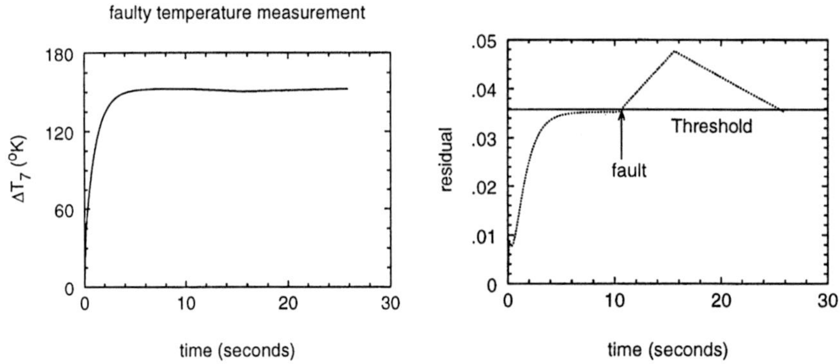

Figure 5.4. The faulty output and residual when a fault occurs in the temperature sensor for T_7

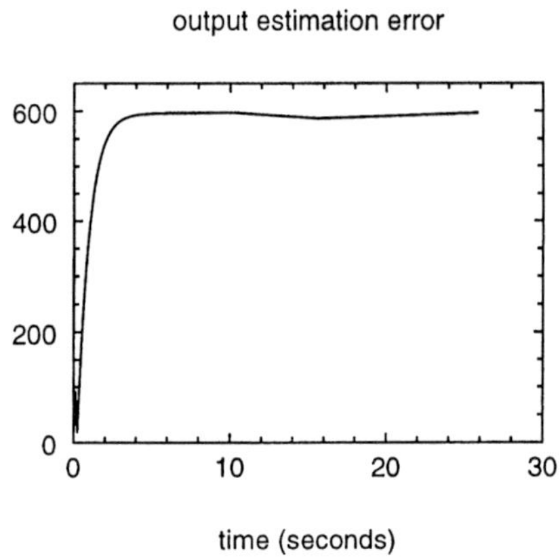

Figure 5.5. The output estimation error when a fault occurs in the temperature sensor for T_7

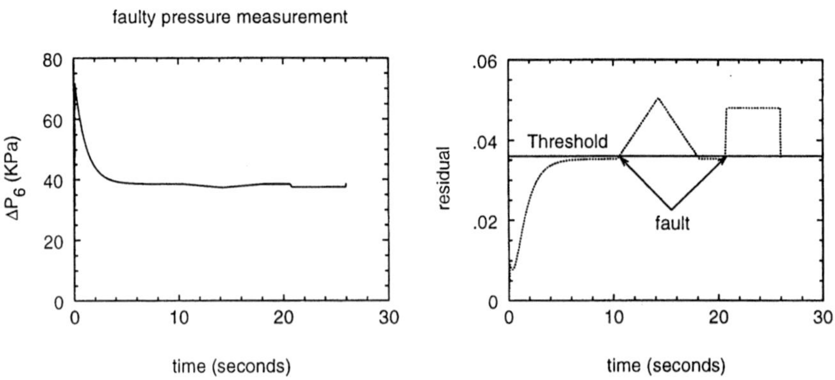

Figure 5.6. Faulty output of the pressure measurement P_6 and corresponding residual

$$B = \begin{bmatrix} 0.000139 & 0.000195 \\ 0.000067 & -0.000005 \\ 0.003188 & 0.000601 \\ 0.007840 & -0.000273 \\ 0.003123 & -0.001516 \end{bmatrix} \quad C = I_{5\times 5} \;\; D = 0_{5\times 2}$$

Modeling errors are represented by the term $Ed(k)$ in the dynamic equation. The term $d_1(k) = Ed(k)$ is now determined via the augmented observer approach, as explained in Section 5.3.1. Assume that the input of the system is $u = [1,\;\; 1]^T$, the "true" system output $\{y_t(k)\}$ is generated using the 17_{th} order continuous-time model, then the data $\{u(k),\; y_t(k)\}$ is fed to an augmented observer to estimate $d_1(k)$. The result is shown in Fig.5.7.

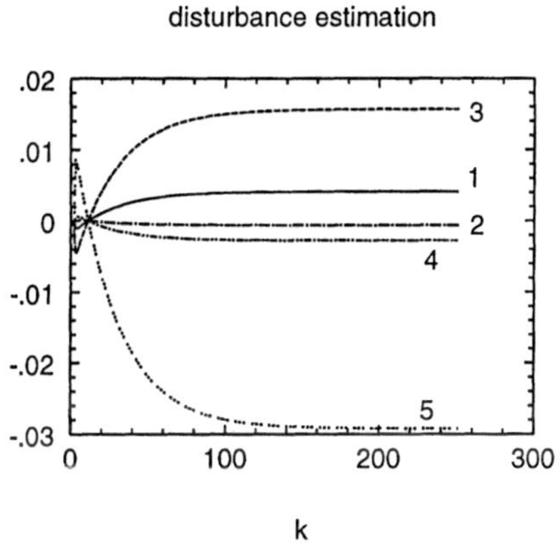

Figure 5.7. The disturbance vector $d_1(k)$ for the step input case

From this diagram, it can be seen that the elements of $d_1(k)$ converge after a short transient. Our interest here is in the direction (distribution) of the term $d_1(k)$, i.e., the relative magnitudes of all elements of this vector. It can also be seen the relative magnitude of all elements of $d_1(k)$ converge. It can then be assumed that:

$$d_1(k) = E_1 d(k)$$

Here, E_1 is a 5×1 vector, it is here used to represent the direction of the $d_1(k)$, and $d(k)$ is a scalar which is the magnitude of the $d_1(k)$. In fact, all directions of $d_1(k)(k = 0, 1, 2, \cdots)$ are slightly different. An optimally representative direction vector E_1^* must be aligned to all the directions of $d_1(k)(k = 0, 1, 2, \cdots)$ "as closely as possible". To obtain a reliable direction, the steady-state disturbances $\{d_1(200), d_1(201), \cdots, d_1(251)\}$ are used to compute this optimal direction. The method of decomposing $d_1(k)$ to E and $d(k)$, using $d_1(k) = Ed(k)$ and with E matrix of rank less than n, is presented in Section 5.3.2. This technique is now used to determine the rank one matrix E_1 as:

$$E_1^* = [0.4126 \;\; -0.0617 \;\; 1.5659 \;\; -0.2776 \;\; -2.9231]^T$$

Normally, the estimation of the disturbance vector $d_1(k)$ will be different for different inputs to the system. In order to check the generality of the direction of the disturbance distribution using the simulation, the system input is changed to $u = [sin(\pi t/3), cos(\pi t/3)]^T$. The estimation of the disturbance signals is shown in Fig.5.8.

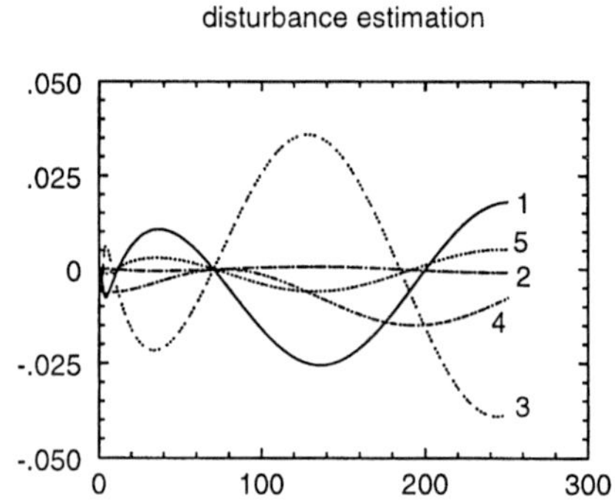

Figure 5.8. The disturbance vector $d_1(k)$ for the sinusoidal input case

Although the magnitude of $d_1(k)$ is time-varying, its direction (the relative magnitude of all elements) is almost constant. Following the procedure described Section 5.3.2, the approximate direction has been obtained as:

$$E_2^* = [0.5334 \quad -0.0768 \quad 1.9658 \quad -0.3698 \quad -3.7068]^T$$

In general, for a complex non-linear system, the operating point will change according to the process inputs and outputs. Hence, it is instructive to consider the system to function at another operating point. In this study, this has been chosen as $95\%N_H$ (or almost full dry power), using the non-linear thermodynamic engine model to generate the linearized parameters. For this case of changed operation, the direction of the disturbances will also be changed. If step inputs are applied to both u_1 and u_2, the approximate direction is obtained as:

$$E_3^* = [1.0511 \quad -0.1545 \quad 4.3087 \quad -0.9646 \quad -7.8283]^T$$

For a sinusoidal input, the approximate direction is:

$$E_4^* = [1.1580 \quad -0.1644 \quad 4.3874 \quad -0.8722 \quad -8.2010]^T$$

Although there are differences between E_1^*, E_2^*, E_3^* and E_4^*, the generalized misalignment angles between them are very small. In fact, the generalized misalignment angles are: $\angle(E_1^*, E_2^*) = 0.3764^o$, $\angle(E_1^*, E_3^*) = 1.5633^o$ and $\angle(E_1^*, E_4^*) = 0.5712^o$. So, it is reasonable to say that the disturbance direction is almost constant (E_1^* is used as a representative in the study) for the system studied here, although the system is a fully non-linear gas turbine model. The results in an interesting basis for further study.

A 5_{th} order discrete-time observer is used to generate the disturbance de-coupling residuals. The first step to complete a disturbance de-coupling (robust residual generation) design is to compute the residual weighting matrix Q (see Section 4.3), such that $QCE = 0$ holds true. This weighting matrix is obtained as:

$$Q = \begin{bmatrix} -0.367 & -0.441 & 0.656 & 0.409 & 0.270 \\ -0.116 & 0.895 & 0.334 & 0.245 & 0.121 \\ 0.879 & -0.050 & 0.350 & -0.034 & 0.316 \end{bmatrix}$$

which ensures that $QCE_1^* = 0$, $QCE_3^* = 0$, $QCE_2^* \approx 0$ and $QCE_4^* \approx 0$. The desired eigenvalues of the observer are $\{0, 0, 0, 0, 0\}$ such that the observer has a *state dead-beat* structure. The desired left eigenvectors of the observer are the rows of the matrix $H = QC = Q$. The observer gain matrix can be derived using eigenstructure assignment. In this example, as all eigenvalues of the observer are zero, the gain matrix is simply derived as $K = A$. Because $QCE_1^* = 0$, $QCE_3^* = 0$, $QCE_2^* \approx 0$, $QCE_4^* \approx 0$ and the rows of the matrix $QC = H$ are the *left* eigenvectors of the observer corresponding to zero-valued eigenvalues, i.e. the robustness conditions hold true, the fault detection scheme is then always robust (such that disturbance de-coupling always holds) when the system works at different operating points and different types of inputs.

The designed robust FDI scheme is applied to detect faulty sensors in the jet engine system. Simulation is based on the 17_{th} order thermodynamic jet engine continuous-time model. A particular emphasis of this assessment study is the power of the method to detect soft or incipient faults which are otherwise unnoticeable in the measurement signals. These attributes are well illustrated in the following graphical time response results. As the FDI scheme has been made robust against modeling errors, the scheme is able to detect incipient faults under conditions of modeling uncertainty. The uncertainty of the jet engine system has been increased further by simulating the effect of random noise generated through a small malfunction in the fuel flow regulator system - to emulate the possibility of a high interference level arising in the electronic system. This has been achieved by adding a zero-mean Gaussian random signal with variance of 1% of demanded fuel-flow, to the fuel flow actuation signal in the model. The inputs to the system are $u = [1, \quad 1]^T$, and initial values are zero. The linear model used has been based on a per-unit scaling of the engine dynamics and hence the final results have been scaled to give meaningful magnitudes.

Fig.5.9 shows the residual norm and the output estimate error norm for both fault-free and faulty cases.

The result in Fig.5.9 shows that the residual is very small in the fault-free case, i.e., disturbance de-coupling is achieved. The output estimation error which represents the non-robust design is very large, even when no faults occur, and this cannot be used to detect faults reliably.

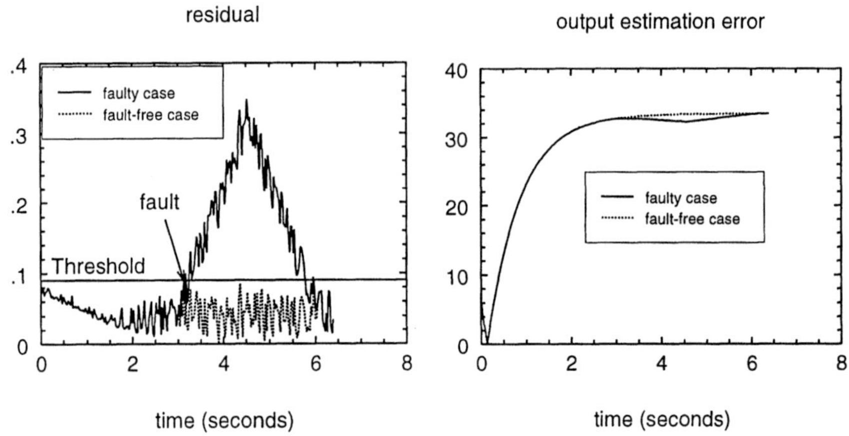

Figure 5.9. The residual $(r(k))$ norm and the output estimation error $(e_y(k))$ norm

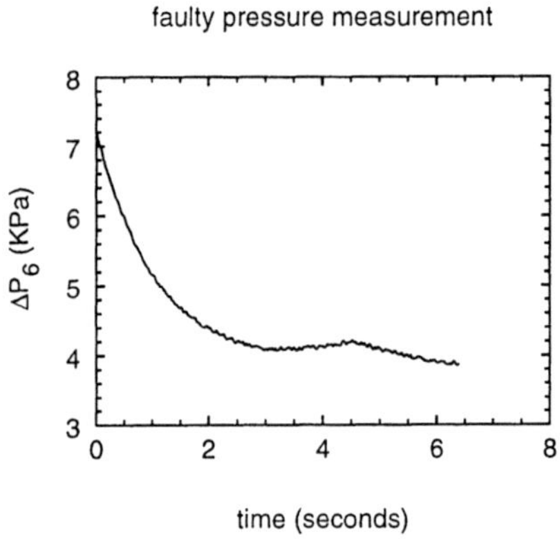

Figure 5.10. Faulty pressure (P_6) measurement

Fig.5.10 shows the faulty output of the pressure sensor P_6; the fault is very small compared with the output, and consequently, which cannot directly be detected in the output. The corresponding residual and the output estimate error for this faulty case are shown in Fig.5.9. It can be seen that the residual has a very significant increase when the fault occurs. Despite the actuation noise, a threshold can easily be placed on the residual signal to declare the occurrence of faults. But, one cannot be sure whether a fault has even occurred in the system when using the information from the output estimate error.

Fig.5.11 shows the fault detection performance of the residual for detecting a parabolic fault in spool speed sensor N_H. The results show the fault can be reliably detected from the residual, but cannot be detected using the output estimation error. This result has proved once again the importance of a robust residual in fault diagnosis.

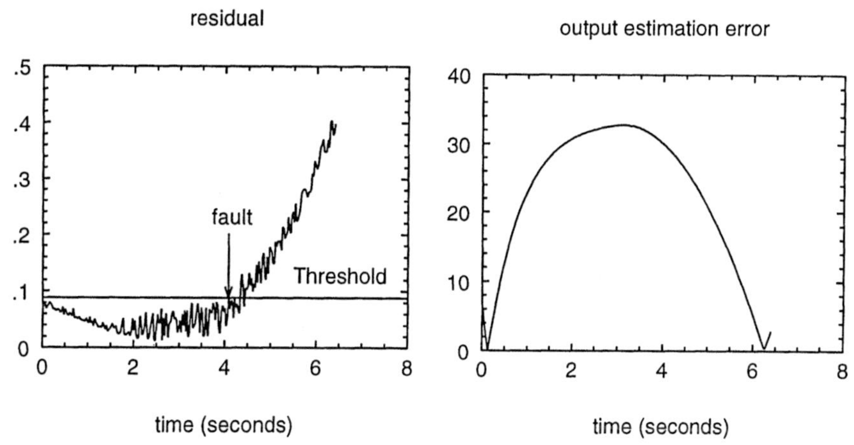

Figure 5.11. The residual norm and the output estimation error norm for the case a parabolic fault on the spool speed sensor for N_H

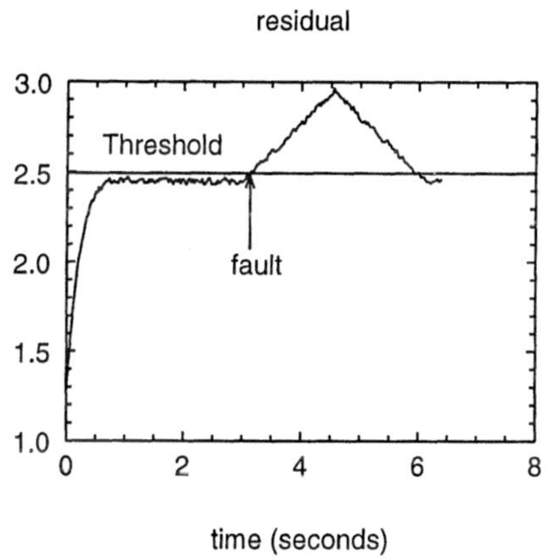

Figure 5.12. Residual when a fault occurs in the temperature measurement T_7 with operating condition changed

In general, for a complex non-linear system, the operating point changes according to the process operation. Hence, it is instructive to consider the system to function at different operating points. A robust FDI scheme should work well for a range of operating points. In order to assess the robustness performance, the scheme is used to detect the fault when the system is working at another operating point (in the presence of demanded changes in high compressor speed N_H), the result is shown in Fig.5.12. Note that, although the magnitude of the residual is changed, the fault can also be easily detected from the significant increase of the residual.

5.6 Conclusion

One critical limitation of the model-based approach to fault diagnosis is that modeling uncertainty is inevitable. For complex systems such as a jet engine, the effects of uncertainty are more pronounced compared with other systems. In order to design robust FDI schemes, we should have a mathematical description of modeling uncertainty. Furthermore, it is necessary to make sure that this description can be handled in a straightforward and systematic manner. Modeling uncertainty can be accounted for using an additional term in the dynamic equation of the system; this additional term has a certain structure. Normally, it is assumed that the distribution of this additional term is known *a priori*. Based on this description, the disturbance de-coupling approach is used to design a robust FDI scheme. For most real systems, the distribution matrix which represents the information about uncertainty is unknown.

This chapter has discussed the methods for determining the disturbance distribution matrix for uncertainty. The main aim has been to bridge theoretical assumptions with practical realty. Two principle methods for determining disturbance distribution matrix have been presented. The first method is the direct determination & optimization method, whose strength is simple and direct and does not require real or simulated system input and output data. Its disadvantage is that it requires some *a priori* information about modeling uncertainty. However, this chapter has presented ways to determine disturbance distribution matrices for a wide range of possible situations. Hence, it can be claimed that the method is *general* in application. The second method is the estimation and de-convolution method. One disadvantage is that it requires that the system has at least n (state dimension) independent measurements. However, for many fault diagnosis problems, e.g., the jet engine example, there are usually a large number of measurements available and the dynamics of the system can be approximated by a relatively low order model. The method can be used for a number of fault diagnosis problems and, as real or simulated system input and output data are used, the results can be affected by the system inputs; different inputs may give arise different distribution matrices. This is a disadvantage of this estimation method. It can be seen that the two methods have compromising properties. One can choose which method is more suitable for a particular problem.

In this chapter, a jet engine example has been used to illustrate the application of the techniques developed. The jet engine is a very complex system and the non-linearities and modeling errors are inevitable. This presents a big challenge for achieving reliable FDI using model-based approaches. Excellent results have been obtained and these indicate the effectiveness of the method for detecting soft (small) and hence incipient faults.

6 ROBUST RESIDUAL GENERATOR DESIGN VIA MULTI-OBJECTIVE OPTIMIZATION AND GENETIC ALGORITHMS

6.1 Introduction

To ensure reliable operation of control systems, hard faults in system components are not tolerable and must be detected before they actually occur. Hopefully, faults are detected during the maintenance stage. However, the situation is different for soft (incipient) faults. Their effect on the system is very small and almost unnoticeable during their incipient stage. They may develop slowly to cause very serious effects on the system, although these incipient faults may be tolerable when they first appear. Hence, the most important issue of reliable system operation is to detect and isolate incipient faults as early as possible. An early indication of incipient faults can give the operator enough information and time to take proper measures to prevent any serious consequence on the system.

Model-based FDI approaches have been demonstrated to be capable of detecting and isolating abrupt and hard faults very quickly and reliably. However, the detection of incipient faults presents a challenge to model-based FDI techniques due to the inseparable mixture between fault effects and modeling uncertainty. Hard or sudden faults normally have a larger effect on the detection residual than the effect of modeling uncertainty. Hence the fault can be detected by placing an appropriate threshold on the residual. However, incipient faults have a lower effect; the effect can even be lower than the response due to modeling uncertainty, so that the detection of soft, slowly developing (or incipient) faults is not very straightforward without any consideration in robustness of FDI algorithms. As discussed in previous chapters, the residual has to be designed to be robust against modeling uncertainty to detect incipient faults.

Although many approaches have been developed, robust FDI is still an *open problem* for further research. One of the most important approaches for robust FDI is the use of disturbance de-coupling principles, which have been studied in Chapters 3 & 4. One should recall that the idea is to treat modeling uncertainty as exogenous disturbances and de-couple their effect from the residual. The main disadvantage is that the distribution of disturbances is required to facilitate designs, although the disturbance itself is assumed unknown. For most uncertain systems, the modeling uncertainty is expressed in terms of modeling errors. Hence, the disturbance de-coupling approach cannot be applied directly. Chapter 5 introduces many ways on representing modeling errors as unknown disturbance with an approximate distribution matrix. In this way, robust FDI is partially achievable. There are some successful applications, however it would be better to relax the restriction on the assumption about modeling uncertainty in the design of robust residual generators. In this chapter, the modeling uncertainty is simply treated as an additive disturbance term in the dynamic equation. There are no requirements to use information about the distribution (structure) of the disturbance or uncertainty, although this information can be used if it is available.

For disturbance de-coupling approaches in FDI, the aim is to completely eliminate the disturbance effect from the residual. However, the complete elimination of disturbance effects may not be possible due to the lack of design freedom. Moreover, it may be problematic, in some cases, because the fault effect may also be eliminated. Hence, an appropriate criterion for robust residual design should take account of the effects of both modeling errors and faults. There is a trade-off between sensitivity to faults and robustness to modeling uncertainty and hence this is an issue of prime concern. Robust residual generation can be then considered as a *multi-objective optimization problem*, i.e. the maximization of fault effects and the minimization of uncertainty effects. The problem of maximizing fault effects and at the same time minimizing disturbance effects was studied by Frank & Wünnenberg (Frank and Wünnenberg, 1989; Wünnenberg, 1990) in the time domain, Frank & Ding (Ding and Frank, 1989; Ding and Frank, 1991; Frank and Ding, 1993; Frank and Ding, 1994) and Qiu and Gertler (1993) in the frequency domain. In their studies, a ratio between disturbance effects and fault effects is minimized. The main problem is that they only considered the cases when the disturbance distribution matrix is known. The multi-objective design in the time-domain for systems with bounded parameter uncertainty and disturbances has been studied by Chen et al. (1993) and is extended in Chapter 7.

Taking from the studies by Chen, Patton and Liu (1994a; 1994b; 1996), this chapter introduces a optimal residual design approach which is based on the combination of multi-objective optimization and genetic algorithms. In this approach the residual is generated via an observer. In order to make the residual become insensitive to modeling uncertainty and sensitive to sensor faults, a number of performance indices are defined to achieve good fault diagnosis performance. Some performance indices are defined in the frequency domain to account for the fact that modeling uncertainty effects and faults occupy different frequency bands. The numerical optimization technique is used to find the observer gain and weighting matrices. To utilize the knowledge of the modeling uncertainty and fault frequency bands, frequency-dependent weighting factors are introduced in the performance indices (cost functions). The main idea of robust FDI is to distinguish faults and the uncertainty effects in residuals, and this is only possible when they are "physically" distinguishable. Otherwise, no matter what mathematical method is chosen, one cannot discriminate between these two effects. Moreover, some information about both faults and disturbances must be available. In the previous chapters of this book, the faults have been assumed to have different distribution directions from those of the uncertainty. In the technique describe in this chapter, the information on frequency distribution ranges of faults, noise and modeling uncertainty (once known) can be incorporated into a robust residual design.

To design robust residuals, a multi-objective optimization problem needs to be solved. This chapter uses the method of inequalities to solve this multi-objective optimization problem. All objectives are reformulated into a set of inequality constraints on performance indices. The genetic algorithm is thus used to search an optimal solution to satisfy these inequality constraints. The use of genetic algorithms obviates the requirement for the calculation of cost function gradients and also increases the possibility of finding global optimization. A flight control example is used in this chapter to illustrate the technique developed. The fault detection performance is examined in the presence of modeling errors. The simulation results show that the fault detection algorithm designed by the introduced method can detect incipient sensor faults very effectively.

6.2 Residual Generation and Performance Indices

6.2.1 Residual generation and responses

Consider the following mathematical description of the monitored system:

$$\left\{ \begin{array}{lcl} \dot{x}(t) & = & Ax(t) + Bu(t) + R_1 f(t) + d(t) \\ y(t) & = & Cx(t) + Du(t) + R_2 f(t) \end{array} \right. \tag{6.1}$$

where $x(t) \in \mathbb{R}^n$ is the state vector, $u(t) \in \mathbb{R}^r$ is the control input vector and $y(t) \in \mathbb{R}^m$ is the measurement vector, $f(t) \in \mathbb{R}^g$ represents the fault vector which is considered as an unknown time function. The matrices A, B, C and D are system parameter matrices and the pair $\{C, A\}$ is assumed observable.

The vector $d(t)$ is the disturbance vector which can also be used to represent modeling errors such as:

$$d(t) = \Delta A x(t) + \Delta B u(t)$$

Note that this form of uncertainty representation is very general as the distribution matrix is not required. The matrices R_1 and R_2 are fault distribution matrices which represent the influence of faults on the system. They can be determined if one has defined which faults are to be diagnosed. For two most common cases: sensor and actuator faults, these matrices are:

$$R_1 = \begin{cases} 0 & \text{sensor faults} \\ B & \text{actuator faults} \end{cases} \qquad R_2 = \begin{cases} I_m & \text{sensor faults} \\ D & \text{actuator faults} \end{cases}$$

The residual generator studied in this chapter, shown in Fig.6.1, is based on a full-order observer. The basic idea is to estimate the system output from the measurements using an observer. The weighted output estimation error is then used as a residual. The flexibility in selecting the observer gain and the weighting matrix provides freedom to achieve good detection performance.

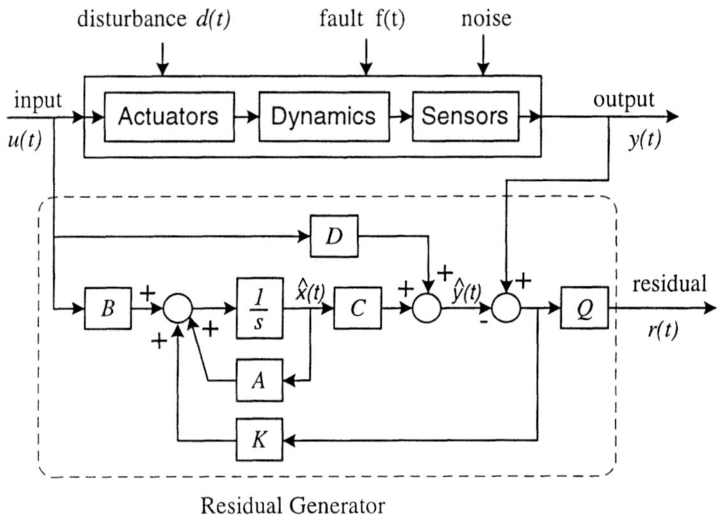

Figure 6.1. Robust residual generation via a full-order observer

The residual generator is described as:

$$\begin{cases} \dot{\hat{x}}(t) &= (A - KC)\hat{x}(t) + (B - KD)u(t) + Ky(t) \\ \hat{y}(t) &= C\hat{x}(t) + Du(t) \\ r(t) &= Q[y(t) - \hat{y}(t)] \end{cases} \tag{6.2}$$

where $r \in \mathbb{R}^p$ is the residual vector, $\hat{x}$ and $\hat{y}$ are state and output estimations. The matrix $Q \in \mathbb{R}^{p \times m}$ is the residual weighting factor which, in most cases, is static but can also be dynamic. When this residual generator is applied to the monitored system described by Eq.(6.1), the state estimation error ($e(t) = x(t) - \hat{x}(t)$), and the residual are governed by the following equations:

$$\begin{cases} \dot{e}(t) &= (A - KC)e(t) + d(t) + R_1 f(t) - KR_2 f(t) \\ r(t) &= QCe(t) + QR_2 f(t) \end{cases} \tag{6.3}$$

The residual response to faults and disturbances is thus:

$$\begin{aligned} r(s) &= Q\{R_2 + C(sI - A + KC)^{-1}(R_1 - KR_2)\}f(s) \\ &\quad + QC(sI - A + KC)^{-1}[d(s) + e(0)] \\ &= G_{rf}(s, K, Q)f(s) + G_{rd}(s, K, Q)[d(s) + e(0)] \end{aligned} \tag{6.4}$$

where $e(0)$ is the initial value of the state estimation error.

6.2.2 *Performance indices in robust residual generation*

Both faults and disturbances affect the residual, and discrimination between these two effects is difficult. To reduce false and missed alarm rates, the effect of faults on the residual should be maximized and the effect of disturbances on the residual should be minimized. One can maximize the effect of the faults by maximizing the following performance index, in the required frequency range $[\omega_1, \omega_2]$:

$$\bar{J}_1(K, Q) = \inf_{\omega \in [\omega_1, \omega_2]} \underline{\sigma}\{QR_2 + QC(j\omega I - A + KC)^{-1}(R_1 - KR_2)\} \tag{6.5}$$

This is equivalent to the minimization of the following performance index :

$$J_1(K, Q) = \sup_{\omega \in [\omega_1, \omega_2]} \overline{\sigma}\{[QR_2 + QC(j\omega I - A + KC)^{-1}(R_1 - KR_2)]^{-1}\} \tag{6.6}$$

where $\underline{\sigma}\{.\}$ and $\overline{\sigma}\{.\}$ denote the minimal and maximal singular values.

Similarly, one can minimize the effects of both disturbance and initial condition by minimizing the following performance index:

$$J_2(K, Q) = \sup_{\omega \in [\omega_1, \omega_2]} \overline{\sigma}\{QC(j\omega I - A + KC)^{-1}\} \tag{6.7}$$

Besides faults and disturbances, noise in the system can also affect the residual. To illustrate this, assume that $\zeta(t)$ and $\eta(t)$ are input and sensor noise signals, the system equations in this case is:

$$\begin{cases} \dot{x}(t) = Ax(t) + Bu(t) + d(t) + R_1 f(t) + \zeta(t) \\ y(t) = Cx(t) + Du(t) + R_2 f(t) + \eta(t) \end{cases} \tag{6.8}$$

It can be seen that the sensor noise as well as faults acting through $R_2 f(t)$ affect the system at the same excitation point and hence affect the residual in the same way. To reduce the noise effect on the residual, the following norm should be minimized.

$$\| Q - QC(j\omega I - A + KC)^{-1}K \|$$

The minimization of the above norm contradicts with the requirement for maximizing the effects of faults on the residual. Fortunately, the frequency ranges of the faults and noise are normally different. For an incipient fault signal, the fault information is contained within a low frequency band as the fault development is slow. However, the noise comprises mainly high frequencies signals. Based on these observations, the effects of noise and faults can be separated by using different frequency-dependent weighting penalties. In this case, the performance index $J_1(K,Q)$ is:

$$J_1(K,Q) = \sup_{\omega\in[\omega_1,\omega_2]} \bar{\sigma}\{W_1(j\omega)[QR_2 + QC(j\omega I - A + KC)^{-1}(R_1 - KR_2)]^{-1}\} \tag{6.9}$$

To minimize the effect of noise on the residual, a new performance index is introduced as:

$$J_3(K,Q) = \sup_{\omega\in[\omega_1,\omega_2]} \bar{\sigma}\{W_3(j\omega)Q[I - C(j\omega I - A + KC)^{-1}K]\} \tag{6.10}$$

In order to maximize the effects of faults at low frequencies and minimize the noise effect at high frequencies, the frequency-dependent weighting factor $W_1(j\omega)$ should have large magnitude in the low frequency range and small magnitude at high frequencies. The frequency effect of $W_3(j\omega)$ should be opposite to $W_1(j\omega)$ and can be chosen as $W_3(j\omega) = W_1^{-1}(j\omega)$.

The disturbance (or modeling error) and input noise affect the residual in the same way. As both effects should be minimized, the performance index J_2 does not necessarily need to be weighted. However, modeling uncertainty and input noise effects may be more serious in one or more frequency bands. The performance index should reflect this fact, and hence a frequency-dependent weighting factor must also be placed on $J_2(K,Q)$, in some situations.

$$J_2(K,Q) = \sup_{\omega\in[\omega_1,\omega_2]} \bar{\sigma}\{W_2(j\omega)QC(j\omega I - A + KC)^{-1}\} \tag{6.11}$$

Now, considering the steady state value of the residual:

$$r(\infty) = QR_2 f(\infty) + QC(A - KC)^{-1}(KR_2 - R_1)f(\infty) - (A - KC)^{-1}d(\infty) \tag{6.12}$$

After the transient period, the residual steady state value plays an important role in FDI. Ideally, it should reconstruct the fault signal. The disturbance effects on residual can be minimized by minimizing the following performance index:

$$J_4(K) = \| (A - KC)^{-1} \|_\infty \tag{6.13}$$

When J_4 is minimized, the matrix K is very large and the norm $\| (A - KC)^{-1}K \|$ approaches to a constant value. This means that the fault effect on the residual has not been changed by reducing the disturbance effect. This is what is required for good FDI performance.

6.2.3 Remarks on performance indices

6.2.3.1 The choice of norms. In the definition of performance indices, the infinity norm of matrices are used. However, other matrix norms (such as the Frobenius norm) are also useful. To examine the function of different matrix norms, let's consider the disturbance effect on the residual.

$$r(s) = G_{rd}(s, K, Q)d(s)$$

It is well known that the residual norm is bounded by:

$$||r(s)|| \leq ||G_{rd}(s, K, Q)||||d(s)||$$

If the infinity norm is used, this inequality becomes:

$$||r(s)||_\infty \leq ||G_{rd}(s, K, Q)||_\infty ||d(s)||_\infty$$

This measures the worst effects, i.e., the largest component of the residual due to the largest component of disturbance will be minimized if $||G_{rd}(s, K, Q)||_\infty$ is minimized.

If the Frobenius norm is used, the corresponding relations is:

$$||r(s)||_2 \leq ||G_{rd}(s, K, Q)||_F ||d(s)||_2$$

This measures the average effects, i.e., the energy of the residual due to the disturbance will be minimized if $||G_{rd}(s, K, Q)||_F$ is minimized. Different norm measures have different characteristics, however if one format of norm for a particular matrix is minimized, other kind of norms for the same matrix are unlikely to be large. This can be proved using the following inequality (Golub and Van Loan, 1989, p.57):

$$\frac{1}{\sqrt{n}}||G_{rd}||_\infty \leq ||G_{rd}||_F \leq \sqrt{np}||G_{rd}||_\infty$$

where p is the row number of G_{rd} and n is column number. From this relation, it can be seen that the Frobenius norm will be bounded if the infinity norm is minimized and verse versa.

6.2.3.2 Disturbance distribution. In this chapter, there is no requirement on the disturbance distribution. However, this information can also be incorporated into performance indices, if it is available. If the disturbance distribution matrix is known, i.e.

$$d(t) = Ed'(t)$$

where E is a known matrix and $d'(t)$ is an unknown vector. In this case, the performance index J_2 can be modified as:

$$J_2(K, Q) = \sup_{\omega \in [\omega_1, \omega_2]} \overline{\sigma}\{W_2(j\omega)QC(j\omega I - A + KC)^{-1}E\} \qquad (6.14)$$

6.2.3.3 Fault isolation. As discussed in Sections 2.7.1 & 3.3, a structured residual set should be generated to isolate faults. The word "structured" here signifies the sensitivity and insensitivity relations that any residual will have, i.e. whether it is designed to be sensitive to a group of faults whilst, insensitive to another group of faults. The faults contained in the vector $f(t)$ can be divided into two groups $f^1(t)$ and $f^2(t)$ and the system equation in this case is:

$$\left\{ \begin{array}{lcl} \dot{x}(t) & = & Ax(t) + Bu(t) + R_1^1 f^1(t) + R_1^2 f^2(t) + d(t) \\ y(t) & = & Cx(t) + Du(t) + R_2^1 f^1(t) + R_2^2 f^2(t) \end{array} \right. \tag{6.15}$$

If the residual is to be designed sensitive to $f^1(t)$ and insensitive to $f^2(t)$, the performance index J_1 should be modified as:

$$J_1(K,Q) = \sup_{\omega \in [\omega_1, \omega_2]} \overline{\sigma}\{W_1(j\omega)[QR_2^1 + QC(j\omega I - A + KC)^{-1}(R_1^1 - KR_2^1)]^{-1}\} \tag{6.16}$$

In addition to the four performance indices defined, a new performance index $J_5(K,Q)$ which is to be minimized should be introduced to make the residual insensitive to $f^2(t)$.

$$J_5(K,Q) = \sup_{\omega \in [\omega_1, \omega_2]} \overline{\sigma}\{W_5(j\omega)[QR_2^2 + QC(j\omega I - A + KC)^{-1}(R_1^2 - KR_2^2)]\} \tag{6.17}$$

If only sensor faults are to be isolated, the design problem is easier to solve. This is because, if the residual is to be sensitive to a group of sensor faults, only the measurements from this set of sensors will be used in the residual generation. The detailed discussion on robust sensor and actuator fault isolation can be found in Section 3.3.

6.3 Parameterization In Observer Design

Four performance indices $J_1(K,Q)$, $J_2(K,Q)$, $J_3(K,Q)$ and $J_4(K,Q)$ have now been defined. To achieve robust FDI (in terms of minimizing false and missed alarm rates), one has to solve a multi-objective optimization problem. One of the parameter sets to be designed is the observer gain matrix K which must guarantee the stability of the observer. This leads to a constrained optimization problem which is difficult to solve. Within the context of control system design, this stability constraint is normally changed to the assignment of eigenvalues in the left hand side of the complex plane (or within the unit disc, for the discrete-time domain). The observer design is a dual of the controller design problem and all techniques in control design can be applied. Here, the eigenstructure assignment method is chosen to give the parameterization of the gain matrix K (Burrows and Patton, 1991; Patton and Liu, 1994; Liu and Patton, 1996; Chen, Patton and Liu, 1994a; Chen, Patton and Liu, 1994b; Chen, Patton and Liu, 1996; Patton, Chen and Liu, 1997; Liu and Patton, 1998). Note that the gain matrix can also be parameterized in other ways. However, the parametric representation in terms of eigenstructure has many advantages, the most important one is that the eigenvalues can be specified in predefined points or regions according to required residual responses.

The eigenvalues of the observer can be real or complex-conjugate. Assume that there are n_r real eigenvalues λ_i $(i = 1, \cdots, n_r)$ and n_c pairs of complex-conjugate eigenvalues $\lambda_{j,re} \pm j\lambda_{j,im}$ $(j = 1, \cdots, n_c)$, and n_r and n_c satisfy the following relation:

$$n_r + 2n_c = n$$

6.3.1 *Real eigenvalue case*

Assume that v_i is the i_{th} right eigenvector of $(A^T - C^T K^T)$ corresponding to the i_{th} eigenvalue λ_i of $(A^T - C^T K^T)$, we then have that:

$$(A^T - C^T K^T)v_i = \lambda_i v_i \tag{6.18}$$

or

$$v_i = -(\lambda_i I - A^T)^{-1} C^T K^T v_i \tag{6.19}$$

To define a design parameter vector w_i as:

$$w_i = K^T v_i \tag{6.20}$$

The eigenvector v_i can expressed via this design parameter vector:

$$v_i = -(\lambda_i I - A^T)^{-1} C^T w_i \tag{6.21}$$

6.3.2 *Complex-conjugate eigenvalue case*

Assume that $v_{j,re} + jv_{j,im}$ is the j_{th} right eigenvector of $(A^T - C^T K^T)$ corresponding to the j_{th} eigenvalue $\lambda_{j,re} + j\lambda_{j,im}$ of $(A^T - C^T K^T)$, we have:

$$(A^T - C^T K^T)(v_{j,re} + jv_{j,im}) = (\lambda_{j,re} + j\lambda_{j,im})(v_{j,re} + jv_{j,im}) \tag{6.22}$$

This equivalent to:

$$\left\{ \begin{array}{lcl} (A^T - C^T K^T)v_{j,re} & = & \lambda_{j,re} v_{j,re} - \lambda_{j,im} v_{j,im} \\ (A^T - C^T K^T)v_{j,im} & = & \lambda_{j,im} v_{j,re} + \lambda_{j,re} v_{j,im} \end{array} \right. \tag{6.23}$$

or

$$\left\{ \begin{array}{lcl} (\lambda_{j,re} I - A^T)v_{j,re} - \lambda_{j,im} v_{j,im} & = & -C^T K^T v_{j,re} \\ \lambda_{j,im} v_{j,re} + (\lambda_{j,re} I - A^T)v_{j,im} & = & -C^T K^T v_{j,im} \end{array} \right. \tag{6.24}$$

To define:

$$A_j = \left[\begin{array}{cc} \lambda_{j,re} I - A^T & -\lambda_{j,im} I \\ \lambda_{j,im} I & \lambda_{j,re} I - A^T \end{array} \right] \qquad C_c = \left[\begin{array}{cc} C^T & 0 \\ 0 & C^T \end{array} \right]$$

and

$$\left\{ \begin{array}{lcl} w_{j,re} & = & K^T v_{j,re} \\ w_{j,im} & = & K^T v_{j,im} \end{array} \right. \tag{6.25}$$

this leads to:

$$\begin{bmatrix} v_{j,re} \\ v_{j,im} \end{bmatrix} = -A_j^{-1} C_c \begin{bmatrix} w_{j,re} \\ w_{j,im} \end{bmatrix} \tag{6.26}$$

To put Eqs. (6.20) & (6.25) together, the parametric representation of the observer gain matrix K is given by:

$$K = [WV^{-1}]^T \tag{6.27}$$

where

$$W = [w_1 \cdots w_{n_r} \ ; \ w_{1,re} \cdots w_{n_c,re} \ ; \ w_{1,im} \cdots w_{n_c,im}] \in \mathbb{R}^{m \times n}$$

is the design parameter matrix whose elements can be determined arbitrarily.

$$V = [v_1 \cdots v_{n_r} \ ; \ v_{1,re} \cdots v_{n_c,re} \ ; \ v_{1,im} \cdots v_{n_c,im}] \in \mathbb{R}^{n \times n}$$

Any column vector of this matrix is a function of the corresponding column vector in the matrix W. All columns in this matrix can be calculated via either Eq.(6.21) or Eq.(6.26).

6.3.3 Eigenvalue specifications

The eigenvalues λ_i $(i = 1, \cdots, n_r)$ and $\lambda_{j,re} \pm j\lambda_{j,im}$ $(j = 1, \cdots, n_c)$ have to be given by the designer prior to the design procedure. In practice, the eigenvalues do not need to be assigned at a specific point in the complex plane. However, we do need to assign eigenvalues in predefined regions to meet stability and response requirements, i.e.

$$\lambda_i \in [L_i, \ U_i] \qquad i = 1, \cdots, n_r$$

for real eigenvalues. For complex-conjugate eigenvalues, the relations will be:

$$\left. \begin{array}{l} \lambda_{j,re} \in [L_{j,re}, \ U_{j,re}] \\ \lambda_{j,im} \in [L_{j,im}, \ U_{j,im}] \end{array} \right\} \quad \text{for} \quad j = 1, \cdots, n_c$$

The assignment of eigenvalues in regions rather than at specific points increases the design freedom. However the inequality constraints on eigenvalues are introduced by doing this. To remove these constraints, a simple transformation for a real eigenvalue can be introduced (Burrows and Patton, 1991):

$$\lambda_i = L_i + (U_i - L_i) sin^2(z_i) \tag{6.28}$$

where $z_i \in \mathbb{R}$ $(i = 1, \cdots, n_r)$ can be freely chosen. Similar to the real eigenvalue case, the real and the imaginary parts of a complex-conjugate eigenvalue pair can be expressed by:

$$\left\{ \begin{array}{lll} \lambda_{j,re} & = & L_{j,re} + (U_{j,re} - L_{j,re}) sin^2(z_j) \\ \lambda_{j,im} & = & L_{j,im} + (U_{j,im} - L_{j,im}) sin^2(z_{j+1}) \end{array} \right. \tag{6.29}$$

where $z_j, z_{j+1} \in \mathbb{R}$ can be determined arbitrarily. Now, the constrained performance indices $J_j(K, Q)$ $(j = 1, 2, 3, 4)$ have been transformed into unconstrained performance indices $J_j(Z, W, Q)$, where W and $Z = [z_1 \cdots z_n] \in \mathbb{R}^{1\times n}$ can be chosen freely. The multi-objective optimization problem for robust FDI is solved in the following sections by combining the method of inequalities and the genetic algorithm.

6.4 Multi-Objective Optimization and the Method of Inequalities

6.4.1 Multi-objective optimization

The use of multi-objective optimization is very common in engineering design problems. Generally speaking, a solution does not exist which minimizes all performance indices simultaneously. A set of parameters which minimizes a particular performance index may let other performance indices become very large and unacceptable. Hence, some compromises and trade-offs must be made in the design. The trade-off is based on the relative importance of objectives. As the number of objectives increases, trade-offs between objectives are likely to become complex and less easily quantifiable. There is, therefore, much reliance on the intuition of the designer and his/her ability to express preferences throughout the optimization cycle. This is relatively easier to be solved using numerical search algorithms, as the designer can alter his preference throughout the optimization cycle and enter them into a numerically tractable and realistic design problem.

Mixed objective strategy: A commonly used approach in multi-objective optimization is the mixed objective approach, for example, Burrows and Patton (1991) applied this approach to control system design. In this approach, all objective functions are mixed together according to different weighting factors. The emphasis on different objectives can be made using different magnitudes of weighting factors. For the optimization problem presented in this chapter, the performance indices can be mixed together in the following ways:

$$J(Z, W, Q) = \sum_{i=1}^{4} \alpha_i J_i(Z, W, Q) \tag{6.30}$$

$$J(Z, W, Q) = \frac{\sum_{i=2}^{4} \alpha_i J_i(Z, W, Q)}{\bar{J}_1(Z, W, Q)} \tag{6.31}$$

$$J = \sup_{\omega\in[\omega_1,\omega_2]} \overline{\sigma} \left\{ \begin{bmatrix} W_1(j\omega)[QR_2 + QC(j\omega I - A + KC)^{-1}(R_1 - KR_2)]^{-1} \\ W_2(j\omega)QC(j\omega I - A + KC)^{-1} \\ W_3(j\omega)Q[I - C(j\omega I - A + KC)^{-1}K] \\ (A - KC)^{-1} \end{bmatrix} \right\} \tag{6.32}$$

where $\alpha_i \geq 0$ $(i = 1, \cdots, 4)$ are weighting factors which should be decided according to the relative importance of objectives. The multi-objective optimization problem is now reformulated into the minimization of the mixed single cost function $J(Z, W, Q)$.

The minimax optimization: The multi-objective optimization problem can also be solved via a minimax optimization procedure in which the largest normalized performance index is to be minimized:

$$\min J(Z, W, Q) = \min\{\max_i \frac{J_i(Z, W, Q)}{C_i}\} \tag{6.33}$$

where C_i $(i = 1, \cdots, 4)$ are the normalizing factors. The preference on different objectives can be achieved by altering the normalizing factors. It is interesting to note that, minimax optimization considers the worst case which is the same as the use of infinity norms.

6.4.2 The method of inequalities

A more attractive approach for solving the multi-objective optimization problem in control system design is the method of inequalities, proposed by Zakian (Zakian and Al-Naib, 1973; Zakian, 1979). The main philosophy behind this approach is to replace the minimization of the performance index by an inequality constraint on the performance index. The simultaneous minimization of all performance indices is normally impossible. However, in engineering design problems, what is one normally required is the restriction of the performance index within a pre-defined region. The optimization problem is posed as the *satisfaction* of a set of inequalities, rather than the *minimization* of some objective functions with inequalities acting as side-constraints. The shift of emphasis from objective functions to a set of inequalities gives a more accurate formal representation of many design problems, and leads to an iterative design procedure in which the designer changes the "trade-off" between conflicting constraints by adjusting the inequalities, rather than some objective functions. This is attractive, because it is usually much easier to understand the physical implications of changes in constraining inequalities than changes in an objective function.

Optimization is still required in the method of inequalities, but the shift of emphasis from minimization to inequality satisfaction means that the usual ideas on the choice of the optimization algorithms have to be revised: speed of convergence in the neighborhood of a minimum becomes much less important than the likelihood of finding at least one feasible point – namely one at which all the inequalities are satisfied.

For the problem presented in this chapter, the multi-objective optimization problem is being reformulated into that of searching for a parameter set $\{Z, W, Q\}$ to satisfy the following inequalities:

$$J_i(Z, W, Q) \leq \varepsilon_i, \qquad \text{for} \qquad i = 1, 2, 3, 4 \tag{6.34}$$

where the real number ε_i represents the numerical bound on the performance index $J_i(Z, W, Q)$ required by the designer. If the minimal value of $J_i(Z, W, Q)$

achieved by minimizing $J_i(Z, W, Q)$ itself is J_i^*, the objective bound must be set as: $\varepsilon_i > J_i^*$. This is based on the fact that a parameter set which minimizes a particular performance index can make other performance indices very large. If $J_i^*(Z_i^*, W_i^*, Q_i^*)$ is the minimal value of $J_i(Z, W, Q)$ achieved at the parameter set $\{Z_i^*, W_i^*, Q_i^*\}$, the following inequalities hold true:

$$J_i(Z_j^*, W_j^*, Q_j^*) \geq J_i^*(Z_i^*, W_i^*, Q_i^*) \tag{6.35}$$

where $j \neq i, j \in \{1, 2, 3, 4\}$; for $i = 1, 2, 3, 4$. As a general rule, the performance boundaries ε_i should be set as:

$$J_i^*(Z_i^*, W_i^*, Q_i^*) < \varepsilon_i \leq \max_{j \neq i, j \in [1,4]} \{J_i(Z_j^*, W_j^*, Q_j^*) \tag{6.36}$$

for $i = 1, 2, 3, 4$. The problem of multi-objective optimization is to find a parameter set to make all performance indices lie in an acceptable region. By adjusting the bounds ε_i, one can place a different emphasis on each of the objectives. If the performance index J_j is important for the problem, one can let ε_j near to J_j^*. If the performance index J_k is less important, one can let ε_k be far away from J_k^*.

Example of the method of inequalities: A simple example show in Fig.6.2 is used to illustrate the use of the method of inequalities.

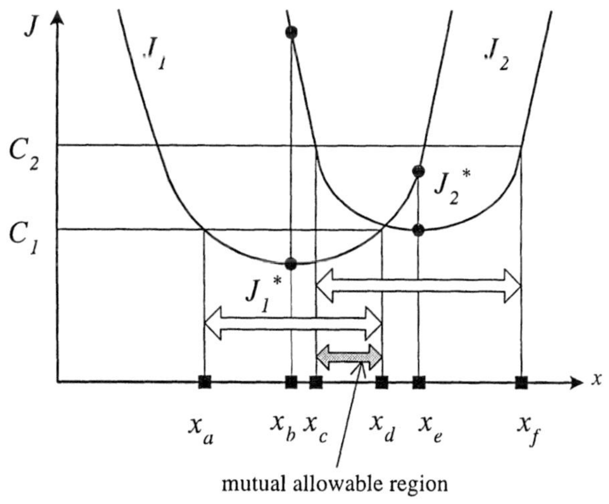

Figure 6.2. An example of the method of inequalities

In this example, two performance indices $J_1(x)$ and $J_2(x)$ are to be minimized. It can be seen that x_b is the minimization point for $J_1(x)$, however this point is not acceptable for $J_2(x)$. The point which minimizes both $J_1(x)$ and $J_2(x)$ does not exist. To solve this problem, the requirement for optimization should be relaxed. In stead of the minimization of $J_1(x)$ and $J_2(x)$, the problem is transformed to the satisfaction of the following inequalities:

$$\begin{cases} J_1(x) & \leq & C_1 \\ J_2(x) & \leq & C_2 \end{cases}$$

Any point in the region $[x_c, x_d]$ can satisfy the design requirements. Within this region, one can also improve the optimization performance further using either the mixed-objective method or the minimax method. If the minimax principle is used, the optimal point will be x_d. If the mixed-objective method is used, the optimal point will be close to the average point of the region $[x_c, x_d]$.

Moving-boundaries algorithm: Zakian (Zakian and Al-Naib, 1973; Zakian, 1979) suggests an algorithm for satisfying the inequalities which he calls the *moving-boundaries algorithm* (Maciejowski, 1989, pp.341–346). The procedure of this algorithm which provides the solution to the problem presented, is given below.

Let us firstly normalize the performance indices as follows:

$$\phi_i(Z, W, Q) = \begin{cases} \frac{J_i(Z,W,Q)}{\varepsilon_i} & \text{for} \quad \varepsilon_i \neq 0 \\ J_i(Z, W, Q) + 1 & \text{for} \quad \varepsilon_i = 0 \end{cases} \tag{6.37}$$

The problem is now to satisfy the following normalized inequalities:

$$\phi_i(Z, W, Q) \leq 1 \tag{6.38}$$

Let $\mathcal{S}_i$ be the set of parameters (Z, W, Q) for which the i_{th} objective is satisfied:

$$\mathcal{S}_i = \{(Z, W, Q) \quad : \quad \phi_i(Z, W, Q) \leq 1\} \tag{6.39}$$

Then the admissible or feasible set of parameters for which all objectives hold is:

$$\begin{aligned} \mathcal{S} &= \mathcal{S}_1 \bigcap \mathcal{S}_2 \bigcap \mathcal{S}_3 \bigcap \mathcal{S}_4 \\ &= \{(Z, W, Q) : \max_{i=1}^{4}\{\phi_i(Z, W, Q)\} \leq 1\} \end{aligned} \tag{6.40}$$

which shows that the search for an admissible parameter set (Z, W, Q) can be pursued via optimization, in particular by solving:

$$\min_{(Z,W,Q)} \{\max_{i=1}^{4}\{\phi_i(Z, W, Q)\} \leq 1 \tag{6.41}$$

Now, let (Z^k, W^k, Q^k) be the values of the parameters at k_{th} step, and define:

$$\mathcal{S}_i^k = \{(Z, W, Q) : \phi_i(Z, W, Q) \leq \Delta^k\} \text{ for } i = 1, \cdots, 4 \tag{6.42}$$

where

$$\Delta^k = \max_{i=1,2,3,4}\{\phi_i(Z^k, W^k, Q^k)\} \tag{6.43}$$

and also define

$$\mathcal{S}^k = \mathcal{S}_1^k \bigcap \mathcal{S}_2^k \bigcap \mathcal{S}_3^k \bigcap \mathcal{S}_4^k \tag{6.44}$$

$$\Sigma^k = \sum_{i=1}^{4} \phi_i(Z^k, W^k) \tag{6.45}$$

Hence $\mathcal{S}^k$ is the k_{th} set of parameters for which all objectives satisfy:

$$\phi_i(Z, W, Q) \leq \Delta^k \quad \text{for} \quad i = 1, \cdots, 4 \tag{6.46}$$

It is clear that $\mathcal{S}^k$ contains both (Z^k, W^k, Q^k) and the admissible set $\mathcal{S}$. Σ^k is a combined measurement of all objectives. The task now is to find a new parameter set which moves objectives towards the final feasible set. The strategy for finding the new parameter set is to minimize the largest performance index, i.e., Δ^k. If the largest performance index Δ^k cannot be improved, the improvement of the combined performance index Σ^k is considered. If one now finds new parameters $(\overline{Z}^k, \overline{W}^k, \overline{Q}^k)$, such that:

$$\overline{\Delta}^k < \Delta^k \tag{6.47}$$

or

$$\overline{\Delta}^k = \Delta^k \quad \text{and} \quad \overline{\Sigma}^k < \Sigma^k \tag{6.48}$$

where $\overline{\Delta}^k$ and $\overline{\Sigma}^k$ are defined similarly to Δ^k and Σ^k, then we accept $(\overline{Z}^k, \overline{W}^k, \overline{Q}^k)$ as the next set of parameters, i.e., $(Z^{k+1}, W^{k+1}, Q^{k+1}) = (\overline{Z}^k, \overline{W}^k, \overline{Q}^k)$. This leads to:

$$\phi_i(Z^{k+1}, W^{k+1}, Q^{k+1}) \leq \phi_i(Z^k, W^k, Q^k), \quad \text{for } i = 1, \cdots, 4 \tag{6.49}$$

and

$$\mathcal{S} \subset \mathcal{S}^{k+1} \subset \mathcal{S}^k \tag{6.50}$$

So that the boundary of the set in which the parameters are located has been moved towards the admissible set, or rarely, has remained unaltered. The process of finding the optimization solution is terminated when both Δ^k and Σ^k cannot be reduced further. But the process of finding an admissible parameter set (Z, W, Q) is terminated when $\Delta^k \leq 1$, i.e., when the boundaries of $\mathcal{S}^k$ have converged to the boundaries of $\mathcal{S}$. The process of the moving-boundaries algorithm is illustrated in Fig.6.3.

If the Δ^k persists in being larger than 1, this may be taken as an indication that the objectives may be inconsistent, whilst their magnitudes give some measure of how closely it is possible to approach the objectives. In this case, some of the inequality constraints should be relaxed until they are satisfied. From a practical viewpoint, the approximate optimal solution is also useful if the absolute optimal solution is not achievable.

The difficult part of the algorithm is the generation of a trial parameter set $(\overline{Z}^k, \overline{W}^k, \overline{Q}^k)$, given (Z^k, W^k, Q^k). Many methods have been proposed since

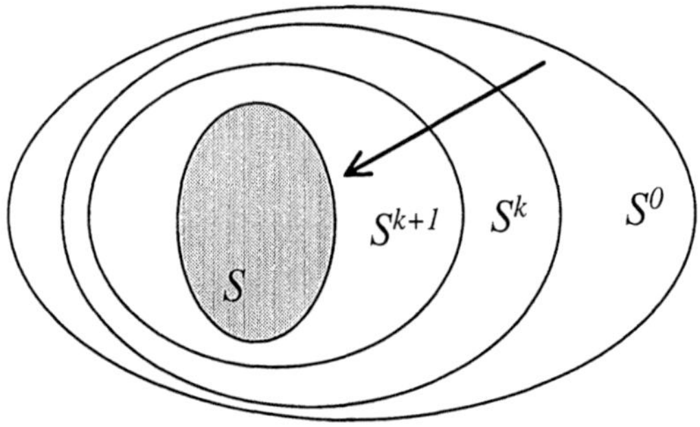

Figure 6.3. The "moving-boundaries" process

Zakian introduced the method of inequalities and a short review can be found in Maciejowski (1989, p.345). It is suggested that the relatively crude direct search methods such as the simplex method can be used to solved this problem. Patton and Liu (1994) suggested a method to generate the trial parameter set via genetic algorithms in the design of robust controllers. This method is extended by Chen et al. (1994a; 1996) to the robust FDI problem. The combination of genetic algorithms with the method of inequalities for solving the multi-objective optimization problem, defined in this chapter is discussed in Section 6.5.

6.5 Optimization via Genetic Algorithms

Most optimization techniques can be classified broadly into calculus-based techniques or direct-search methods. In recent years, the direct-search techniques, which are problem-independent, have been widely used in optimization. Unlike calculus-based methods (gradient descent, etc.), direct search algorithms do not require the use of derivatives. Consequently, it eases the analytical analysis in the calculation of derivatives and it is less likely for direct search algorithms to get "trapped" into local minima. Gradient-descent methods, on the other hand, calculate the slope of the objective surface at the current position in all directions and move in the direction with the most negative slope. This works well when the objective surface is relatively smooth, with few local minima. However, real-world data are often multimodal and contaminated by noise which can further distort the objective surface.

6.5.1 Introduction to genetic algorithms

The most important direct search algorithm in optimization is the genetic algorithm (GA) which was invented to mimic some of the processes observed in natural evolution. The technique was pioneered by Holland and his associates in the 1970's (Holland, 1975), and in the last six years has been receiving growing interest in both research and application (Goldberg, 1989; Frenzel, 1993). GAs are parameter search procedures based upon the mechanics of natural genetics. All natural species survive by adapting themselves to the environment. This natural adaption is the underlying theme of GAs. GAs search combines a Darwinian survival of the fittest strategy to eliminate unfit characteristics and uses random information exchange, with exploitation of knowledge contained in old solutions, to effect a search mechanism with surprising power and speed.

Genetic algorithms are different from other optimization techniques in many ways, notably they are:

- GAs constitute a *parallel* search of the solution space, as opposed to a point-by-point search in gradient-descent methods. By using a population of trial solutions, the genetic algorithm can effectively explore many regions of the search space simultaneously. Optimization methods more usually provide iterative progress (global or local) solution, based on a single region in the parameter space; the region may only include a local minimum and another region must then be used to locate a global minimum. This is one of the reasons why GAs are less sensitive to local minima.

- GAs manipulate *representations (or codings)* of the parameter set, rather than the parameters themselves.

- GAs do not require derivative information or other auxiliary knowledge concerning problems to be solved. The only problem-specific requirement is the ability to evaluate the trial solutions on objective function, and the relative fitness levels influence the directions of search.

- GAs use probabilistic rather than deterministic transition rules.

A genetic algorithm is an exploratory procedure that is able to locate close-to-global optimal solutions to complex problems. It maintains a set of trial solutions (often called individuals), and forces them to "evolve" towards an acceptable solution. The procedure starts with an initial random population and employing survival-of-the-fittest and exploiting old knowledge in the gene pool. Each generation's ability to solve the problem should be improved. The computational structure of a genetic algorithm is shown in Fig.6.4.

The main stages involved in GAs discussed in Frenzel (1993) and Davis (1991) are shown in below:

Representation (or coding): The parameter set is represented by a coding scheme which can be recognized by computers. These representations are normally referred to as *chromosomes.* The most common coding scheme is the use of binary strings, where selections of the string represent encoded parameters. The number of digits assigned to strings will determine the numerical accuracy.

Evaluation: To evaluate the objective fitness of the current chromosomes in each generation. Each chromosome in the population is decoded and evaluated on how well it solves the problem. The fitness measure is used in the next step to determine how many offspring will be generated from any particular chromosome.

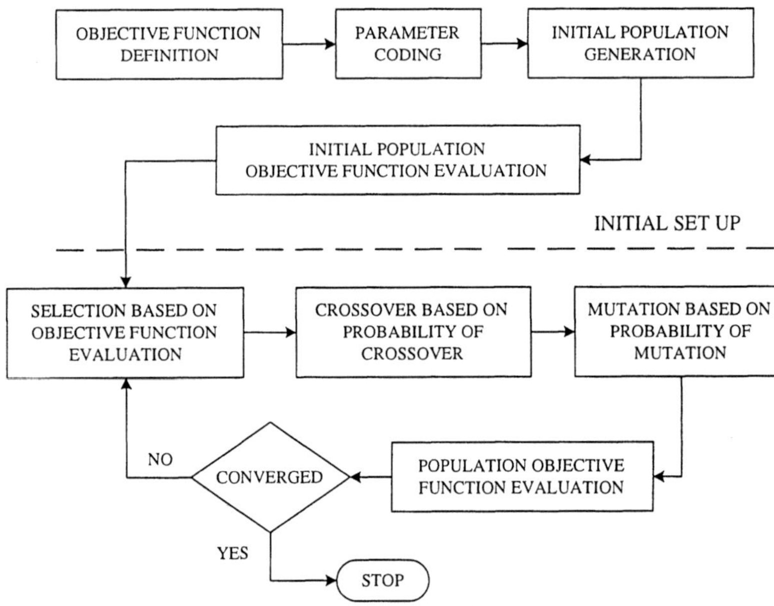

Figure 6.4. Computational structure of genetic algorithms

Reproduction: In this stage, a new population is created based on the evaluation of the current one. For every chromosome in the current population, a number of exact copies are generated with the best chromosomes producing the most copies. This is the step that allows GAs to take advantage of a survival-of-the-fittest strategy. There are several ways to calculate the number of offspring that each chromosome will be allocated. The two most popular methods are referred to as *ratioing* and *ranking*.

In ratioing, each individual reproduces in proportion to its fitness. So, an individual whose fitness is ten times better than another will produce ten times the number of offspring. This way as superior chromosomes emerge they can guide the population quickly. The disadvantage is that if a superior individual surfaces early and dominates the population, then the population may converge prematurely on a possible suboptimal (or local minimal) solution.

For the ranking method, the number of offspring each chromosome will generate is determined by how it ranks in the population. For example, the top 20% of the population might generate two offspring each, the bottom 20% of the population generate no offspring, and the rest generate just one offspring apiece. Using this method, no one chromosome can overpower the population in a single generation. Also, no matter how close the actual fitness values are, there is always constant pressure to improve. The primary disadvantage of ranking is speed because better chromosomes are not capable of guiding the population easily. This forces good solutions to develop more slowly.

Recombination: The reproduction creates a population whose member are currently the best solution for the problem, however many of the chromosomes are identical and no-one is different from the previous generation. The reproduction simply produces multiple copies of existing chromosomes. Recombination combines chromosomes from the population and produces new chromosomes that, while they did not exist in the previous generation, maintain many of features of the previous generation. In natural evolution, recombination and reproduction occur in the same step. However, in GAs they are often separated to facilitate experimentation with different methods. The most important method for recombination is *crossover* in which two individuals are randomly selected from the population and, governed by a specified crossover probability or rate, subsection of the two chromosomes are swapped about a randomly chosen crossover point. During recombination, GAs exploit knowledge of the gene pool by allowing good chromosomes to combine with chromosomes that are not as good. This is based on the assumption that each individual, no matter how good it appears, does not contain the complete answer to the problem. The answer is contained in the population as a whole, and the best solution can only be found by combining chromosomes.

Mutation: This step in creating a new generation is motivated by the possibility that the initial population did not contain all of the information necessary to solve the problem. Moreover, it is possible that the individuals that produce no offspring may have had some information that is essential to the solution. The injection of new information into the population is called mutation. One of methods to implement mutation is to change randomly a fixed number of bits every generation based upon a specified mutation probability.

Elitism: It is possible that the best member of the population may fail to produce offspring in the next generation. The *elitist* strategy fixes this potential source of loss by copying the best member of each generation into the succeeding generation. The elitist strategy may increase the speed of domination of a population by a super individual, but on balance it appears to improve genetic algorithm performance. More specifically, the elitist can improve the speed of convergence, but it could give a local minimum due to the domination of a super individual. The use of the elitist strategy depends on problems, if the performance index has many local minimums, it is not good idea to use it.

GENETIC ALGORITHM PARAMETERS: The best values for mutation rate, crossover percentage, and other parameters are problem specific. It is even possible to find the best values using genetic algorithms! However, certain generalizations can be made (Frenzel, 1993). If the population is too small, relative to the size of the search space, it will be difficult to effectively search the entire region. Furthermore, large mutation rates tend to disrupt the steady improvement resulting from crossover and reproduction. Researchers have found that a population of 30 individuals, a crossover probability of 60%, and a mutation probability of 3% seems to be a good starting point (Frenzel, 1993).

6.5.2 Procedure of genetic algorithms in satisfying performance inequalities

In the implementation of genetic algorithms, it is not necessary to include all main stages given above. There are many variations in the implementation. Some stages may need to be modified to best suit particular problems. The genetic algorithm is used here to search the optimal solutions in the moving-boundaries process of satisfying performance inequalities. The procedure of the optimal search via GA is first suggested by Liu and Patton (1996) and later modified to suit robust FDI design by Chen et al. (1994a; 1996). This optimization procedure includes the following steps:

Step 1: Chromosomal representation. Each solution in the population is represented as a real number string rather than as a binary string. For $W \in \mathbb{R}^{m\times n}$, $Z \in \mathbb{R}^{1\times n}$ and $Q \in \mathbb{R}^{p\times m}$, the chromosomal representation may be expressed as an array:

$$P = [Z, w_1^T, \cdots, w_n^T, q_1^T, \cdots, q_m^T]$$

This kind of chromosomal representation has *two* advantages. One is that it guarantees that the domain expertise embodied in the representation will be preserved. The other is that the algorithm to be developed will feel natural to the designer.

Step 2: Generation of the initial population. N (an odd number) sets of parameter string P for the initial population are randomly generated.

Step 3: Evaluation of the performance functions. Evaluate the performance function $\phi_i(P_j)$ $(i = 1, 2, 3, 4)$ for all N sets of the parameter P_j and determine:

$$\begin{aligned} \Delta_j &= \max\{\phi_1(P_j), \phi_2(P_j), \phi_3(P_j), \phi_4(P_j)\} \\ \Sigma_j &= \phi_1(P_j) + \phi_2(P_j) + \phi_3(P_j) + \phi_4(P_j) \end{aligned}$$

for $j = 1, 2, \cdots, N$.

Step 4: Selection. According to the fitness of the performance functions for each set of parameters, cull the $(N-1)/2$ weaker members of the population and reorder the sets of the parameters. The fitness of the performance functions is measured by:

$$F_j = \frac{1}{\Delta_j}, \quad \text{for} \quad j = 1, 2, \cdots, N$$

Step 5: Cross-over. Perform the Cross-over using an average cross-over function to produce the $(N-1)/2$ offsprings. The average cross-over operator takes two parents which are selected in step 4 and produces one child that is the result of averaging the corresponding fields of the two parents. In other words, the average Cross-over function is defined as:

$$P_{C_j} = \frac{P_{j+1} + P_j}{2}, \quad \text{for} \quad j = 1, 2, \cdots, \frac{N-1}{2}$$

Step 6: Mutation. A real number mutation operator, called *real number creep*, is used. The function we are optimizing is a continuous one with hills and valleys. If we are on a good hill, we want to jump around on it, to move nearer to the top. Real number creep can have that effect. What it does is to sweep along the chromosome, creeping any value up or down a small random amount. The maximum amount that this operator can alter the value of a field is a parameter of the operator. Hence it is the probability of altering any field. The mutation operation is defined as:

$$P_{M_j} = P_{C_j} + d_m \xi_j, \quad \text{for} \quad j = 1, 2, \cdots, \frac{N-1}{2}$$

where d_m is the maximum value to be altered and $\xi_i \in [-1, 1]$ is a random variable with zero mean.

Step 7: Elitism. The elitist strategy copies the best parameter set into the succeeding parameter sets. It prevents the best parameter set from loss in the succeeding parameter sets. It may increase the speed of domination of a population by a super individual, but on balance it appears to improve genetic algorithm performance. The best parameter set P_b is defined as one satisfying:

$$\Sigma_b = \min_l \{\Sigma_l : \Sigma_l \le \Sigma_m - \alpha(\Delta_l - \Delta_m), \text{and} \Delta_l \le \Delta_m + \delta\}$$

where

$$\Delta_m = \min\{\Delta_1, \Delta_2, \Delta_3, \Delta_4\}$$

Σ_m and Σ_l are corresponding to Δ_m and Δ_l, $\alpha > 1$ and δ is a positive number, which are given by the designer, for example $\alpha = 1.1$ and $\delta = 0.1$.

Step 8: New offsprings. Add the $(N-1)/2$ new offsprings to the population which are generated in a random fashion. Actually, the new offsprings are formed by mutating the best parameter set P_b with a probability, i.e.

$$P_{N_j} = P_b + d_n \xi_j, \quad \text{for} \quad j = 1, 2, \cdots, \frac{N-1}{2}$$

where d_n is the maximal value to be altered and $\xi_j \in [-1, 1]$ is a random with zero mean. Thus, the next population is formed by the parameter set P_{M_j} $(j = 1, 2, \cdots, (N-1)/2)$, P_{N_j} $(j = 1, 2, \cdots, (N-1)/2)$ and P_b.

Step 9: Termination checking. Continue the cycle initiated in Step 4 until convergence is achieved. The population is considered to have converged when

$$\Delta_j - \Delta_b \le \varepsilon, \quad \text{for} \quad j = 1, 2, \cdots, N$$

where ε is a positive number.

Take the best solution in the converged generation and place it in a second "initial generation". Generate the other $N-1$ parameter sets in this second initial generation at random and begin the cycle again until a satisfactory solution is obtained or Δ_b and Σ_b cannot be reduced any further.

The design software for implementing the algorithm mentioned above is not included in this book. However with some modifications, the design algorithm for combining the method of inequality and genetic algorithms included in the Eigenstructure Assignment Toolbox (Liu and Patton, 1998) can be used for robust FDI.

6.6 Detection of Incipient Sensor Faults in Flight Control Systems

As modern aircraft and onboard equipment become more and more complex, the probability of potential faults increases. One of the biggest challenges in the design of flight control systems is a requirement for the flight of the aircraft to recover safely from structural damage and/or system faults. Regardless of whether the aircraft is equipped with a special control reconfiguration capability, reliable fault diagnostic information is extremely important to the pilot. Prompt presentation of fault information to the pilot could enable him to take accommodating action to the malfunction, using system redundancy. Sensors are the most important components for flight control and aircraft safety due to its roles in flight control and navigation. Any sensor fault must be detected as early as possible to prevent serious accident. The problem of detecting and isolating faults in flight control systems has been studied for many years (Deckert, Desai, Deyst and Willsky, 1977; Deckert, Desai, Deyst and Willsky, 1978; Bundick, 1985; Weiss and Hsu, 1985; Bundick, 1991), and model-based approaches have been demonstrated to be capable of detecting and isolating faults very quickly and reliably. The main challenge is the detection and isolation of incipient faults in the presence of modeling uncertainty and noise. To diagnose incipient faults, a FDI systems have to be made robust against modeling uncertainty and noise. The technique presented in this chapter is used to design robust residuals to diagnose incipient sensor faults in a flight control system.

The flight control system example considered here is the lateral control system of a remotely-piloted aircraft (Mudge and Patton, 1988). The linearized lateral dynamics are given by the state space model matrices:

$$A = \begin{bmatrix} -0.277 & 0 & -32.9 & 9.81 & 0 \\ -0.1033 & -8.525 & 3.75 & 0 & 0 \\ 0.3649 & 0 & -0.639 & 0 & 0 \\ 0 & 1 & 0 & 0 & 0 \\ 0 & 0 & 1 & 0 & 0 \end{bmatrix}, \quad B = \begin{bmatrix} -5.432 & 0 \\ 0 & -28.64 \\ -9.49 & 0 \\ 0 & 0 \\ 0 & 0 \end{bmatrix}$$

$$C = \begin{bmatrix} 0 & 1 & 0 & 0 & 0 \\ 0 & 0 & 0 & 1 & 0 \\ 0 & 0 & 0 & 0 & 1 \end{bmatrix} \qquad D = 0_{3\times 2}$$

where the state vector and control input are:

$$\begin{bmatrix} v \\ p \\ r \\ \phi \\ \psi \end{bmatrix} = \begin{bmatrix} \text{sideslip} \\ \text{roll rate} \\ \text{yaw rate} \\ \text{bank angle} \\ \text{yaw angle} \end{bmatrix} \qquad \begin{bmatrix} \tau \\ \xi \end{bmatrix} = \begin{bmatrix} \text{rudder} \\ \text{aileron} \end{bmatrix}$$

The system is unstable and needs to be stabilized. Since the purpose of the example is to illustrate the fault detection capability, the system is simply stabilize using a state feedback controller provided by Liu and Patton (1996). The FDI system in the flight control system is illustrated in Fig.6.5 in which the input signal to actuators and the output from sensors are available for fault detection and isolation. Note that the control reconfiguration issues are not considered in this chapter, although they are very important.

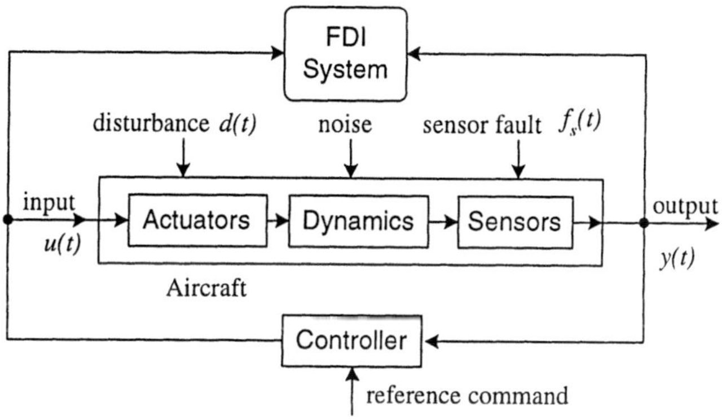

Figure 6.5. FDI in flight control systems

An observer is designed to generate residual signal for FDI. To make the residual have the required response, the observer eigenvalues are constrained within the following regions:

$$-5 \leq \lambda_1 \leq -0.2 \qquad -15 \leq \lambda_2 \leq -3$$
$$-10 \leq \lambda_{3,re} \leq -2 \qquad 0.2 \leq \lambda_{3,im} \leq 4$$
$$-30 \leq \lambda_5 \leq -8$$

Note that the eigenvalue λ_4 is the conjugate of the eigenvalue λ_3, i.e., $\lambda_4 = \lambda_3^*$. The weighting penalty factors for the performance functions J_1 and J_3 are chosen as:

$$W_1(j\omega) = \frac{500}{(j\omega + 10)(j\omega + 50)}, \qquad W_3(j\omega) = \frac{1}{W_1(j\omega)}$$

which places emphasis on J_1 at low frequencies and J_3 at high frequencies. By minimizing J_1 and J_3, the fault effect can be maximized and the noise effect can be minimized. To simplify the optimization procedure, the residual weighting matrix is set as $Q = I_3$. Table 6.1 lists the performance indices for different observer gains. In this table, K_i^* $(i = 1, 2, 3, 4)$ is the observer gain matrix which minimizes J_i $(i = 1, 2, 3, 4)$. It can be seen that a design which minimizes a particular performance function makes all other performance functions unacceptably large. Hence, multi-objective optimization must be used to reach a reasonable compromise. In order to use the method of inequalities to solve this problem, a set of performance index bounds ε_i $(i = 1, 2, 3, 4)$ are chosen as shown in the Table 6.1.

Table 6.1. Performance indices for different designs

	J_1	J_2	J_3	J_4
K_1^*	189.58	2.5949	24.8288	0.00935
K_2^*	3865.26	0.07576	23.415	0.00798
K_3^*	3274.55	0.11232	22.40	0.00798
K_4^*	3.9×10^6	10700	34600	2×10^{-7}
Bounds	2000	0.16	22.5	0.006
$K_{optimal}$	1950.7	0.1492	22.420	0.00512
K_{place}	2800.39	0.1784	22.668	0.00965

The genetic algorithm is used to search for solutions which satisfy all performance index boundaries. The optimal observer gain matrix found is:

$$K_{optimal} = \begin{bmatrix} -189.1419 & 0.8083 & 18.8392 \\ 17.9317 & -0.7936 & -0.7943 \\ 15.4684 & 2.8543 & 7.6140 \\ -0.7606 & 6.9329 & 0.1537 \\ -1.2303 & 0.2329 & 9.8678 \end{bmatrix}$$

with corresponding eigenvalues:

$$\{-1.5371, -4.7045, -3.4973 \pm 2.1194i, -19.9994\}$$

The performance indices under this gain are shown in Table 6.1. This design is an acceptable compromise. To demonstrate the effectiveness of the developed method, an observer gain matrix K_{place} using the MATLAB routine *place*, to assign eigenvalues at: $\{-0.5, -14, -4.8 \pm 1.6i, -20\}$ is also designed. The performance indices for this design are also shown in the Table 6.1.

The simulation is used to assess the performance of the observer-based residual generator in the detection of incipient sensor faults. The control commands for both inputs are set as a unit sinusoid function. The sensor noise comprises a random summation of multi-frequency signals with all frequencies larger than $20rad/s$. In the simulation, all aerodynamic coefficients have been perturbed by $\pm 10\%$. The fault is a slowly developing signal whose shape is shown in Fig.6.6.

The simulated fault is added to the roll rate sensor. To illustrate the small nature of the incipient fault, Fig.6.7 shows the plot of both faulty and normal measurements of the roll rate p. It can be seen that the fault is hardly noticeable in the measurement and cannot be detected easily, without the assistance of the residuals.

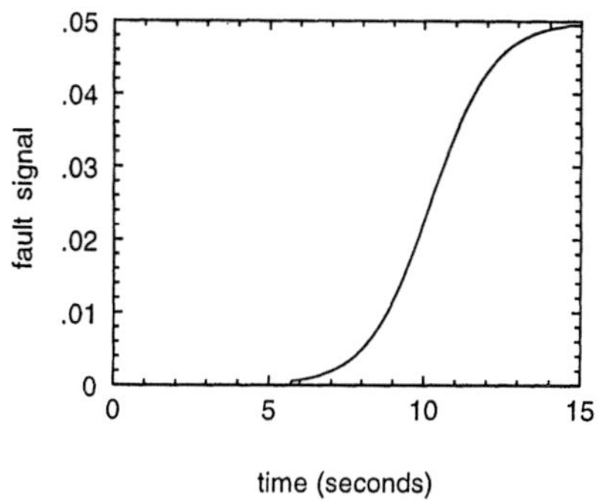

Figure 6.6. The fault signal shape

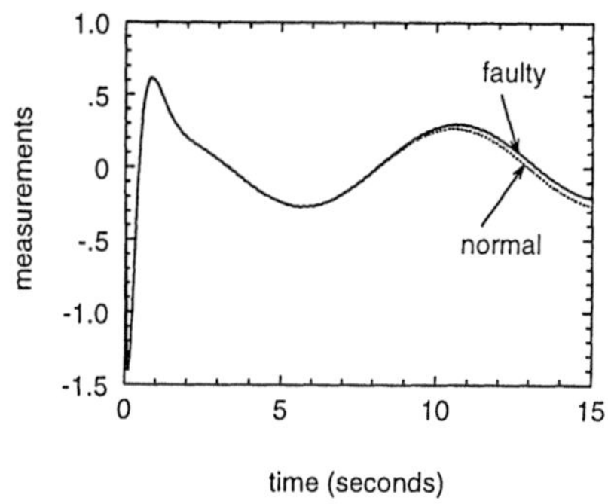

Figure 6.7. The faulty and normal measurements for roll rate p

Fig.6.8 shows the residual response for the case when a fault occurs in the roll rate sensor. The residual responses for other faulty cases are similar to the response shown in Fig.6.8. The residual response demonstrates that the residual changes very significantly after a fault occurs in one of the sensors. Hence, the residual can be used to detect incipient sensor faults reliably even in the presence of modeling errors and noise. To reduce the effect of noise further, the residual signal has been filtered by a low-pass filter.

Note that this example only considers the robust residual generation for fault detection, as it is believed that the design of an optimal residual is the most important task to be considered. Fault isolation can be achieved by designing structured residual sets. For the system considered in this chapter, one can design four different observer-based residual generators to generate four residual vectors. The four observers are driven by different subsets of measurements, namely, $\{p, \phi, \psi\}$, $\{r, \phi, \psi\}$, $\{p, r, \psi\}$ and $\{p, \phi, r\}$ (Patton and Kangethe, 1989). This chapter has only considered the design of one of these observers, although the principle is valid for the design of the other observers.

6.7 Conclusions

This chapter has described a systematic approach to the design of optimal residuals which satisfy a set of objectives. These objectives are essential for

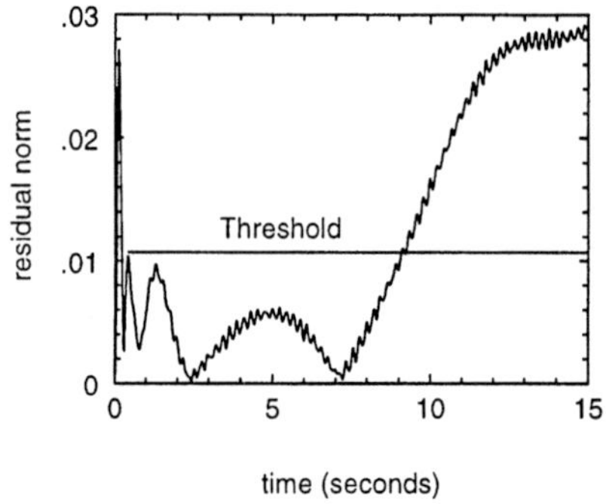

Figure 6.8. The residual norm when a fault occurs in the roll rate sensor

achieving robust diagnosis of incipient faults. Some performance indices are expressed in the frequency domain which can take account of the frequency distribution of different factors that affect the residuals. It is the first time such a consideration has been addressed and solved in a fault diagnosis design. It has been proved that the frequency-dependent weighting factors incorporated into performance indices play an important role in the optimal design. They are problem-dependent and must be chosen very carefully. The multi-objective optimization problem has been reformulated into one of satisfying a set of inequalities on the performance indices. The genetic algorithm has been used to search the optimal solution to satisfy these inequalities on the performance indices. The method has been applied to the design of an observer-based residual generator for detecting incipient sensor faults in a flight control system and the simulation results show the effectiveness of the method. Considering the extreme difficulty in enhancing the fault diagnosis performance under modeling uncertainty and noise, any improvement in the robustness of residual design is very useful. The scope of application of this work extends to all systems with possible incipient faults.

7 ROBUST RESIDUAL GENERATION USING OPTIMAL PARITY RELATIONS

7.1 Introduction

In Chapters 3–6, robust observer-based residual generators have been discussed. This chapter focuses on the problem of robust residual generation via optimal parity relations. The parity relation is one of the most commonly accepted approaches for generating residuals. To achieve robustness for this approach, Chow and Willsky (1984) reformulated the design of parity relations for robust residual generation as a minmax optimization problem. The optimal objective they defined specifies robustness with respect to a particular operating point, thereby allowing the possibility of adaptively choosing the best parity relations. However, the main drawback of their method is that it leads to an extremely complex optimization problem for which there is no analytical solution. Lou et al. (1986) proposed an alternative method to find "optimally robust parity relations" for generating robust residuals. They used multiple models to describe the modeling uncertainty due to parameter variations so that the residual becomes minimally sensitive to system parameters variation. The introduction of the multiple model description in parity relation design and the provision of an analytical strategy for solving the optimization problem are the main contributions of Lou et al. (1986). However, the optimal objective they proposed seems inappropriate, because they only considered the minimization of effects of parameter variations. A residual designed to be insensitive to modeling uncertainty may also be insensitive to faults. An appropriate performance index for robust residual design should take account of both effects of modeling uncertainty and faults. Following this philosophy, Wünnenberg and Frank (Wünnenberg and Frank, 1988; Frank, 1990; Wünnenberg and Frank, 1990; Wünnenberg, 1990) studied the design of optimal parity relations by adopting a modified performance index which is the ratio of the modeling uncertainty response effect to that of the fault effect. However, the modeling uncertainty description they used was the unknown input (or disturbance) description which, as discussed Chapters 2–5, cannot be used to represent a wide range of uncertain situations without any modification and approximation. This disappointing feature was due to the lack of application study even in a simple academic exercise or simulation setting.

Following the study by Chen et al. (1993) and Chen (1995), this chapter re-examines the design of optimal parity relations for robust residual generation by considering the modeling uncertainty due to both parameter variations and disturbances. To generate robust residuals, two objective functions for the design of parity relations are defined. The optimization objectives are the minimization of effects due to the modeling uncertainty and the maximization of fault effects. Together these lead to a multi-objective optimization problem which is solved by forming a "mixed" objective function optimization problem. This objective function (performance index) represents the trade-off between two design objectives, its solution is obtained using the matrix theory of generalized eigenstructure and singular value decomposition. The method used in this chapter utilizes advantages offered by studies of Lou et al. (1986) and Wünnenberg and Frank (Wünnenberg and Frank, 1988; Frank, 1990; Wünnenberg, 1990). An example is used to illustrate the method proposed, and the results show that the method is very effective for robust residual generation.

7.2 Performance Indices for Optimal Parity Relation Design

The basic principle of the parity relation approach for residual generation has been presented in Section 2.8.2. Here, the parity relation for dynamic systems with modeling uncertainty is examined. Consider the discrete-time system model with the following description:

$$\left\{ \begin{array}{rcl} x(k+1) & = & A_t x(k) + B_t u(k) + E_t^1 d(k) + R_t^1 f(k) \\ y(k) & = & C_t x(k) + D_t u(k) + E_t^2 d(k) + R_t^2 f(k) \end{array} \right. \tag{7.1}$$

where $u(k) \in \mathbb{R}^r$ is the input vector, $y(k) \in \mathbb{R}^m$ is the output vector and $x(k) \in \mathbb{R}^n$ is the state vector, $f(k) \in \mathbb{R}^g$ denotes a fault vector which may contain actuator, component or sensor faults, $d(k) \in \mathbb{R}^q$ is the unknown input (or disturbance) vector. $\{A_t, B_t, C_t, D_t, E_t^1, E_t^2, R_t^1, R_t^2\}$ are system model matrices with appropriate dimensions. These matrices are not known precisely due to the modeling uncertainty and the subscript "t" denotes variation. These matrices have nominal values as: $\{A, B, C, D, E^1, E^2, R^1, R^2\}$, although their exact values are unknown.

As pointed out in Section 2.8.2, the redundancy relations can be constructed by collecting a batch of data with window length s as follows:

$$\underbrace{\begin{bmatrix} y(k-s) \\ y(k-s+1) \\ \vdots \\ y(k) \end{bmatrix}}_{Y(k)} - H_t \underbrace{\begin{bmatrix} u(k-s) \\ u(k-s+1) \\ \vdots \\ u(k) \end{bmatrix}}_{U(k)} = W_t x(k-s)$$

$$+ L_t \underbrace{\begin{bmatrix} d(k-s) \\ d(k-s+1) \\ \vdots \\ d(k) \end{bmatrix}}_{D(k)} + M_t \underbrace{\begin{bmatrix} f(k-s) \\ f(k-s+1) \\ \vdots \\ f(k) \end{bmatrix}}_{F(k)} \tag{7.2}$$

where

$$H_t = \begin{bmatrix} D_t & 0 & \ldots & 0 \\ C_t B_t & D_t & \ldots & 0 \\ \vdots & \vdots & \ddots & \vdots \\ C_t A_t^{s-1} B_t & C_t A_t^{s-2} B_t & \ldots & D_t \end{bmatrix} \in \mathbb{R}^{(s+1)m \times (s+1)r} \tag{7.3}$$

and

$$W_t = \begin{bmatrix} C_t \\ C_t A_t \\ \vdots \\ C_t A_t^s \end{bmatrix} \in \mathbb{R}^{(s+1)m \times n} \tag{7.4}$$

and the matrix M_t is constructed by replacing $\{D_t, B_t\}$ with $\{R_t^2, R_t^1\}$ in Eq.(7.3), similarly the matrix L_t is constructed by replacing $\{D_t, B_t\}$ with $\{E_t^2, E_t^1\}$ in Eq.(7.3).

To simplify the notation, Eq.(7.2) can be rewritten as:

$$Y(k) = H_t U(k) + W_t x(k-s) + L_t D(k) + M_t F(k) \tag{7.5}$$

According to Chow and Willsky (1984) and Lou et al. (1986) and from Section 2.8.2, a residual signal is defined as:

$$r(k) = v^T [Y(k) - HU(k)] \tag{7.6}$$

where $v \in \mathbb{R}^{(s+1)m}$ is the residual generating vector which can also be a matrix (see Section 2.8.2). The matrix H is the nominal value of H_t and can be constructed by replacing $\{A_t, B_t, C_t, D_t\}$ with their nominal values in Eq.(7.3).

Eq.(7.6) is the *computational form* of a residual generator which shows the residual signal as a function of measured inputs and outputs of the monitored system. Substituting Eq.(7.5) into Eq.(7.6), we have:

$$\begin{aligned} r(k) &= v^T[W_t x(k-s) + (H_t - H)U(k) + L_t D(k) + M_t F(k)] \\ &= v^T[W_t \quad (H_t - H) \quad L_t] \begin{bmatrix} x(k-s) \\ U(k) \\ D(k) \end{bmatrix} + v^T M_t F(k) \\ &= v^T Z_t X(k) + v^T M_t F(k) \end{aligned} \tag{7.7}$$

where

$$Z_t = [W_t \quad (H_t - H) \quad L_t] \;\in\; \mathbb{R}^{(s+1)m \times (n+(s+1)r+(s+1)q)}$$

$$M_t \;\in\; \mathbb{R}^{(s+1)m \times (s+1)g}$$

In order to detect faults, we should make the residual signal $r(k)$ become zero for the fault-free case and non-zero for the faulty case; this requires that:

$$v^T Z_t = 0 \tag{7.8}$$

$$v^T M_t \neq 0 \tag{7.9}$$

Normally, Z_t and M_t are unknown and time-varying, so that Eq.(7.8) cannot hold true for a wide range of modeling uncertainty. Here the uncertainty is considered as bounded, i.e. the parameter variations are contained within a pre-defined bound, e.g.

$$\begin{aligned} A - \Delta A \le A_t \le A + \Delta A \quad &; \quad B - \Delta B \le B_t \le B + \Delta B \\ C - \Delta C \le C_t \le C + \Delta C \quad &; \quad D - \Delta D \le D_t \le D + \Delta D \\ E^1 - \Delta E^1 \le E_t^1 \le E^1 + \Delta E^1 \quad &; \quad E^2 - \Delta E^2 \le E_t^2 \le E^2 + \Delta E^2 \\ R^1 - \Delta R^1 \le R_t^1 \le R^1 + \Delta R^1 \quad &; \quad R^2 - \Delta R^2 \le R_t^2 \le R^2 + \Delta R^2 \end{aligned}$$

where $A_1 \le A_2$ means that all elements of the matrix A_1 is not larger than the corresponding element in the matrix A_2. The real system matrix $A_t, \cdots$ can be any values within the pre-defined bounds. This statement is absolutely correct, however it does not provide any aids for design. To achieve a realistic design, let us consider $\{A_t, B_t, C_t, D_t, E_t^1, E_t^2, R_t^1, R_t^2\}$ in a finite set of possibilities, say $\{A_i, B_i, C_i, D_i, E_i^1, E_i^2, R_i^1, R_i^2\}$ $(i = 1, 2, \cdots, N)$ within their bounds. In practice, this might involve choosing representative points out of the actual continuous range of parameter values, reflecting any desired weighting on the likelihood or importance of particular sets of parameters. This finite selection corresponds to a multiple model system representation. In this situation, a set of corresponding matrices Z_i and M_i $(i = 1, \cdots, N)$ are obtained, and an ideal residual generation vector v should satisfy the following equations:

$$v^T Z_i = 0 \;;\quad i = 1, 2, \cdots, N \tag{7.10}$$

$$v^T M_i \neq 0 \;;\quad i = 1, 2, \cdots, N \tag{7.11}$$

The above equations can be rewritten as:

$$v^T Z = 0 \tag{7.12}$$
$$v^T M \neq 0 \tag{7.13}$$

where:

$$\begin{aligned} Z &= [Z_1, \ Z_2, \ \cdots, \ Z_N] \in \mathbb{R}^{(s+1)m \times N(n+(s+1)r+(s+1)q)} \\ M &= [M_1, \ M_2, \ \cdots, \ M_N] \in R^{(s+1)m \times N(s+1)g} \end{aligned}$$

The condition for a solution of Eq.(7.12) to be exist is that:

$$rank(Z) \leq (s+1)m - 1 \tag{7.14}$$

When this condition is satisfied, a solution v^* for Eq.(7.12) exists. If this solution also satisfies Eq.(7.13), it can be used to form an optimal parity relation for generating robust residuals. However, the above condition cannot normally be satisfied for cases when the parameter variations are very significant. For such cases, it is necessary to find a rank deficient matrix Z^* which is a close approximation to the matrix Z, i.e.

$$\min \|Z - Z^*\|_F \quad \text{subject to} \quad rank(Z^*) \leq (s+1)m - 1 \tag{7.15}$$

It can be proved that the above optimization problem is equivalent to the following (Lou et al., 1986):

$$\min J_1 = \min \left\{ \sum_{i-1}^{N} \| v^T Z_i \|^2 \right\} = \min \left\{ v^T Z Z^T v \right\} \quad \text{s.t.} \ v^T v = 1 \tag{7.16}$$

A solution to this problem can only minimize the sensitivity to modeling uncertainty, it cannot guarantee the maximal sensitivity to faults. Hence, to achieve an optimally robust design, it is necessary to introduce another design objective as follows:

$$\max J_2 = \max \left\{ \sum_{i=1}^{N} \| v^T M_i \|^2 \right\} = \max \left\{ v^T M M^T v \right\} \quad \text{s.t.} \ v^T v = 1 \tag{7.17}$$

A mutually optimal solution v^* for the above two optimization problems can used to generate robust residuals which are insensitive to modeling uncertainty. This is because we have already taken the modeling uncertainty (in the form of multiple models) into account in the problem formulation.

7.3 Optimally Robust Parity Relations Design via Multi-Objective Optimization

Section 7.2 shows that the robust residual design is achievable by solving two optimization problems. This is a multi-objective optimization problem and the simultaneous optimal solution may not exist. As discussed in Section 6.4, the multi-objective optimization can be solved by the method of inequalities, combined with a proper numerical search algorithm. However, this chapter considers analytical solutions for multi-objective optimization.

7.3.1 Solving optimization problems via SVD

The optimization problem defined in the last section is similar to the optimization problem studied in Section 5.2.5 and can also be solved via Singular Value Decomposition (SVD). Let the SVD of Z be:

$$Z = \Phi[diag\{\sigma_1, \sigma_2, \cdots, \sigma_z\}, \quad 0]\Theta^T$$

Φ and Θ are orthogonal matrices, $\sigma_1 \leq \sigma_2 \leq \cdots \leq \sigma_z$ are singular values of Z. As shown in Lou et al. (1986), the vector v which minimizes J_1 lies in a subspace spanned by the matrix:

$$P = [\phi_1, \cdots, \phi_{k_1}] \tag{7.18}$$

where $\phi_1, \cdots, \phi_{k_1}$ are first k_1 columns of Φ and k_1 $\{= (s+1)m - rank(Z^*)\}$ is a pre-defined constant which is the possible independent solution number of the vector v of the optimization problem $\min J_1$. In this situation, the minimum of J_1 is:

$$J_1^* = \sum_{i=1}^{k_1} \sigma_i$$

Similarly, the vector v which maximizes J_2 lies in a subspace spanned by the matrix:

$$Q = [\bar{\phi}_1, \cdots, \bar{\phi}_{k_2}] \tag{7.19}$$

where $\bar{\phi}_1, \cdots, \bar{\phi}_{k_1}$ are last k_2 columns of the orthogonal matrix $\bar{\Phi}$ and $M = \bar{\Phi}\bar{\Sigma}\bar{\Theta}^T$.

The optimal solution v^* for minimizing J_1 and maximizing J_2 is:

$$v^* \in span\{P\} \cap span\{Q\} \tag{7.20}$$

Note that this solution is relevant to constants k_1 and k_2, and different optimal solutions can be obtained by changing these two constants.

Remark: In the study by Lou et al. (1986), only the multiple model uncertainty was considered in optimal parity relations design. The design problem is to find the vector v which satisfies the following equations:

$$v^T H = 0; \quad \text{and} \quad v^T H_i = 0, \quad \text{for} \quad i = 1, \cdots, N \tag{7.21}$$

which equivalent to

$$v^T \Xi = v^T [H \; H_1 \; \cdots \; H_N] = 0 \tag{7.22}$$

If $rank(\Xi) = n$, there is no solution for the above equation. To find an optimally robust parity relation, a low rank matrix Ξ^* needs to be found to approximate Ξ^*. Alternatively, the optimal vector v should minimizes the performance index

$$J_H = v^T \Xi \Xi^T v \tag{7.23}$$

subject to $v^T v = 1$. This problem is similar to the problem of minimizing the performance index J_1 which can be solved by SVD.

7.3.2 *Solutions for multi-objective optimization*

The simultaneous optimal solution for the multi-objective optimization does not exist, if there is no intersection between the solution spaces P and Q, i.e. $span\{P\} \cap span\{Q\} = \{0\}$. For most problems, this would be the case. Hence, a compromise should be made, i.e. one needs to find a solution which gives an acceptable design although it does not optimize both performance indices. Three methods are presented here to produce a compromised optimal solution.

7.3.2.1 Multi-objective optimization via optimal projection. As a vector v which minimizes J_1 lies in $span\{P\}$, an acceptable vector v should be near to the subspace $span\{P\}$. Hence, the distance between the vector v to this subspace can be used as a measure to evaluate the satisfactory degree of the vector v to the performance index J_1. A mixed performance index J which accounts for both J_1 and J_2 is defined as:

$$\begin{aligned} J &= \alpha \parallel v - v_P^* \parallel^2 + \beta \parallel v - v_Q^* \parallel^2 = \alpha \parallel v - P_1 v \parallel^2 + \beta \parallel v - Q_1 v \parallel^2 \\ &\quad \text{s.t.} \quad \alpha + \beta = 1 \quad \text{and} \quad v^T v = 1 \end{aligned} \tag{7.24}$$

where v_P^* and v_Q^* are projections of the vector v onto subspaces $span\{P\}$ and $span\{Q\}$ respectively, and

$$P_1 = P(P^T P)^{-1} P^T \quad ; \quad Q_1 = Q(Q^T Q)^{-1} Q^T$$

The weighting factors α and β can be adjusted to satisfy different design goals. For example, if a low missed-detection rate is required one can increase β, on the other hand if a low false detection rate is required, one can increase α. This mixed performance index formulation can be extended to include more terms (sub-indices), e.g., the residual response to noise etc. The robust residual design can be achieved by solving the following optimization problem:

$$\begin{aligned} \min J &= \min \left\{ \parallel v - P_1 v \parallel^2 + \parallel v - Q_1 v \parallel^2 \right\} \\ &= \min \left\{ v^T [\alpha (I - P_1)^T (I - P_1) + \beta (I - Q_1)^T (I - Q_1)] v \right\} \\ &\quad \text{s.t.} \quad \alpha + \beta = 1 \quad \text{and} \quad v^T v = 1 \end{aligned} \tag{7.25}$$

Once again, this problem can be solved via the Singular Value Decomposition of the matrix $[\sqrt{\alpha}(I - P_1)^T \ \ \sqrt{\beta}(I - Q_1)^T]$.

7.3.2.2 A two-stage procedure for solving multi-objective optimization problem. As pointed outed, an optimal solution of minimizing J_1 should lie in the subspace spanned by the matrix P given by Eq.(7.18), i.e. an optimal solution is given by:

$$v = P v_1 \tag{7.26}$$

where $v_1 \in \mathbb{R}^{k_1}$ is an unknown vector and subject to the constraint $v_1^T v_1 = 1$. Substituting v into the J_2 given by Eq.(7.17), we have

$$\max J_2 = \max \left\{ v_1^T (P^T M)(P^T M)^T v_1 \right\} \quad \text{s.t.} \quad v_1^T v_1 = 1 \tag{7.27}$$

This optimization problem can also be solved using the SVD of the matrix $P^T M$. Once v_1 has been obtained, the optimal residual generator v which minimizes the sensitivity to modeling uncertainty and maximizes the sensitivity to faults can be determined by Eq.(7.26).

7.3.2.3 Multi-objective optimization by minimizing a mixed performance index. One of methods to solve the multi-objective optimization problem is to optimize a new cost function J which accounts for both J_1 and J_2. A solution for minimizing J cannot minimize J_1 at the same time as maximizing J_2. However, it could lead to a reasonable solution for robust residual design. A sensible mixture of performance indices is their ratio (or relative magnitude), i.e.

$$J = \frac{J_1}{J_2} = \frac{v^T Z Z^T v}{v^T M M^T v} \tag{7.28}$$

Hence, the robust residual design is achievable by minimizing J. This problem can be solved by introducing the *matrix pencil* concept (Gantmacher, 1959, Vol.I, pp.310-326) as follows:

Definition 7.1 *Given two quadratic forms:*

$$J_1 = v^T Z Z^T v, \qquad J_2 = v^T M M^T v$$

the equation:

$$\det(Z Z^T - \lambda M M^T)$$

is called the characteristic equation of the regular matrix pencil $v^T Z Z^T v - \lambda v^T M M^T v$. The roots of this characteristic equation, denoted by:

$$\lambda_1 \leq \lambda_2 \leq \cdots \leq \lambda_{(s+1)m}$$

are called the **generalized eigenvalues** *of the matrix pencil.*

Since $(Z Z^T - \lambda_i M M^T)$ is singular, there exists a nontrivial solution vector w_i of the following equation:

$$(Z Z^T - \lambda_i M M^T) w_i = 0$$

where w_i is called the generalized eigenvector (or principal vector) of the pencil.

Lemma 7.1 *If $W = [w_1, w_2, \cdots, w_{(s+1)m}]$ is the generalized eigenvector matrix of a regular pencil $v^T Z Z^T v - \lambda v^T M M^T v$, the transformation $v = W\rho$ can be applied to $v^T Z Z^T v$ and $v^T M M^T v$ simultaneously to yield (Gantmacher, 1959, Vol.I, p.314):*

$$J_1 = \sum_{i=1}^{(s+1)m} \lambda_i \rho_i^2 \qquad J_2 = \sum_{i=1}^{(s+1)m} \rho_i^2$$

Proof: See Gantmacher (1959), Vol.I, pp.310-314. ◊ QED

Theorem 7.1 *The performance index J defined by (7.28) is bounded by:*

$$\lambda_1 \leq J = \frac{v^T Z Z^T v}{v^T M M^T v} \leq \lambda_{(s+1)m} \tag{7.29}$$

and

$$J = \begin{cases} \lambda_1 & when \quad v = w_1 \\ \lambda_{(s+1)m} & when \quad v = w_{(s+1)m} \end{cases} \tag{7.30}$$

Proof: Using the results given in the Lemma 7.1, we have

$$J = \frac{\lambda_1 \rho_1^2 + \lambda_2 \rho_2^2 + \cdots + \lambda_{(s+1)m} \rho_{(s+1)m}^2}{\rho_1^2 + \rho_2^2 + \cdots + \rho_{(s+1)m}^2}$$

It follows that:

$$\begin{aligned} \lambda_1 &= \frac{\lambda_1 \rho_1^2 + \lambda_1 \rho_2^2 + \cdots + \lambda_1 \rho_{(s+1)m}^2}{\rho_1^2 + \rho_2^2 + \cdots + \rho_{(s+1)m}^2} \\ &\leq \frac{\lambda_1 \rho_1^2 + \lambda_2 \rho_2^2 + \cdots + \lambda_{(s+1)m} \rho_{(s+1)m}^2}{\rho_1^2 + \rho_2^2 + \cdots + \rho_{(s+1)m}^2} = J \end{aligned}$$

If:

$$\rho = [1, 0, \cdots, 0]$$

we get:

$$v = w_1, \quad \text{and} \quad J = \lambda_1$$

The other side of the inequality ($J \leq \lambda_{(s+1)m}$) can be proved similarly.
◊ QED

From this theorem, the solution which minimizes J can be obtained via the calculation of generalized eigenvalue-eigenvectors of the matrix pencil. The MATLAB function "eig" can be used to find the generalized eigenvectors and eigenvalues.

Three methods for solving the multi-objective optimization problem have now been given. The advantage of the first method is that it can easily satisfy different design goals (low missed-detection rate or low false detection rate) by adjusting weighting factors α and β. However, the solution procedure involves two optimization steps and is very complicated. The third method (Section 7.3.2.3) has the opposite advantages and disadvantages compared with the first method (7.3.2.1). Although the way of mixing design criteria in the thrid method is the same as that given by Wünnenberg and Frank (Wünnenberg and Frank, 1988; Frank, 1990; Wünnenberg and Frank, 1990; Wünnenberg, 1990), the way of handling modeling uncertainty is completely different. The technique developed here can be applied to systems with both modeling errors and unknown disturbances, whilst the technique developed by Wünnenberg and Frank can only be used to tackle disturbances. Hence the technique developed here has wider application.

7.4 A Numerical Illustration Example

A problem of designing robust residual for a four-dimensional system operating at a set-point with two actuators and three sensors is now considered. This example is a modification to the example in Chow and Willsky (1984). The system matrices are:

$$A = \begin{bmatrix} 0.5 & -0.7 & 0.7 & 0.0 \\ 0.0 & 0.8 & 0.6\gamma & 0.0 \\ -1.0 & 0.0 & 0.0 & 0.1 \\ 0.0 & 0.0 & -\gamma & 0.4 \end{bmatrix}, \quad B = \begin{bmatrix} 0 & 0 \\ 1 & 0 \\ 0 & 1 \\ 0 & 0 \end{bmatrix}, \quad C = \begin{bmatrix} 0 & 0 & 1 & 0 \\ 0 & 1 & 0 & 0 \\ 0 & 0 & 0 & 1 \end{bmatrix}$$

$$D = 0_{3\times 2}$$

Consider the situation when the fault occurs in the first sensor, the corresponding fault distribution matrices are:

$$R_1 = 0_{4\times 1} \qquad R_2 = \begin{bmatrix} 1 \\ 0 \\ 0 \end{bmatrix}$$

Except for two elements in the A matrix, all parameters are known exactly. The modeling uncertainty is denoted by the parameter γ whose nominal value is $\gamma_0 = 0.15$ and the bound is $\gamma \in [0.1, \;\; 0.2]$. Taking the parity relations of order $s = 2$, the residual is generated by the parity relation:

$$r(k) = v^T \begin{bmatrix} y(k-2) \\ y(k-1) \\ y(k) \end{bmatrix} - v^T H \begin{bmatrix} u(k-2) \\ u(k-1) \\ u(k) \end{bmatrix} \tag{7.31}$$

By choosing the uncertain parameter γ with representative values of 0.1, 0.125, 0.15, 0.175, 0.2 within the uncertainty bound, 5 sets of model matrices are obtained. The residual generation vector v is designed by method described in Section7.3.2.3 given in the last section.

$$v = \begin{bmatrix} 0.2449 \\ 0.0375 \\ 0.1022 \\ -0.1749 \\ -0.6686 \\ -0.3415 \\ 0.3498 \\ 0.3945 \\ 0.2367 \end{bmatrix} \quad \text{and} \quad (v^T H)^T = \begin{bmatrix} -0.3530 \\ -0.1749 \\ 0.3945 \\ 0.3498 \\ 0 \\ 0 \end{bmatrix}$$

For this design, the values of the objective functions are:

$$J_1 = 1.6e^{-29}, \quad J_2 = 1.0648, \quad \frac{J_1}{J_2} = 1.505e^{-29}$$

i.e., an almost perfect robust design has been achieved. Now the simulation is used to assess the fault detection performance of the designed residual signal. The design is carried out at the nominal point ($\gamma = 0.15$), but the simulation is carried out at a non-nominal point ($\gamma = 0.1875$). Each control input is a unit step function with a small level of additive Gaussian white noise. Two faults have been added to sensor 1; one is a ramp up and ramp down signal and the other is a step signal.

Fig.7.1 shows faulty measurements. It can be seen that the fault is very small and is hardly noticeable from the output. However, it is very easy to detect from the robust residual Fig.7.2. We have carried out a number simulations in which the uncertain parameter is assigned within its bound of $\gamma \in [0.1, \quad 0.2]$, and the results (shown in Fig.7.2) are almost identical except for some fluctuation due to noise.

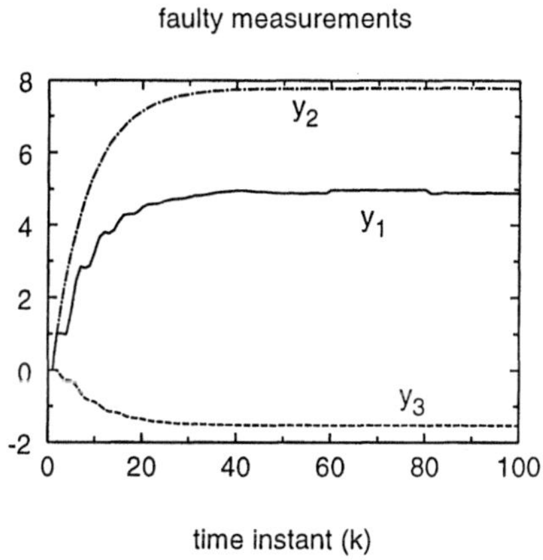

Figure 7.1. Faulty measurement when $\gamma = 0.1875$

Fig.7.2 also shows the result when the uncertain parameter is taken outside its bound, say $\gamma = -0.1$. It is interesting to see that the result is almost the same as the case when the uncertain parameter lies within its bound. To extend this idea further, if we let $\gamma = 1$, the system is unstable for this setting. Very surprisingly, the residual signal (Fig.7.2) is almost the same as the others. This shows that the residual is *robust* over a wide range of parameter variations.

7.5 Discussion on Designing Optimal Parity Relations

7.5.1 Robust fault isolation

Robust fault isolation can be achieved using *robust structured residual sets.* A robust structured residual is robust against modeling uncertainty and sensitive to a group of faults, whilst insensitive to another group of faults. If the fault vector $f(k)$ is re-grouped as two sub-vectors $\overline{f}(k)$ and $\underline{f}(k)$, the faults and

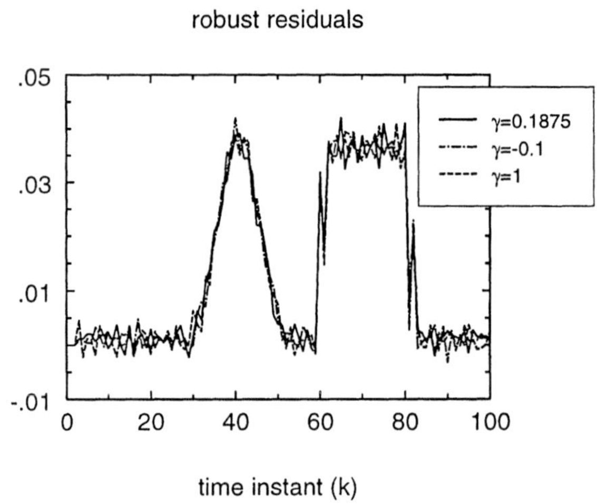

Figure 7.2. Robust residuals for different operating points

associated distribution matrices are:

$$R_t^1 f(k) = [\overline{R}_t^1 \;\; \underline{R}_t^1] \begin{bmatrix} \overline{f}(k) \\ \underline{f}(k) \end{bmatrix} \quad ; \quad R_t^2 f(k) = [\overline{R}_t^2 \;\; \underline{R}_t^2] \begin{bmatrix} \overline{f}(k) \\ \underline{f}(k) \end{bmatrix}$$

In this case, the system equation can be rewritten as:

$$\begin{cases} x(k+1) = A_t x(k) + B_t u(k) + [E_t^1 \;\; \underline{R}_t^1] \begin{bmatrix} d(k) \\ \underline{f}(k) \end{bmatrix} + \overline{R}_t^1 \overline{f}(k) \\ y(k) = C_t x(k) + D_t u(k) + [E_t^2 \;\; \underline{R}_t^2] \begin{bmatrix} d(k) \\ \underline{f}(k) \end{bmatrix} + \overline{R}_t^2 \overline{f}(k) \end{cases}$$

If a structured residual is to be designed to be insensitive to faults grouped in the vector $\underline{f}(k)$, this vector can be treated in the same way as a disturbance vector in an optimal parity relation design. The performance indices should be modified correspondingly.

The design of robust directional residual vectors using optimal parity relations can be found in Gertler and Kunwer (1993).

7.5.2 Probability distribution of multiple models

The probability of the system works at a certain operating point may be larger than that at other operating points. This fact should be taken into consideration in the design of optimal parity relations. The performance indices are thus modified accordingly, to place different emphases on the different model descriptions:

$$\min J_1 = \min \left\{ \sum_{i=1}^{N} \| p_i v^T Z_i \|^2 \right\} \quad \text{s.t. } v^T v = 1 \tag{7.32}$$

$$\min J_2 = \min \left\{ \sum_{i=1}^{N} \| p_i v^T M_i \|^2 \right\} \quad \text{s.t.} \quad v^T v = 1 \tag{7.33}$$

where p_i is the probability that the system operates at the i_{th} model ($i = 1, 2, \cdots, N$), and

$$\sum_{i=1}^{N} p_i = 1$$

7.5.3 Orthogonal parity relations

This is an approach proposed by Gertler and colleagues (Gertler, Fang and Luo, 1990; Gertler and Luo, 1989; Gertler, Luo, Anderson and Fang, 1990; Gertler and Singer, 1990; Gertler, 1991; Gertler and Kunwer, 1993; Gertler, 1998) to design robust and (or) isolable residual sets. The method is based on the z-transformed input-output relationship of the monitored system, i.e.

$$\Psi_y(z)y(z) = \Psi_u(z)u(z) + \Psi_d(z)d(z) + \Psi_f(z)f(z) \tag{7.34}$$

where $\Psi_y(z)$, $\Psi_u(z)$, $\Psi_d(z)$ and $\Psi_f(z)$ are known z-polynomial matrices. A primary residual vector can be directly obtained by rearranging the above equation as follow:

$$r'(z) = \begin{cases} \Psi_y(z)y(z) - \Psi_u(z)u(z) & \text{computational form} \\ \Psi_d(z)d(z) + \Psi_f(z)f(z) & \text{evaluation form} \end{cases} \tag{7.35}$$

This primary residual can be used to detect faults, however it does not have robust and isolable properties. To design robust and (or) isolable residuals, the primary residual should be transformed as:

$$r(z) = T(z)r'(z) \tag{7.36}$$

where $T(z)$ is a z-polynomial matrix to be designed for achieving required robust and isolable properties. The response of this transformed residual to faults and disturbances is:

$$r(z) = T(z)\Psi_d(z)d(z) + T(z)\Psi_f(z)f(z) \tag{7.37}$$

To make the residual insensitive to the disturbance $d(z)$, the transformation matrix $T(z)$ should be made orthogonal to $\Psi_d(z)$, this is the basic principle of the orthogonal parity relation approach for robust residual generation. Similarly, the residual can be designed to be insensitive to the i_{th} fault component, if $T(z)$ is made to be orthogonal to the i_{th} column of $\Psi_f(z)$. If sufficient design freedom is available, a totally robust and isolable residual can be designed if the matrix $T(z)$ satisfies:

$$T(z)D(z) = 0 \quad ; \quad T(z)\Psi_f(z) = I$$

This approach is, in principle simple, however it is not easy to implement because the numerical operation of polynomial matrices is not an easy task.

Moreover, this approach is only effective to uncertainty caused by unknown disturbances and cannot be directly applied to robust design against to modeling errors (Gertler, 1991; Gertler and Kunwer, 1993; Gertler, 1998).

7.5.4 Design of robust parity relations via optimization

The design of robust residuals can be treated as an optimization problem in which fault effects should be maximized and modeling uncertainty effects should be minimized. This philosophy has been adopted in many research studies, for example, Staroswiecki *et al* (Staroswiecki, Cassar and Cocquempot, 1993a; Staroswiecki, Cassar and Cocquempot, 1993b) have defined a multi-objective optimization problem in robust parity relation design and the solution for this optimization has also be presented. However, they assumed that faults and/or disturbances are either pulse or step functions in the calculation of residual sensitivity cost functions, this limits the application domain of their approach. The approach presented in this chapter does not make any assumptions concerning fault and disturbance functions and hence has a wider application domain.

Kinnaert (1993a; 1993b; 1996) formulated the robust parity relation design as a constrained optimization problem, the aim being to construct a number of parity relations, as follows:

$$r_i(k) = w_i^T \begin{bmatrix} Y(k) \\ U(k) \end{bmatrix} \quad ; \quad i = 1, 2, \cdots, g$$

Note that this residual definition is just a rearrangement of the definition given in Eq.(7.6). The performance index and constraints are evaluated using the expectation value of the residual under different hypothesis as follows:

$$\min_{w_i} \lim_{k\to\infty} \mathcal{E}\{r_i^2(k/\text{no fault})\}$$

subject to:

$$\begin{cases} \lim_{k\to\infty} \frac{\mathcal{E}\{r_i^2(k/\text{no fault})\}}{\mathcal{E}\{r_i^2(k/\text{fault i})\}} \le \delta_{ii} < 1 \\ \lim_{k\to\infty} \frac{\mathcal{E}\{r_i^2(k/\text{fault j})\}}{\mathcal{E}\{r_i^2(k/\text{fault i})\}} \le \delta_{ij} < 1 \ ; \ j \in \{1, \cdots, i-1, i+1, \cdots, g\} \\ w_i^T w_i = 1 \end{cases}$$

where $\mathcal{E}\{\cdot\}$ denotes the expectation operator, δ_{ij} are design parameters to be decided by designer. The first constraint is to assure the robustness and the second constraint is to guarantee isolability. If the statistical properties of measurement noise, disturbances and faults are known *a priori* the optimization problem can be rewritten as:

$$\min_{w_i} w_i^T \Phi_0 w_i$$

subject to:

$$\begin{cases} w_i^T(\Phi_0 - \delta_{ii}\Phi_i) \leq 0 \\ w_i^T(\Phi_j - \delta_{ij}\Phi_i) \leq 0 \quad ; \; j \in \{1, \cdots, i-1, i+1, \cdots, g\} \\ w_i^T w_i = 1 \end{cases}$$

where Φ_i $(i = 1, \ldots, g)$ are related to the statistical properties of measurement noise, disturbances and faults and a complex computation procedure is presented in Kinnaert (1993a; 1993b; 1996). It can be seen that this is a constrained optimization problem with a quadratic cost function under non-convex quadratic inequality constraints. It is only possible to find a numerical solution for this optimization problem through complicated search algorithms.

The main disadvantage of Kinnaert's approach is that it requires the statistical properties of measurement noise, disturbances and faults which are normally unavailable. Another disadvantage is that the optimization procedure is very complex and there are no analytical solutions. With the cost of great complexity, there is no evidence to show that it can give diagnostic performance better than the approach presented in this chapter.

7.5.5 *Closed-loop optimal parity relations*

Wu and Wang (1993; 1995) suggested an approach to designing robust residuals based on parity checking on the output estimation errors. The approach involved two stages: the first stage is to estimate the system output and generate the output estimation error via a full-order state observer, the second is to construct parity relations using the output estimation error. As described in Sections 2.8.1 & 6.2.1, when a full-order observer is applied to a system without faults and modeling uncertainty, the state estimation error $e(k) = x(k) - \hat{x}(k)$ and the output estimation error $e_y(k) = y(k) - \hat{y}(k)$ are driven by the following equation:

$$\begin{cases} e(k+1) &= (A - KC)e(k) \\ e_y(k) &= Ce(k) \end{cases} \tag{7.38}$$

where K is the observer gain matrix. The output estimation error $e_y(k)$ can be used directly as a residual vector, however Wu and Wang (1993; 1995) construct the residual as:

$$r(k) = v^T \begin{bmatrix} e_y(k-s) \\ e_y(k-s+1) \\ \vdots \\ e_y(k) \end{bmatrix} \tag{7.39}$$

where the vector v^T satisfies the following equation:

$$v^T \begin{bmatrix} C \\ C(A-KC) \\ C(A-KC)^2 \\ \vdots \\ C(A-KC)^s \end{bmatrix} = 0 \tag{7.40}$$

It can be proved that the residual generated by Eq.(7.39) is equivalent to the residual generated by Eq.(7.6) when the observer gain matrix is zero, i.e. $K = 0$ (Wu and Wang, 1993; Wu and Wang, 1995). This shows once again that the parity relation is a special case of the observer-based residual generator in which the dynamic feedback is zero.

Wu and Wang (1993; 1995) demonstrated an optimization procedure to find K and v for achieving residual robustness against modeling uncertainty. Because there is more design freedom (i.e. the choice of K) in the closed-loop parity relation design, the robustness and sensitivity performances can be better than those of the original parity relation design. However, the extra price to pay is the increased complexity in implementation. Wang and Wu (1993) applied the closed-loop parity relations to fault diagnosis of closed-loop control systems. They have shown that the feedback controller can also be modified to achieve maximal diagnostic sensitivity to faults. This is consistent with the idea given by Wu (1992) in which the effect of a fault in the residual is sensitized by means of feedback controller design. This also shows that the fault diagnosis scheme and the robust controller should be designed together to achieve maximal closed-loop reliability and performance.

7.6 Summary

In this chapter, the problem of finding optimally robust residuals for systems with bounded parameter variations and unknown disturbances has been discussed. This parity relation design problem has been formulated into a multi-objective optimization problem, yielding a robust residual which is maximally sensitive to faults, whilst minimally sensitive to modeling uncertainty (including modeling errors and unknown disturbances). Three methods for solving this multi-objective optimization problem have been proposed. The simplest method is to mix performance indices as a single optimization index according to the design objective, which is solved using the generalized eigenvalue-eigenvector concept. As the robust criterion has been given quantitatively, the residual designed using different parity relations can be ordered according to robustness. Both modeling errors (in term of parameter variations) and disturbances have been considered in the robust residual design procedure, the technique developed can be used to diagnose incipient faults in a wide range of systems with modeling uncertainty. This principle has been well illustrated using a numerical example. Some other developments in designing optimal parity relations for robust FDI have been commented, and these developments are also compared with the technique introduced in this Chapter.

8 FREQUENCY DOMAIN DESIGN AND H_∞ OPTIMIZATION METHODS FOR ROBUST FAULT DIAGNOSIS

8.1 Introduction

As described in previous chapters, there are many ways, such as the unknown input observer, eigenstructure assignment, optimally robust parity relations, for eliminating or minimizing disturbance and modeling error effects on residual and hence for achieving robustness in FDI . Whilst these techniques are different, one feature is common among them, the original frameworks of these methods were developed for ideal systems or with a special uncertainty structure and then efforts have been made to include non-ideal or more general uncertainties. In contrast, H_∞ optimization is a robust design method with the original motivation firmly rooted in the consideration of various uncertainties, especially the modeling errors. H_∞ optimization has been developed from the very beginning with the understanding that no design goal of a system can be perfectly achieved without being compromised by an optimization in the presence of uncertainty. Hence this technique is very suitable for tackling uncertainty issues. After two decades of development, it is now playing a leading role in tackling the robustness problem in control systems. It is reasonable to seek the application of these examines in other areas, including the robust design of FDI systems. This chapter studies the frequency domain approaches, including H_∞ optimization, for robust fault diagnosis.

Unlike robust control design, the use of frequency domain techniques for robust FDI has not received enough attention until recently. Patton et al. (1986)

first discussed the possibility of using frequency distribution information to design FDI algorithms, however they did not give further guidance as to how this could be achieved. Ding and Frank (1989) proposed a frequency domain optimal observer design method for robust FDI by using a complicated frequency domain optimization procedure. After Viswanadham, Taylor and Luce (1987) developed the factorization approach in residual design, the frequency domain approach for robust FDI started to show some promises. Viswanadham and Minto (1988) discussed the possibility of solving the robust FDI problem using factorization approach. Following this line of enquiry, Ding and Frank (1990) reformulated the frequency domain residual generation method via factorization of the system transfer matrix, however the robustness issue is not their primary concern in design. Ding and Frank (1991) then attempted to solve the robust FDI problem using the inner-outer factorization method. More advanced developments were reported by Frank (1991a), Frank and Ding (1993), Ding et al. (1993), Frank and Ding (1994), García, Köppen-Seliger and Frank (1995), Frank and Ding (1997) and Ding and Guo (1997). Qiu and Gertler (1993) made some important modifications in robust FDI design by using the factorization-based H_∞-optimization technique.

The factorization-based H_∞-optimization technique is useful in solving robust control as well as robust FDI problems. However, the more elegant and advanced H_∞-optimization methods are based on the use of the Algebraic Riccati Equation (ARE) (Doyle, Glover, Khargonekar and Francis, 1989; Shaked and Theodor, 1992b; Zhou, Doyle and Golver, 1996). Mangoubi et al. (1992) first solved the robust FDI estimation problem by using the Riccati equation approach through the use of H_∞ and μ robust estimator synthesis methods developed by Appleby et al. (1991). A direct formulation of the FDI problem as a robust H_∞ filter design problem was given in Edelmayer, Bokor and Keviczky (1994; 1996; 1997a). This approach results in solving a Riccati equation for a standard H_∞ filter problem. Edelmayer, Bokor and Keviczky (1997b) further developed this approach for linear time-varying systems. The ARE-based approach for robust FDI is also studied by Yao et al. (1994). Mangoubi, Appleby, Verghese and Vander Velde (1995) combined H_∞ robust FDI design with statistical methods for the detection of faults. To deal with modeling errors as well as disturbances in robust FDI design, Niemann and Stoustrup (1996) introduced modeling error blocks into the standard H_∞ observer design framework. The weighting factors are then introduced in the standard problem formulation to find the optimal FDI solution. This approach is further extended to non-linear systems where the nonlinearity is treated in the same way as a modeling error block (Stoustrup and Niemann, 1998). Another approach of tackling modeling errors has been proposed recently by Eich and Oehler (1997) where the μ model validation method is used.

The majority of studies discussed so far involve the use of a slightly modified H_∞ filter for the residual generation. That is to say the design objective is to minimize the effect of disturbances and modeling errors on the estimation error and subsequently on the residual. However, the robust residual generation problem is different from the robust estimation problem because it does not only require the disturbance attenuation. The residual has to be remain sensitive to faults whilst the effect of disturbance is minimized. Sauter et al. (1997) studied this problem where the fault sensitivity is enhanced by applying an optimal post-filter to the "primary residual". The problem of enhancing fault sensitivity while increasing robustness against disturbances and modeling errors was studied extensively in by Sadrnia, Chen and Patton (1997a; 1997b; 1997c). The essential idea is to reach an acceptable compromise between disturbance robustness and fault sensitivity. In the beginning, an observer with very small disturbance sensitivity bound is designed using the Riccati equation approach. Then, the fault sensitivity is checked. If the fault sensitivity is too small, then the disturbance robustness requirement should be relaxed. That is to say to design another optimal observer with an increased disturbance sensitivity bound. This procedure is likely to be repeated several times. The final goal is to find a design which provides the maximum ratio between fault sensitivity and disturbance sensitivity.

An important contribution in the use of the standard H_∞ technique for robust FDI was made by Stoustrup *et al* (Kilsgaard, Rank, Niemann and Stoustrup, 1996; Stoustrup et al., 1997; Stoustrup and Grimble, 1997; Niemann and Stoustrup, 1997). In their investigation, the robust controller and robust FDI system are designed simultaneously. One of advantages of this integrated design approach is that it can be easily formulated into a standard H_∞ problem. Moreover, the modeling errors can be incorporated directly into the standard problem formulation and solved using robust control techniques such as μ-synthesis. Instead of generating a diagnostic residual, the fault estimation problem is tackled in their study. The estimated fault can be used to FDI as well as control reconfiguration. Both nominal and robust design cases are investigated and the conclusion is that the separation principle is only valid for the nominal case.

The reason for using H_∞ in connection with the FDI problem is due to the robustness aspect with respect to modeling uncertainties. However, from an FDI point of view, it is more natural to use an l_∞ optimization of the FDI filter where the magnitude of the residual due to disturbances and modeling uncertainties is bounded. This will make the detection of fault signal more readily comparable to the case where an H_∞ has been applied. The application of l_∞ for the design of integrated control and FDI systems was studied by Faitakis and Kantor (1996).

In the similar way to robust control, the optimal filtering problem can be solved in the linear matrix inequality (LMI) formulation (Gahinet, Nemirovski, Laub and Chilali, 1995; Park and Kailath, 1997; Palhares and Peres, 1998). The application of LMI in robust FDI observer design was studied by Hou and Patton (1996) and Patton and Hou (1997). The most important feature in their study is the introduction of fault sensitivity into the robust FDI observer design. The worst case fault sensitivity is used as a constraint for the optimization problem which is solved through the use of LMI methods.

This chapter starts with the factorization approach for robust FDI design, then proceeds to the standard H_∞ problem formulation. Finally, the LMI approach for robust FDI observer design is introduced.

8.2 Robust Fault Detection via Factorization Approach

8.2.1 Residual generator design using factorization

The design of a residual generator using factorization was first proposed by Viswanadham, Taylor and Luce (1987) and later extended by Ding and Frank (1990) and the advanced development can be found in Frank and Ding (1994) and Kinnaert and Peng (1995). To introduce this approach, let us start with the system model. A system with faults and disturbances can be described by the state space model as:

$$\begin{cases} \dot{x}(t) &= Ax(t) + Bu(t) + R_1 f(t) + E_1 d(t) \\ y(t) &= Cx(t) + Du(t) + R_2 f(t) + E_2 d(t) \end{cases} \tag{8.1}$$

where $x(t) \in \mathbb{R}^n$ is the state vector, $y(t) \in \mathbb{R}^m$ is the output vector, $u(t) \in \mathbb{R}^r$ is the known input vector and $d(t) \in \mathbb{R}^q$ is the unknown input (or disturbance) vector, $f(t) \in \mathbb{R}^g$ represents the fault vector which is considered as an unknown time function. A, B, C, D, E_1, E_2, R_1 and R_2 are known matrices with appropriate dimensions. Note that only the disturbance uncertainty is considered in this system model. Other kinds of uncertainties such as modeling errors can be treated as approximate disturbances for robust FDI design using techniques introduced in Chapter 5.

The input-output model is thus given by:

$$y(s) = G_u(s)u(s) + G_f(s)f(s) + G_d(s)d(s) \tag{8.2}$$

where the transfer function matrices are:

$$G_u(s) = C(sI - A)^{-1}B + D \tag{8.3}$$
$$G_f(s) = C(sI - A)^{-1}R_1 + R_2 \tag{8.4}$$
$$G_d(s) = C(sI - A)^{-1}E_1 + E_2 \tag{8.5}$$

According to the notation used in robust control, the transfer function matrices (which are proper real-rational matrices) can be denoted as:

$$G_u(s) = \left[\begin{array}{c|c} A & B \\ \hline C & D \end{array}\right] \;;\; G_f(s) = \left[\begin{array}{c|c} A & R_1 \\ \hline C & R_2 \end{array}\right] \;;\; G_d(s) = \left[\begin{array}{c|c} A & E_1 \\ \hline C & E_2 \end{array}\right] \tag{8.6}$$

The standard robust control notation RH_∞ is used throughout this chapter. This notation stands for the set of all real-rational transfer function matrices that are stable and proper. Alternatively speaking, any transfer matrix in RH_∞ is realizable using a stable linear system.

Definition 8.1 *(Vidyasagar, 1985; Zhou et al., 1996): Two transfer matrices $M(s)$ and $N(s)$ in RH_∞ are right coprime over RH_∞ if they have the same* *number of columns and if there exist transfer matrices $X_r(s)$ and $Y_r(s)$ in RH_∞ such that*

$$[X_r(s) \;\; Y_r(s)] \begin{bmatrix} M(s) \\ N(s) \end{bmatrix} = X_r(s)M(s) + Y_r(s)N(s) = I$$

Similarly, two transfer matrices $\tilde{M}(s)$ and $\tilde{N}(s)$ in RH_∞ are left coprime over RH_∞ if they have the same number of rows and if there exist transfer matrices $X_l(s)$ and $Y_l(s)$ in RH_∞ such that

$$\begin{bmatrix} \tilde{M}(s) & \tilde{N}(s) \end{bmatrix} \begin{bmatrix} X_l(s) \\ Y_l(s) \end{bmatrix} = \tilde{M}(s)X_l(s)M(s) + \tilde{N}(s)Y_l(s) = I$$

Now let $G_u(s)$ be a proper real-rational matrix. A *right-coprime factorization (rcf)* of $G_u(s)$ is a factorization $G_u(s) = N(s)M^{-1}(s)$ where $N(s)$ and $M(s)$ are right coprime over RH_∞. Similarly, a *left-coprime factorization (lcf)* has the form $G_u(s) = \tilde{M}^{-1}\tilde{N}(s)$ where $\tilde{M}(s)$ and $\tilde{N}(s)$ are left coprime over RH_∞.

Lemma 8.1 *(Vidyasagar, 1985; Zhou et al., 1996): For any proper real rational matrix $G_u(s)$ $(m \times r)$, there always exists a double (left and right) coprime factorization given by:*

$$G_u(s) = N(s)M^{-1}(s) = \tilde{M}^{-1}(s)\tilde{N}(s) \tag{8.7}$$

where $N(s)$ $(m \times r)$, $M(s)$ $(r \times r)$, $\tilde{M}(s)$ $(m \times m)$ and $\tilde{N}(s)$ $(m \times r)$ are right and left coprime RH_∞ matrices of $G_u(s)$, respectively. For this double coprime factorization there exist RH_∞ transfer matrices $X_r(s)$, $Y_r(s)$, $X_l(s)$ and $Y_l(s)$ satisfying Bezout identity (Vidyasagar, 1985):

$$\begin{bmatrix} X_r(s) & Y_r(s) \\ -\tilde{N}(s) & \tilde{M}(s) \end{bmatrix} - \begin{bmatrix} M(s) & -Y_l(s) \\ N(s) & X_l(s) \end{bmatrix} = I \tag{8.8}$$

Of course implicit in these definitions is the requirement that both $M(s)$ and $\tilde{M}(s)$ be square and nonsingular. The eight transfer matrices mentioned in the above Lemma can be determined by Lemma 8.2.

Lemma 8.2 *(Nett et al., 1984; Zhou et al., 1996): Suppose $G_u(s)$ is a proper real-rational matrix and*

$$G_u(s) = \left[\begin{array}{c|c} A & B \\ \hline C & D \end{array}\right] \tag{8.9}$$

is a stabilizable and detectable realization. Let K_c and K be such that $A + BK_c$ and $A - KC$ are both stable, then eight transfer matrices in the double coprime factorization can be determined as follows:

$$M(s) = \left[\begin{array}{c|c} A + BK_c & B \\ \hline K_c & I \end{array}\right] \;\; ; \;\; N(s) = \left[\begin{array}{c|c} A + BK_c & B \\ \hline C + DK_c & D \end{array}\right] \tag{8.10}$$

$$\tilde{M}(s) = \left[\begin{array}{c|c} A-KC & K \\ \hline -C & I \end{array}\right] \quad ; \quad \tilde{N}(s) = \left[\begin{array}{c|c} A-KC & B-KD \\ \hline C & D \end{array}\right] \qquad (8.11)$$

$$X_r(s) = \left[\begin{array}{c|c} A-KC & B-KD \\ \hline -K_c & I \end{array}\right] \quad ; \quad Y_r(s) = \left[\begin{array}{c|c} A-KC & K \\ \hline -K_c & 0 \end{array}\right] \qquad (8.12)$$

$$X_l(s) = \left[\begin{array}{c|c} A+BK_c & K \\ \hline C+DK_c & I \end{array}\right] \quad ; \quad Y_l(s) = \left[\begin{array}{c|c} A+BK_c & K \\ \hline -K_c & 0 \end{array}\right] \qquad (8.13)$$

With the coprime factorization, the residual generator can be designed accordingly to the following Theorem.

Theorem 8.1 (Residual Generator Design via Factorization) *Let the left-coprime factorization of the transfer matrix $G_u(s)$ be:*

$$G_u(s) = \tilde{M}^{-1}(s)\tilde{N}(s) \qquad (8.14)$$

then the diagnostic signal residual $r \in \mathbb{R}^p$ can be generated using the following frequency domain residual generator (Ding and Frank, 1990):

$$r(s) = Q(s)(\tilde{M}(s)y(s) - \tilde{N}(s)u(s)) \qquad (8.15)$$

where $Q(s)$ $(p \times m)$ is a RH_∞ transfer function matrix, defined as a weighting matrix which can be static as well as dynamic.

Proof: Substituting Eq.(8.2) into Eq.(8.15) and assuming the disturbance vector (d) is zero-valued, we have

$$r(s) = Q(s)(\tilde{M}(s)G_u(s)u(s) + \tilde{M}(s)G_f(s)f(s) - \tilde{N}(s)u(s))$$

Due to the factorization (Eq.(8.14)), we have $\tilde{M}(s)G_u(s) = \tilde{N}(s)$. Finally,

$$r(s) = Q(s)\tilde{M}(s)G_f(s)$$

That is to say the signal r is only affected by the fault f. According to the residual definition given in Section 2.5, the signal r is a residual signal. The residual generator described by Eq.(8.15) complies with the general residual generator structure design by Eq.(2.13) in Section 2.5. ◊ **QED**

The residual generator described by the frequency domain formula (8.15) is illustrated by Fig. 8.1.

To implement this residual generator, the transfer matrices $\tilde{M}(s)$ and $\tilde{N}(s)$ have to be expressed in terms of the system state space model as shown in Lemma 8.2. That is to say

$$\tilde{M}(s) = \left[\begin{array}{c|c} A-KC & K \\ \hline -C & I \end{array}\right] := I - C(sI - A + KC)^{-1}K \qquad (8.16)$$

$$\tilde{N}(s) = \left[\begin{array}{c|c} A-KC & B-KD \\ \hline C & D \end{array}\right] := D + C(sI - A + KC)^{-1}(B - KD) \quad (8.17)$$

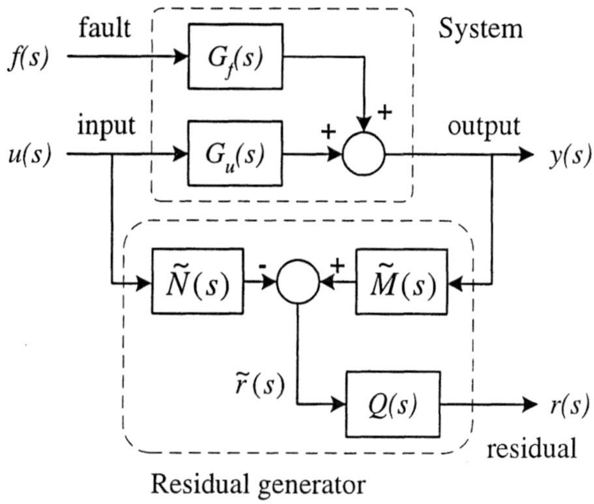

Figure 8.1. Frequency domain residual generator

with K ensuring the stability of the matrix $A - KC$. If we compare these two transfer matrices with the residual transfer matrices of a full-order observer-based residual generator given by Eq. (2.45) in Section 2.8.1, it can be seen that they are almost identical. This demonstrates that the residual generator designed in the frequency domain is totally equivalent to the observer-based residual generator. The two design methods are just two different tools, but they achieve the same goal. Actually, Ding et al. (1994) have shown that an observer can be designed using frequency domain factorization. To explain the equivalence further in terms of residual generator implementation, let us examine the "primary residual" which is the residual generated without the weighting transfer matrix (or post-filter) $Q(s)$,

$$\tilde{r}(s) = \tilde{M}(s)y(s) - \tilde{N}(s)u(s) = [\tilde{M}(s) \quad -\tilde{N}(s)] \begin{bmatrix} y(s) \\ u(s) \end{bmatrix} \tag{8.18}$$

The transfer matrix of this primary residual generator is:

$$G_{\tilde{r},uy}(s) = [\tilde{M}(s) \quad -\tilde{N}(s)] = \left[\begin{array}{c|c} A - KC & [K \quad B - KD] \\ \hline -C & [I \quad -D] \end{array}\right] \tag{8.19}$$

The state space realization of this transfer matrix is thus:

$$\begin{bmatrix} \dot{\hat{x}} \\ \tilde{r} \end{bmatrix} = \begin{bmatrix} A - KC & K & B - KD \\ -C & I & -D \end{bmatrix} \begin{bmatrix} \hat{x} \\ y \\ u \end{bmatrix} \tag{8.20}$$

Or, this can be implemented conventional through a full-order observer with K being the gain matrix.

$$\left\{ \begin{array}{rcl} \dot{\hat{x}}(t) & = & A\hat{x}(t) + Bu(t) + K[y(t) - C\hat{x}(t) - Du(t)] \\ \tilde{r}(t) & = & y(t) - [C\hat{x}(t) + Du(t)] \end{array} \right. \tag{8.21}$$

The state space implementation only gives us the primary residual. It can be seen that the primary residual is the difference between the actual and estimated outputs. To improve the residual quality, this primary residual is post-processed through the weighting transfer matrix $Q(s)$. The design of $Q(s)$ for achieving required FDI performance is discussed in the following Sections 8.2.2 & 8.2.3.

8.2.2 Perfect fault detection and isolation and perfect disturbance de-coupling

To achieve perfect fault detection (PFD), a residual signal should only be affected by faults, i.e.

$$\textbf{PFD:} \qquad \left\{ \begin{array}{ll} r(t) = 0 & \text{if } f(t) = 0 \\ r(t) \neq 0 & \text{if } f(t) \neq 0 \end{array} \right. \tag{8.22}$$

Note that this definition is defined after the transients of the residual generator have been settled down. When there are no disturbances and modeling errors, the perfect fault detection can easily be achieved by any residual generation method satisfying the detectability condition. For systems with disturbances and modeling errors, PFD cannot be always achieved. In this situation, the optimal (or approximately perfect) fault detection is tackled by robust residual generation methods.

To achieve perfect fault isolation (PFI), the residual vector should have the same dimension as the fault vector and each residual component should only be affected by one fault, i.e.

$$r(t) = \begin{bmatrix} r_1(t) \\ \vdots \\ r_g(t) \end{bmatrix} \in \mathbb{R}^g$$

$$\textbf{PFI:} \quad \text{for } i = 1, \cdots, g \quad \left\{ \begin{array}{ll} r_i(t) = 0 & \text{if } f_i(t) = 0 \\ r_i(t) \neq 0 & \text{if } f_i(t) \neq 0 \end{array} \right. \tag{8.23}$$

If PFI conditions are satisfied, PFD is also achieved. That is to say, PFI implies PFD. If there are no disturbances and modeling errors, perfect fault isolation can be achieved by the dedicated observer scheme (Frank, 1987; Patton et al., 1989) or using the *single-sensor/actuator parity relation* (Massoumnia and Vander Velde, 1988; Tsai and Chou, 1993; Peng et al., 1997). In contrast to PFD, PFI cannot always be achieved even if there are no disturbances and modeling errors.

Now let us discuss prefect fault detection and isolation (PFDI) for systems with disturbances and faults using the factorization approach.

Lemma 8.3 *(Frank and Ding, 1994): The perfect fault detection and isolation is achievable if and only if there exist a RH_∞ parameterization (or weighting) matrix $Q(s)$ so that:*

$$Q(s)\tilde{M}(s)G_f(s) = T(s) = diag(t_1(s), \cdots, t_g(s)) \in RH_\infty \tag{8.24}$$

and

$$Q(s)\tilde{M}(s)G_d(s) = 0 \tag{8.25}$$

This Lemma can be easily verified using the definition of PFDI and the residual generator defined by the factorization formulation (Eq.(8.15). The condition (8.25) denotes perfect disturbance de-coupling, whilst the condition (8.24) signifies PFI which also implies PFD.

The following Theorem proposed by Frank and Ding (1994) gives conditions for verifying the PFDI possibility using the system transfer matrices $G_f(s)$ and $G_d(s)$.

Theorem 8.2 *(Frank and Ding, 1994): For a system described by Eq. (8.2), the PFDI is achievable if and only if*

$$rank\{[G_f(s) \quad G_d(s)]\} = rank\{G_f(s)\} + rank\{G_d(s)\} \tag{8.26}$$

and

$$rank\{G_f(s)\} = g \quad (\textit{the number of independent faults}) \tag{8.27}$$

Proof: see Frank and Ding (1994).

Remarks:

(1) Condition (8.27) implies that the number of independent measurements must not smaller than the number of independent faults in order to achieve PFDI.

(2) Condition (8.26) implies that the number of independent measurements must not smaller than the number of independent faults plus the number of independent disturbances in order to achieve PFDI.

(3) Theorem 8.2 can be interpreted physically. That is to say the condition for achieving PFDI is that the disturbance and fault have totally decoupled effected on the measurement.

8.2.2.1 PFDI for systems without disturbances. The conditions for PFDI were proved by Frank and Ding (1994), however they did not give the algorithm for solving this problem. To find the solution for this problem, let us first consider PFDI without disturbances. In this case, only condition (8.27) is required and the residual response is:

$$r(s) = Q(s)G_f(s)f(s) \tag{8.28}$$

The transfer matrix $G_f(s)$ can be described by a left-coprime factorization as:

$$G_f(s) = \tilde{M}_f^{-1}(s)\tilde{N}_f(s) \tag{8.29}$$

where:

$$\tilde{M}_f(s) = I - C(sI - A + KC)^{-1}K = \tilde{M}(s) \tag{8.30}$$

$$\tilde{N}_f(s) = R_2 + C(sI - A + KC)^{-1}(R_1 - KR_2) \tag{8.31}$$

Substituting Eq.(8.29) into Eq.(8.28), we get

$$r(s) = Q(s)\tilde{N}_f(s)f(s) \tag{8.32}$$

To achieve PFDI for systems without disturbances, we should have:

$$Q(s)\tilde{N}_f(s) = T(s) = diag(t_1(s), \cdots, t_g(s)) \tag{8.33}$$

Since $\tilde{M}(s)$ is a nonsingular matrix, the condition (8.27) is equivalent to

$$rank\{\tilde{N}_f(s)\} = q$$

This implies that $\tilde{N}_f(s)$ is left invertible and the left inverse of this matrix is:

$$\tilde{N}_f^+(s) = [\tilde{N}_f^T(s)\tilde{N}_f(s)]^{-1}\tilde{N}_f^T(s)$$

and the post-filter transfer matrix $Q(s)$ can be determined by:

$$Q(s) = \tilde{N}_f^+(s)diag(t_1(s), \cdots, t_g(s)) \tag{8.34}$$

The proper choice of $t_1(s)$, $\cdots$, $t_g(s)$ will guarantee the matrix $Q(s)$ is a stable and realizable transfer matrix, i.e. a RH_∞ matrix. A more advanced design algorithm of solving PFDI problem without disturbances can be found in Kinnaert and Peng (1995).

When a solution $Q(s)$ cannot be found for the Eq. (8.33), the PFDI is not achievable. In this case, we should consider an optimal FDI problem by minimizing the following performance index:

$$J_f = \|T(s) - Q(s)\tilde{N}_f(s)\|_\infty \tag{8.35}$$

where $\|\cdot\|_\infty$ denotes a H_∞–norm which is defined as:

$$\|G(s)\|_\infty = \sup_\omega \bar{\sigma}\{G(j\omega)\} \tag{8.36}$$

where $\bar{\sigma}\{G(j\omega)\}$ denotes the largest singular value of $G(j\omega)$.

The minimization of the performance index (8.35) is a standard model-matching problem in robust control and the solution can be found in literature such as Francis (1987), Maciejowski (1989), Gao and Antsaklis (1989) and Hung (1993).

8.2.2.2 Perfect disturbance de-coupling. Now let us examine the disturbance de-coupling condition (8.25) alone. Similar to $G_u(s)$ and $G_f(s)$, the transfer matrix $G_d(s)$ can be described by a left-coprime factorization as:

$$G_d(s) = \tilde{M}_d^{-1}(s)\tilde{N}_d(s) \tag{8.37}$$

where:

$$\tilde{M}_d(s) = I - C(sI - A + KC)^{-1}K = \tilde{M}(s) \tag{8.38}$$
$$\tilde{N}_d(s) = E_2 + C(sI - A + KC)^{-1}(E_1 - KE_2) \tag{8.39}$$

Substituting Eq.(8.37) into Eq.(8.25), we get the disturbance de-coupling condition

$$Q(s)\tilde{N}_d(s) = 0 \tag{8.40}$$

With $rank\{\tilde{N}_d(s)\} = rank\{G_d(s)\}$ (because $\tilde{M}_d(s)$ is a nonsingular matrix) and the rank condition (8.26), we know that $rank\{\tilde{N}_d(s)\} \leq m - rank\{G_f(s)\}$, i.e. $\tilde{N}_d(s)$ is a row rank-deficient matrix. For such a matrix, there always exists a matrix $Q_1(s)$ which satisfies the following equation:

$$Q_1(s)\tilde{N}_d(s) = 0 \tag{8.41}$$

Now, the post-filter transfer matrix for achieving disturbance de-coupling is thus given by:

$$Q(s) = Q_2(s)Q_1(s) \tag{8.42}$$

where $Q_2(s)$ is a arbitrary matrix which makes sure that $Q(s)$ is a RH_∞ matrix. Actually, any matrix in the left null space of $\tilde{N}_d(s)$ satisfies the equation (8.41). The technique of determining the left null space of $\tilde{N}_d(s)$ can be found in textbooks on multivariable control, e.g. Vardulakis (1991).

When the perfect disturbance de-coupling cannot be achieved, we should consider the use of optimal disturbance de-coupling. That is, to solve the following minimization problem.

$$\min J_d = \inf_{Q(s)} \|Q(s)\tilde{N}_d(s)\|_\infty \tag{8.43}$$

This problem can be treated as a special case of the model-matching problem.

8.2.2.3 Discussions on PFDI.

Remarks:

(1) The algorithms discussed here require the manipulation of transfer matrices. This is greatly facilitated through the use of symbolic analysis software such as Maple or Mathematica. Further research is needed to find design algorithms which are based on the use of numerical manipulation of state space model matrices for the ease of implementation.

(2) Conditions (8.24) & (8.25) in Lemma 8.3 are solved here separately. To achieve PFDI, these two conditions should be solved together. The effective design algorithm for simultaneously solving these two conditions has not been reported yet in the literature.

(3) The simultaneous solution of conditions (8.24) & (8.25) cannot always be found due to lack of design freedom. That is to say there is not enough number of independent measurements. Therefore, PFDI is not always attempted in practice.

The most important aspect of FDI is disturbance de-coupling because the fault isolation problem can be solved by other ways such as directional and structured residuals discussed in Section 2.7. To design structured residual sets, some faults can be treated as the same way as disturbances. We can achieve this by re-writing the system input-output model as:

$$y(s) = G_u(s)u(s) + G_f^1(s)f^1(s) + G_f^2(s)f^2(s) + G_d(s)d(s) \tag{8.44}$$

where $f^1(s)$ and $f^2(s)$ are fault vectors with each contains some of all faults to be detected. If we want to design a residual which is sensitive to $f^1(s)$ and insensitive to $f^2(s)$, we can treat $f^2(s)$ as the disturbance in the residual generator design. In this case, Eq.(8.44) can be re-written as

$$\begin{aligned} y(s) &= G_u(s)u(s) + G_f^1(s)f^1(s) + [G_f^2(s) \;\; G_d(s)] \begin{bmatrix} f^2(s) \\ d(s) \end{bmatrix} \\ &= G_u(s)u(s) + G_f^1(s)f^1(s) + \bar{G}_d(s)\bar{d}(s) \end{aligned}$$

where $\bar{d}(s)$ is the "disturbance" which is to be decoupled from the residual.

From the discussion above, we can conclude that the optimal FDI performance such as disturbance de-coupling can be achieved by a properly designed weighting (post-filter) transfer matrix $Q(s)$. The first step in designing an optimal residual generator is to design a full-order observer to generate the primary residual $\tilde{r}$. The only requirement for this observer is stability and perhaps some eigenvalue constraints for achieving fast transient response. After this step, the observer can be fixed. The second step is to find the weighting transfer matrix $Q(s)$ for satisfying robustness and isolability conditions. This seems to be "contradiction" with the eigenstructure assignment (EA) approach discussed in Chapter 4. In the EA approach, the gain matrix must be able to assign required left/right eigenvectors for the purpose of disturbance de-coupling. This "contradictory" argument will be clarified in follows. First of all, let us examine the relationship between two sets of left-coprime factorization of the same transfer matrix.

Lemma 8.4 *(Ding and Guo, 1997): Given*

$$G_u(s) = \tilde{M}_1^{-1}(s)\tilde{N}_1(s) = \tilde{M}_2^{-1}(s)\tilde{N}_2(s) \tag{8.45}$$

where

$$\tilde{M}_1(s) = I - C(sI - A + K_1C)^{-1}K_1 \in RH_\infty \quad (8.46)$$
$$\tilde{M}_2(s) = I - C(sI - A + K_2C)^{-1}K_2 \in RH_\infty \quad (8.47)$$
$$(8.48)$$

and K_1 and K_2 are matrices that ensure the stability of matrices $A - K_1C$ and $A - K_2C$. Then, there exists a RH_∞ matrix $Q_0(s)$ which makes:

$$Q_0(s)\tilde{M}_1(s) = \tilde{M}_2 \quad (8.49)$$

and, furthermore this matrix can be expressed as

$$Q_0(s) = I + C(sI - A + K_2C)^{-1}(K_1 - K_2) \quad (8.50)$$

From this Lemma, it can be seen that dynamics of a diagnostic observer can be totally changed by using a weighting transfer matrix. Therefore, any residual generator (including observer-based) can be parameterized by the weighting transfer matrix $Q(s)$. This is summarized by Theorem 8.3.

Theorem 8.3 *(Ding and Guo, 1997): Assume that the matrix K_0 ensures the stability of matrix $A - K_0C$. Then all residual generators can be parameterized by*

$$\begin{aligned} r(s) &= Q(s)(\tilde{M}_0(s)y(s) - \tilde{N}_0(s)u(s)) \\ &= Q(s)\tilde{M}_0(s)(G_d(s)d(s) + G_f(s)f(s)) \quad (8.51) \end{aligned}$$

where $\{\tilde{M}_0(s), \tilde{N}_0(s)\}$ is a left-coprime factorization pair of the transfer matrix $G_u(s) = C(sI - A)^{-1} + D$ defined by

$$\tilde{M}_0(s) = I - C(sI - A + K_0C)^{-1}K_0 \quad (8.52)$$
$$\tilde{N}_0(s) = D + C(sI - A + K_0C)^{-1}(B - K_0D) \quad (8.53)$$

and $Q(s)$ is the parameterized matrix belonging to RH_∞.

This theorem is a statement that the residual performance is independent of the observer gain matrix. This is because any required performance can be achieved by the weighting matrix $Q(s)$. As an example, suppose the gain matrix K_{EA} and the static weighting matrix Q_{EA} of a disturbance decoupled diagnostic observer has been designed using the eigenstructure assignment approach. A simple stable observer with the gain matrix K_S is used to generate the primary residual. According to Lemma 8.4 and Theorem 8.3, this primary residual should be post-processed by the dynamic weighting matrix $Q_S(s)$ in order to make the processed residual has the same FDI performance as provided by the diagnostic observer designed by the EA approach.

$$Q_S(s) = Q_{EA}[I + C(sI - A + K_{EA}C)^{-1}(K_S - K_{EA})]$$

From Eq.(8.51) in the Theorem 8.3, we can see that the main goal of the first design stage (generating primary residual using a full-order observer) is to eliminate the input effect on the residual. The optimal residual performance is achieved in the second design stage using the weight matrix $Q(s)$. That is to say the system input $u(t)$ can be neglected in the design of optimal residual for FDI. This philosophy will be adopted in Sections 8.3 & 8.4. It should be pointed out that the system input can only be neglected when there are no modeling errors in $G_u(s)$.

8.2.3 *Design of optimal residuals*

From the discussion in the previous part of this section, we know that the residual response of the residual generator (8.2) to the system with fault and disturbance (Eq. (8.15)) is given by:

$$r(s) = Q(s)\tilde{N}_f(s)f(s) + Q(s)\tilde{N}_d(s)d(s) \tag{8.54}$$

where RH_∞ transfer matrices $\tilde{N}_f(s)$ and $\tilde{N}_d(s)$ are defined by Eq.(8.2) and Eq.(8.37). Ideally, the disturbance effect should be totally de-coupled from the residual, i.e. we should design the weighting matrix $Q(s)$ which satisfies the perfect disturbance de-coupling condition:

$$Q(s)\tilde{N}_d(s) = 0 \tag{8.55}$$

subject to

$$Q(s)\tilde{N}_f(s) \neq 0 \tag{8.56}$$

The condition (8.55) is only achievable when the number of independent disturbances is smaller than the number of independent measurements. It should be pointed out that the same condition is also true for achieving perfect disturbance de-coupling in the time domain design (see Chapters 3 & 4).

When perfect disturbance de-coupling is not achievable, we have to find some optimal approximation, in the sense of satisfying certain design criteria. Simply, the fault effect on the residual has to be made substantially larger than the disturbance effect on the residual. This can be achieved by maximizing the following performance index

$$J = \frac{||Q(s)\tilde{N}_f(s)||}{||Q(s)\tilde{N}_d(s)||} \tag{8.57}$$

where $||\cdot||$ denotes a norm of the matrix concerned. The use of different norms will result different design methods and also different performance indices. A H_2–norm measures the amplification of a transfer matrix which maps inputs that are either fixed or have a fixed power spectrum into the output, whilst a H_∞–norm measures the amplification of a transfer matrix that maps the input with finite energy into the output (Frank and Ding, 1994; Maciejowski, 1989). Since we do not make any assumption about the fault f and disturbance d in

FDI apart from finite energy, the H_∞–norm is better suited for our optimization problem.

Note that the optimal robust fault detection performance can also be defined in time domain. This has been studied in Chapter 7. The performance index adopted in Chapter 7 uses a ratio of disturbance effect to fault effect. Here we use the inverse ratio: fault effect to disturbance effect, however they are actually equivalent as in Chapter 7 a minimization problem is considered.

If the L_2 norm is used in Eq.(8.57) and the weight matrix $Q(s)$ is a one-dimensional vector $q(s)$, the optimization problem becomes

$$J_{opt} = \max_{q(s)} \frac{||q(s)\tilde{N}_f(s)||_2}{||q(s)\tilde{N}_d(s)||_2} \tag{8.58}$$

It can be shown that the maximization of J leads to the solution of a generalized eigenvalue-eigenvector problem in the frequency domain (Frank, 1991a; Frank and Ding, 1997):

$$v(j\omega)[\tilde{N}_f(j\omega)\tilde{N}_f^T(-j\omega) - \lambda(\omega)\tilde{N}_d(j\omega)\tilde{N}_d^T(-j\omega)] = 0 \tag{8.59}$$

If $\bar{v}(j\omega)$ is the generalized eigenvector corresponding to the largest generalized eigenvalue $\bar{\lambda}(\omega)$, the solution of the frequency domain optimization problem (8.58) finally yields (Frank, 1991a; Frank and Ding, 1997):

$$q_{opt}(s) = \bar{v}(s)\ , \qquad J_{opt} = \sup_{\omega}(\bar{\lambda}(\omega)) \tag{8.60}$$

A detailed discussion of the generalized eigenvalue and eigenvector problem can be found in Section 7.3.2.3.

As an alternative approach, consider the following optimization problem:

$$\max J = \sup_{Q(s)} \frac{||Q(s)\tilde{N}_f(s)||_\infty}{||Q(s)\tilde{N}_d(s)||_\infty} \tag{8.61}$$

which was solved by Ding *et al* (Ding et al., 1993; Frank and Ding, 1994; Frank and Ding, 1997) and Qiu and Gertler (1993) using H_∞ theory.

Frank and Ding (1994) demonstrated that the optimization problem (8.61) is equivalent to

$$\max J = \sup_{Q(s)} ||Q(s)\tilde{N}_f(s)||_\infty \tag{8.62}$$

subject to

$$||Q(s)\tilde{N}_d(s)||_\infty \leq 1 \tag{8.63}$$

in the sense that the maximum values in both cases are equal. Frank and Ding (1994) pointed out that the optimal solution $Q(s)$ of the problem (8.61) is not unique. The optimal solution to problem (8.62), in fact, provides a nominal solution for the problem (8.61).

For the case when $\tilde{N}_d(s)$ has no zeros on the imaginary axis, the optimal solution was given by Frank and Ding (1997).

Theorem 8.4 *(Frank and Ding, 1997): When $\tilde{N}_d(s)$ has no zeros on the imaginary axis, the optimal solution for the problem (8.62) is*

$$Q_{opt}(s) = Q_0(s)G_{do}^{-1}(s) \ , \quad J_{opt} = ||G_{do}^{-1}(s)G_{fo}(s)|| \qquad (8.64)$$

where $G_{do}(s)$ and $G_{fo}(s)$ are outer matrices of the inner-outer factorization (Frank and Ding, 1994; Francis, 1987) of $\tilde{N}_d(s)$ and $\tilde{N}_f(s)$ respectively, i.e.

$$\tilde{N}_d(s) = G_{do}(s)G_{di}(s) \qquad (8.65)$$
$$\tilde{N}_f(s) = G_{fo}(s)G_{fi}(s) \qquad (8.66)$$

The matrix $Q_0(s)$ in Eq.(8.64) satisfies:

$$\begin{cases} Q_0^T(-j\omega)Q_0(j\omega) \leq I \ , & \text{for all } \omega \\ Q_0^T(-j\omega_0)Q_0(j\omega_0) = I \end{cases} \qquad (8.67)$$

where

$$\omega_0 \ : \ ||G_{do}^{-1}(s)G_{fo}(s)||_\infty = \max_\omega \left\{\bar{\sigma}\left[G_{do}^{-1}(j\omega)G_{fo}(j\omega)\right]\right\} = \bar{\sigma}\left[G_{do}^{-1}(j\omega_0)G_{fo}(j\omega_0)\right] \qquad (8.68)$$

This optimal solution has a significant physical meaning (Frank and Ding, 1994; Frank and Ding, 1997). In fact, the matrix-valued division $G_{do}^{-1}(s)G_{fo}(s)$ compares the difference between the transfer matrices $\tilde{N}_d(s)$ and $\tilde{N}_f(s)$, and its H_∞–norm defines a measure of this difference. To reach the maximal difference, which is achievable at the frequency ω_0, a bandpass filter $Q_0(s)$ is used. As a result, this strategy allows the exploitation of all kinds of available frequency information no matter whether this is inherent in the unknown input signal d or within the structure of the path from d to y. In other words, it allows the involvement of frequency spectrum analysis, which has become a standard and powerful tool in practice, to improve the degree of robustness of the residual generator (Frank and Ding, 1997).

The case when $\tilde{N}_d(s)$ has zeros on the imaginary axis, the optimal solution of the problem (8.62) is relatively difficult to find. The technique developed by Frank and Ding (1994) is introduced here.

Theorem 8.5 *(Frank and Ding, 1994): If $\tilde{N}_d(s)$ has zeros on the imaginary axis and its extended inner-outer factorization (Frank and Ding, 1994; Francis, 1987) is given by:*

$$\tilde{N}_d(s) = G_{do}(s)G_{dz}(s)G_{di}(s) \qquad (8.69)$$

with $G_{do}(s)$ is outer, $G_{di}(s)$ is inner and $G_{dz}(s)$ having all the zeros on the imaginary of $\tilde{N}_d(s)$ as its zeros, then the optimal solution of the problem (8.62) is

$$Q_{opt}(s) = Q_0(s)G_{do}^{+}(s) \ , \quad J_{opt} = ||G_{do}^{+}(s)G_{dz}(s)|| \qquad (8.70)$$

where the matrix $Q_0(s)$ in Eq.(8.70) satisfies:

$$\begin{cases} G_{dz}^T(-j\omega)Q_0^T(-j\omega)Q_0(j\omega)G_{dz}(j\omega) \leq I \ , & \text{for all } \omega \\ G_{dz}^T(-j\omega_0)Q_0^T(-j\omega_0)Q_0(j\omega_0)G_{dz}(j\omega_0) = I \end{cases} \qquad (8.71)$$

The proof of this theorem is very involved. Frank and Ding (1994) provided a procedure for proving this theorem. This proof also serves as the design procedure for finding optimal solution $Q_0(s)$. The design algorithm of Frank and Ding (1994) is given here.

Step 1: Find the extended inner-outer factorization of $\tilde{N}_d(s)$:

$$\tilde{N}_d(s) = G_{do}(s)G_{dz}(s)G_{di}(s)$$

where $rank\{G_{dz}(s)\} = k$ (which is the number of zeros of $\tilde{N}_d(s)$ on imaginary axis).

Step 2: Determine the Smith-McMillan form of the transfer matrix $G_{dz}(s)$:

$$G_{dz}(s) = U(s)G_s(s)V(s)$$

where $V(s)$, $U(s)$ are RH_∞-invertible and

$$G_s(s) = diag\{g_1(s), \cdots, g_k(s)\}$$

with

$$g_i(s) = \hat{g}_i(s)\prod_j (s^2 + \alpha_{ij})\prod_j \frac{1}{s+\beta_{ij}}\prod_j \frac{s}{s+\gamma_{ij}} \in RH_\infty\ , \quad ,i = 1, \cdots, k$$

where $\hat{g}_i^{-1}(s) \in RH_\infty$.

Step 3: Set $Q_2(s)$ according to

$$Q_2(s) = V^{-1}(s)Q_1(s)U^{-1}(s)$$

where

$$Q_1(s) = diag\{q_1(s), \cdots, q_k(s)\}$$

with

$$q_i(s) = \prod_{j\neq i}^{k} g_j(s)$$

Step 4: Determine $Q_3(s)$ using the formula:

$$Q_3(s) = q(s)I \in RH_\infty$$

where

$$q(s) = \hat{q}(s)\prod_{i=1}^{k}\left(\prod_j q_{ij}^{(1)}(s)\prod_j q_{ij}^{(2)}(s)\prod_j q_{ij}^{(3)}(s)\right)$$

$$\hat{q}(s) = \left(\prod_{i=1}^{k}\hat{g}_i(s)\right)^{-1}$$

$$q_{ij}^{(1)} = k_{ij1} \frac{1}{s^2 + \beta_{ij1}^{(1)} + \beta_{ij2}^{(1)}}$$

$$q_{ij}^{(2)}(s) = k_{ij2} \frac{s + \beta_{ij1}^{(2)}}{s + \beta_{ij2}^{(2)}}$$

$$q_{ij}^{(3)}(s) = k_{ij3} \frac{s + \beta_{ij1}^{(3)}}{s + \beta_{ij2}^{(3)}}$$

and the parameters in $q_{ij}^{(1)}(s)$, $q_{ij}^{(2)}(s)$ and $q_{ij}^{(3)}(s)$ are determined from the following equations:

$$0 < k_{ij1} < 1$$

$$\beta_{ij2}^{(1)} = \sqrt{(1 - k_{ij1}^2)\omega_0^4 + k_{ij1}^2 \alpha_{ij}^2}$$

$$\beta_{ij1}^{(1)} = \sqrt{2(\beta_{ij2}^{(1)} - (1 - k_{ij1}^2)\omega_0^2 + k_{ij1}^2 \alpha_{ij})}$$

$$k_{ij2}^2 > \beta_{ij}^2 - 2\omega_0^2$$

$$\beta_{ij1}^{(2)} = \sqrt{\frac{\beta_{ij}^2(2\omega_0^2 - k_{ij2}^2 - \beta_{ij}^2) - \omega_0^4}{k_{ij2}^2}}$$

$$\beta_{ij2}^{(2)} = \sqrt{2\omega_0^2 + k_{ij2}^2 - \beta_{ij}^2}$$

$$k_{ij3} < 1$$

$$\beta_{ij2}^{(3)} = \sqrt{\frac{(1 - k_{ij3}^2)\omega_0^4}{\gamma_{ij}^2}}$$

$$\beta_{ij1}^{(3)} = \sqrt{\frac{\gamma_{ij}^2 + (\beta_{ij2}^{(3)})^2 - 2(1 - k_{ij3}^2)\omega_0^2}{k_{ij3}^2}}$$

Step 5: Finally, the optimal weighting transfer matrix is:

$$Q(s) = Q_3(s) Q_2(s) G_{do}^{-1}(s)$$

The design algorithm of Frank and Ding (1994), presented here, is very complicated. It is not easy for an engineer to understand without a deep knowledge of robust control. One could argue the real usefulness of such a complicated algorithm without any assistance of a design software. Unfortunately, the design software has not been seen in the open literature yet. Actually, we can approach the problem from a numerical optimization perspective.

The problem of finding an optimal weighting transfer matrix $Q(s)$ to maximize the performance index (8.61) should be possible to solve by using a numerical search algorithm. For example, the genetic algorithm described in Chapter 6 is an excellent numerical search algorithm. The only problem remained is to parameterize the weighting transfer matrix $Q(s)$ because the numerical search can only manipulate numerical value rather symbolic variables such as s in $Q(s)$. The transfer matrix $Q(s)$ should be realizable, therefore a straightforward parameterization is the state space realization, i.e

$$Q(s) = \left[\begin{array}{c|c} A_Q & B_Q \\ \hline C_Q & D_Q \end{array}\right]$$

with the constraint condition of A_Q being a stable matrix. The dimension of A_Q can be treated as a variable which is to be found by the optimization. The way of parameterization of $Q(s)$ for the numerical optimization is a problem worthy of further investigation.

The problem of maximization of fault effects and minimization of disturbance effects is solved in Chapter 6 using a combination of the method of inequalities and the genetic algorithm. This multi-objective problem is also tackled in Chapter 7, however the performance indices are mixed together to form a single-objective optimization problem (8.61).

Remarks:

(1) Qiu and Gertler (1993) pointed out that the performance index defined in Eq. (8.61) has an upper bound (given in Eq. (8.64)) as well as a lower bound. Both bounds are independent of the weighting matrix $Q(s)$. No matter what design method is used, the *achievable* FDI performance is determined by the system property. The optimal design is to find a residual generator which actually achieves the optimal performance.

(2) The maximization of the performance index defined in Eq. (8.61) does not necessarily guarantee the best FDI performance (Qiu and Gertler, 1993) because the optimization is only achieved at the optimal frequency ω_0. The effect of fault (measured by $\bar{\sigma}(Q\tilde{N}_f)$) may be smaller than the effect of disturbance (measured by $\bar{\sigma}(Q\tilde{N}_f)$) at other frequencies even if J has been maximized. This, however, does not claim that the performance (8.61) is not useful. It may provide, if used with caution, a guideline to improve the optimization procedure. However, both lower and upper bounds of the performance index should be checked during the design (Qiu and Gertler, 1993).

(3) To guarantee the optimal FDI performance over the frequency range which we are interested in, the frequency range which used in the optimal design should be chosen carefully.

(4) The global optimization, i.e. to achieve an optimum over the whole frequency range rather than at the optimal frequency, may not be always possible. To overcome this difficulty, Qiu and Gertler (1993) suggested that the so-called "locally robust" fault detection scheme by introducing frequency weighting on fault and disturbance effects. The performance index is defined as

$$J = \frac{\|Q(s)W_f(s)\tilde{N}_f(s)\|_\infty}{\|Q(s)W_d(s)\tilde{N}_d(s)\|_\infty} \tag{8.72}$$

where $W_f(s)$ and $W_d(s)$ are RH_∞ weighting transfer matrices.

(5) The optimal residual generator design method discussed here considers only the disturbance. To tackle the robustness problem against modeling errors, Frank and Ding (1994) suggested to solve it at the residual evaluation stage by choosing a suitable fixed/adaptive threshold. The problem can, of course, be solved by representing modeling errors as a disturbance with an approximate distribution matrix using the methods introduced in Chapter 5.

8.3 Robust Residual Generation through Standard H_∞ Filtering Formulations

8.3.1 Robust residual generation with disturbance attenuation

Let us first consider the disturbance attenuation problem, i.e. to design a residual generator where the effects of disturbances on the residual are minimized. That can be achieved by designing an optimal H_∞ filter whose disturbance sensitivity is minimized. To tackle this problem, let us use the following system state space model:

$$\begin{cases} \dot{x}(t) &= Ax(t) + E_1 d(t) \\ y(t) &= Cx(t) + E_2 d(t) \end{cases} \tag{8.73}$$

or the input-output model

$$y(s) = G_d(s)d(s) \tag{8.74}$$

where

$$G_d(s) = \left[\begin{array}{c|c} A & E_1 \\ \hline C & E_2 \end{array}\right] \tag{8.75}$$

Faults are not considered in models (8.73) & (8.74), this is because only the disturbance attenuation is considered in a robust FDI filter design here. The control input is also not considered in models (8.73) & (8.74), this is because the control input does not affect the residual response when there are no modeling errors in the system transfer matrix $G_u(s)$. This problem has been discussed in Section 8.2.2. Let us recall the general residual generator formulation defined in Section 2.5:

$$r(s) = H_y(s)y(s) + H_u(s)u(s) \quad ; \quad s.t. \quad H_y(s)G_u(s) + H_u(s) = 0 \tag{8.76}$$

If we assume that $u(s) = 0$ and design the transfer matrix $H_y(s)$, then the transfer matrix $H_u(s)$ can be determined as $H_u(s) = -H_y(s)G_u(s)$. In summary, the control input can be neglected in the residual generator design if there are no modeling errors in the transfer matrix $G_u(s)$.

To generate a residual, we need to estimate an auxiliary signal $z(t)$. The residual signal is the difference between the real and estimated values of this auxiliary signal. Since we need to compare the estimation with the real value of this auxiliary signal, the real value of $z(t)$ should be available. Therefore, we can define the weighted output as an auxiliary signal which is to be estimated by H_∞ filter.

$$z(t) = My(t) \tag{8.77}$$

The residual signal is thus:

$$r(t) = z(t) - \hat{z}(t) \tag{8.78}$$

The design requirement for robust residual generation is to minimize the effect of the disturbance on the residual, i.e. to minimize the following performance index.

$$J_d := \|G_{rd}(s)\|_\infty = \sup_{0<\|d\|_2<\infty} \frac{\|r\|_2}{\|d\|_2} \tag{8.79}$$

where $G_{rd}(s)$ is the transfer matrix from disturbance to residual. The idea of estimating the auxiliary $z(t)$ is illustrated by Fig. 8.2. Where the evaluation signal $\tilde{z}(t)$ which defined by the following equation is used to measure the estimation quality.

$$\tilde{z}(t) = z(t) - \hat{z}(t) \tag{8.80}$$

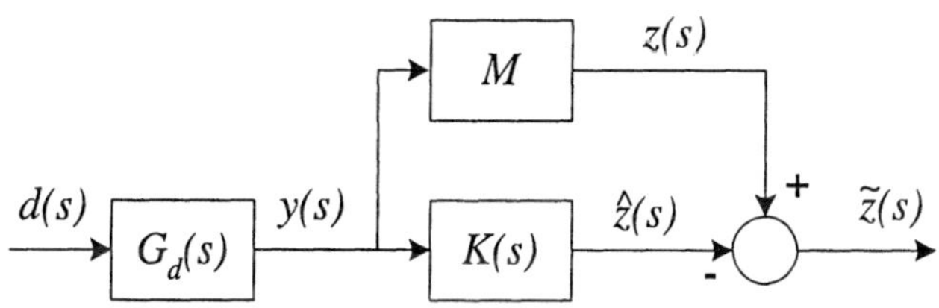

Figure 8.2. Formulation of disturbance attenuation

The system illustrated by Fig. 8.2 can be reformulated into a standard H_∞ problem as given in Fig. 8.3.

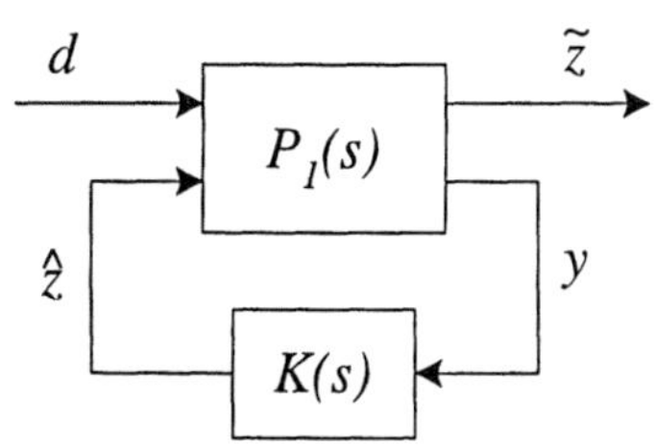

Figure 8.3. Standard problem formulation of disturbance attenuation

The "equivalent" transfer matrix $P_1(s)$ for the standard problem of Fig. 8.3 is given by:

$$P_1(s) = \begin{bmatrix} MG_d(s) & -I \\ G_d(s) & 0 \end{bmatrix} = \left[\begin{array}{c|cc} A & E_1 & 0 \\ \hline MC & ME_2 & -I \\ C & E_2 & 0 \end{array}\right] \tag{8.81}$$

where the transfer matrix $P_1(s)$ is defined as:

$$\begin{bmatrix} \tilde{z} \\ y \end{bmatrix} = P_1(s) \begin{bmatrix} d \\ \hat{z} \end{bmatrix} \tag{8.82}$$

The sensitivity transfer matrix for this standard H_∞ formulation is given by:

$$G_{\tilde{z}d}(s) = LFT(P_1(s), K(s)) = MG_d(s) - K(s)G_d(s) \tag{8.83}$$

where LFT denotes the linear fraction transformation. The standard H_∞ filtering problem is to find a filter $K(s) \in RH_\infty$ such that:

$$\|G_{\tilde{z}d}(s)\|_\infty < \gamma \tag{8.84}$$

where γ (> 0) is a design parameter named as the performance bound. The filtering problem can be regarded as a special H_∞ problem. Compared with the control problems there is no internal stability requirement in the filtering problem. Therefore, standard H_∞ control techniques can be used to solved the problem (8.84). For example, the solution for a general H_∞ filtering problem can be found in Zhou et al. (1996, pp.462-464), Grimble and Elsayed (1990), Shaked and Theodor (1992a) and Yaesh and Shaked (1992). Note that the problem (8.84) can also be treated as a model-matching problem.

A simplified version for the solution of H_∞ robust residual generation, developed by Edelmayer, Bokor and Keviczky (1994; 1996; 1997a), is introduced here.

Theorem 8.6 *(Edelmayer, Bokor and Keviczky, 1994; Edelmayer, Bokor and Keviczky, 1996; Edelmayer, Bokor and Keviczky, 1997a): When $E_2 = 0$ (i.e. no disturbance in the output equation), (A, E_1) is a stabilizable pair and (C, A) is a detectable pair, then the optimal filter $K(s)$ which satisfies (8.84) is given by:*

$$K(s) = \left[\begin{array}{c|c} A - YC^TC & YC^T \\ \hline MC & 0 \end{array}\right] \tag{8.85}$$

where Y is the positive definite solution of the following algebraic Riccati equation:

$$AY + YA^T - Y\left(C^TC - \frac{1}{\gamma^2}(MC)^T(MC)\right)Y + E_1E_1^T = 0 \tag{8.86}$$

The design parameter γ determines the disturbance attenuation performance. The smaller this design parameter, the better the performance of disturbance attenuation. However, if this parameter is too small, the solution of (8.86) may not exist. Therefore, the design procedure starts with sufficiently large γ_0. Then, the performance bound is gradually reduced until a solution for (8.86) can not be found. The minimum design parameter γ_{min} is achievable performance bound. This procedure is known as *γ–iteration.*

The optimal filter (8.85) can be implemented using the state space equations in which the control input (see Eq. (8.1)) can also be included.

$$\begin{cases} \dot{\hat{x}}(t) &= (A - K_oC)\hat{x}(t) + (B - K_oD)u(t) + K_oy(t) \\ \hat{z}(t) &= MC\hat{x}(t) + MDu(t) \\ r(t) &= My(t) - \hat{z}(t) \end{cases} \tag{8.87}$$

where the observer gain matrix is $K_o = YC^T$.

The design problem discussed here gives a solution of the observer gain matrix K_o for achieving optimal disturbance de-coupling. However, it does not give any indication as to how the residual weighting matrix M can be determined. Comparing this with the eigenstructure assignment studied in Chapter 4 in which the observer gain and residual weighting matrices are designed simultaneously in order to achieve disturbance de-coupling. If we have a system satisfying the disturbance de-coupling condition, the eigenstructure assignment can be used to design a residual with disturbances totally decoupled. If the H_∞ filtering method is used, the disturbances are attenuated not decoupled from the residual. For systems of this kind, the eigenstructure assignment is definitely the better choice. For systems that do not satisfy disturbance decoupling conditions, the approximate eigenstructure assignment can be used. However, the residual performance can be neither quantified or guaranteed. The H_∞ filtering technique, on other hand, can give a quantitative upper band on the disturbance attenuation performance.

8.3.2 *Fault estimation*

To detect faults reliably, the residual should be designed to have maximum sensitivity against faults. To concentrate on the fault sensitivity, let us first ignore the disturbance. The system is described by the following state space model:

$$\begin{cases} \dot{x}(t) &= Ax(t) + R_1f(t) \\ y(t) &= Cx(t) + R_2f(t) \end{cases} \tag{8.88}$$

or the input-output model

$$y(s) = G_f(s)f(s) \tag{8.89}$$

where

$$G_f(s) = \left[\begin{array}{c|c} A & R_1 \\ \hline C & R_2 \end{array}\right] \tag{8.90}$$

Using the same argument as in Section 8.3.1, the control input is not considered here. The following residual generator can be designed to generate the residual signal $r(t)$.

$$r(s) = H_y(s)y(s) \tag{8.91}$$

Reliable detection can be ensured if we can solve the optimization problem:

$$\max_{H_y(s)} \left\{ \inf_{0<\|f\|_2<\infty} \frac{\|r\|_2}{\|f\|_2} \right\} \tag{8.92}$$

If the transfer matrix between the residual and the fault is $G_{rf}(s)$, the optimization problem (8.92) is roughly equivalent to:

$$\max_{H_y(s)} \left\{ \inf_{\omega} \underline{\sigma} \left[G_{rf}(j\omega) \right] \right\} \tag{8.93}$$

where $\underline{\sigma}(\cdot)$ denotes the smallest singular value.

The maximization problem (8.93) cannot be easily formulated in an H_∞ setting. To solve this problem, a new performance index should be introduced. If we can make the residual as close to the fault as possible, then the residual can provide all information about the fault. That is to say the fault sensitivity has been maximized. A new performance index is given here:

$$J_f := \|G_{rf}(s) - I\|_\infty = \sup_{0<\|f\|_2<\infty} \frac{\|r - f\|_2}{\|f\|_2} \tag{8.94}$$

The minimization of the performance index in Eq. (8.94) actually solved the fault estimation problem. The idea of fault estimation is illustrated in Fig. 8.4 where $\hat{z}(t)$ is an estimation of fault and $\tilde{z}(t)$ is the signal which is used to evaluate the estimation quality.

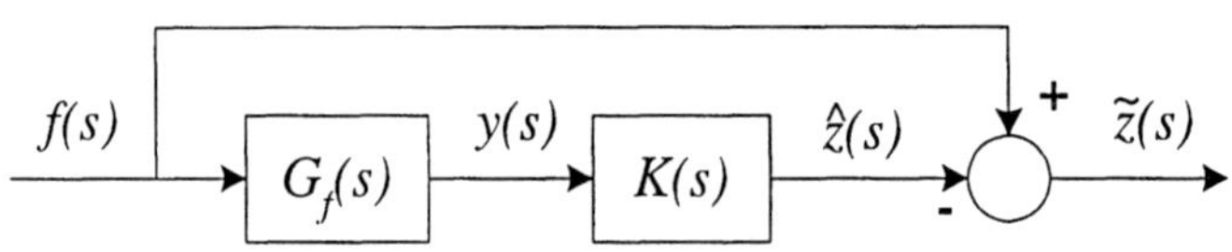

Figure 8.4. Formulation of fault estimation

The system illustrated by Fig. 8.4 can be reformulated into a standard H_∞ problem as given in Fig. 8.5.

The "equivalent" transfer matrix $P_2(s)$ for the standard problem of Fig. 8.5 is given by:

$$P_2(s) = \begin{bmatrix} I & -I \\ G_f(s) & 0 \end{bmatrix} = \left[\begin{array}{c|cc} A & R_1 & 0 \\ \hline 0 & I & -I \\ C & R_2 & 0 \end{array}\right] \tag{8.95}$$

where the transfer matrix $P_2(s)$ is defined as:

$$\begin{bmatrix} \tilde{z} \\ y \end{bmatrix} = P_2(s) \begin{bmatrix} f \\ \hat{z} \end{bmatrix} \tag{8.96}$$

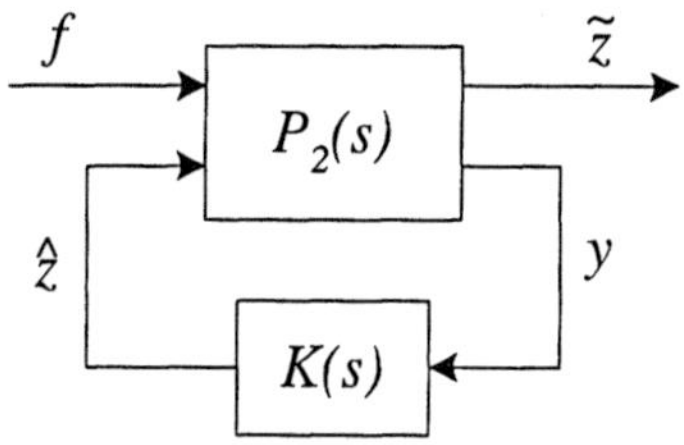

Figure 8.5. Standard problem formulation of fault estimation

The sensitivity transfer matrix for this standard H_∞ formulation is given by:

$$G_{\tilde{z}f}(s) = LFT(P_2(s), K(s)) = I - K(s)G_f(s) \tag{8.97}$$

The standard H_∞ fault estimation is thus to find a filter $K(s) \in RH_\infty$ such that:

$$||G_{\tilde{z}f}(s)||_\infty < \gamma \tag{8.98}$$

The problem can be solved via the standard H_∞ techniques.

Sometimes, it is *not practical* to estimate the fault itself. We can then consider the estimation of a filtered version of the fault which can give us some indication about the fault itself. The problem is thus changed into the estimation of $\bar{f}(t)$ which is the filtered version of the fault, i.e

$$\bar{f}(s) = T(s)f(s) \tag{8.99}$$

where $T(s)$ is a RH_∞ transfer matrix which is normally set as diagonal just like the one used in Eq. (8.24). The filtered fault estimation can then be defined as the minimization of the following performance index:

$$J_f := ||G_{rf}(s) - T(s)||_\infty = \sup_{0<||f||_2<\infty} \frac{||r - \bar{f}||_2}{||f||_2} \tag{8.100}$$

This optimization problem can be formulated according to Fig. 8.6 where $\hat{z}(t)$ is an estimation of fault and $\tilde{z}(t)$ is the signal which is used to evaluate the estimation quality.

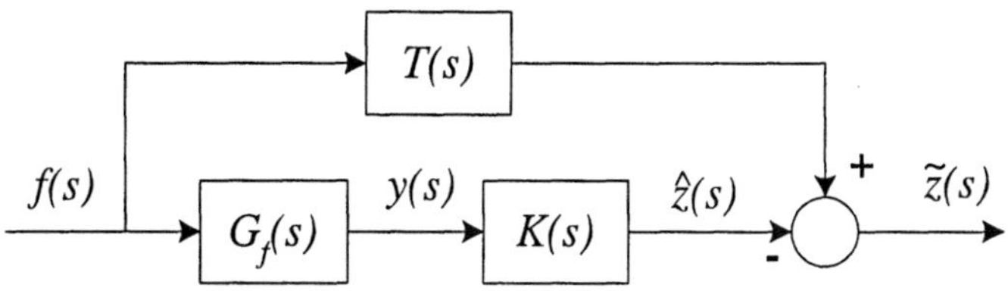

Figure 8.6. Formulation of filtered fault estimation

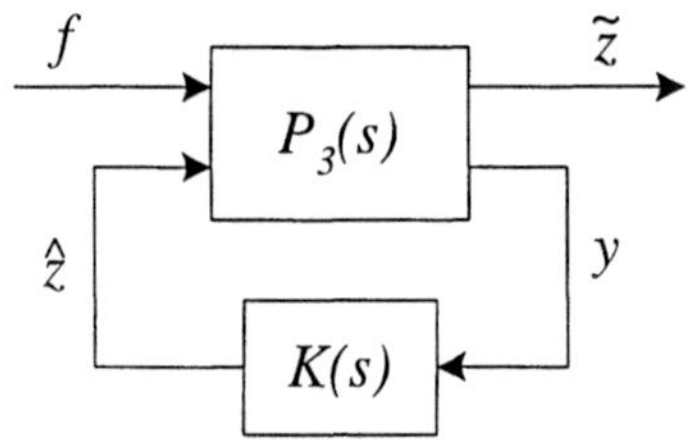

Figure 8.7. Standard problem formulation of filtered fault estimation

The system illustrated by Fig. 8.6 can be reformulated into a standard H_∞ problem as given in Fig. 8.7.

If the state space realization of transfer matrix $T(s)$ is

$$T(s) = \left[\begin{array}{c|c} A_T & B_T \\ \hline C_T & D_T \end{array}\right] \tag{8.101}$$

the "equivalent" transfer matrix $P_3(s)$ for the standard problem of Fig. 8.7 is given by:

$$P_3(s) = \left[\begin{array}{cc} T(s) & -I \\ G_f(s) & 0 \end{array}\right] = \left[\begin{array}{cc|cc} A_T & 0 & B_T & 0 \\ 0 & A & R_1 & 0 \\ \hline C_T & 0 & D_T & -I \\ 0 & C & R_2 & 0 \end{array}\right] \tag{8.102}$$

The sensitivity transfer matrix for this standard H_∞ formulation is given by:

$$G_{\tilde{z}f}(s) = LFT(P_3(s), K(s)) = T(s) - K(s)G_f(s) \tag{8.103}$$

The estimation of the filtered fault within H_∞ formulation is to find a filter $K(s) \in RH_\infty$ such that:

$$||G_{\tilde{z}f}(s)||_\infty < \gamma \tag{8.104}$$

The problem can be solved by the standard H_∞ techniques. Note that the problem can also be regarded as a model-matching problem and solved via methods developed in robust control theory.

8.3.3 Fault estimation with disturbance attenuation

The disturbance attenuation and fault estimation problems have been treated separately so far. In practice, they have to be considered together. That is to say, we need to estimate the result for systems with disturbances. A system with both fault and disturbance terms is modeling by the state space model:

$$\left\{ \begin{array}{rcl} \dot{x}(t) & = & Ax(t) + R_1 f(t) + E_1 d(t) \\ y(t) & = & Cx(t) + R_2 f(t) + E_2 d(t) \end{array} \right. \tag{8.105}$$

or alternatively by the input-output model:

$$y(s) = G_f(s)f(s) + G_d(s)d(s) \tag{8.106}$$

where

$$G_f(s) = \left[\begin{array}{c|c} A & R_1 \\ \hline C & R_2 \end{array}\right] \quad ; \quad G_d(s) = \left[\begin{array}{c|c} A & E_1 \\ \hline C & E_2 \end{array}\right] \tag{8.107}$$

As discussed in Section 8.3.1, the control input does not affect to the residual if there are no modeling errors in the transfer matrix between control input and the system output. Therefore, the control input can be ignored here. Our task here is to find an optimal estimation of the filtered fault when the system has both the fault and disturbance. This can be formulated according to the scheme given in Fig. 8.8.

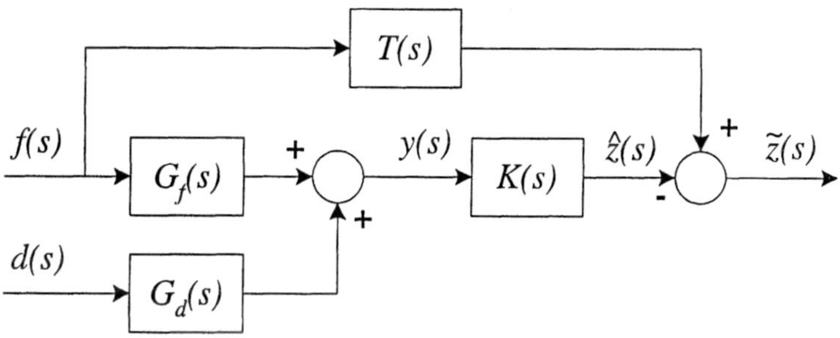

Figure 8.8. Formulation of filtered fault estimation with disturbance attenuation

The objective of the problem is to minimize the following performance index

$$J_f := \|G_{rd_1}(s) - T(s)\|_\infty = \sup_{0<\|d_1\|_2<\infty} \frac{\|r - \bar{f}\|_2}{\|d_1\|_2} \tag{8.108}$$

where d_1 is the generalized disturbance vector which is defined as:

$$d_1 = \begin{bmatrix} f \\ d \end{bmatrix} \tag{8.109}$$

The problem of estimating the fault with the disturbance attenuation property can be reformulated in a standard H_∞ setting as illustrated in Fig. 8.9.

The "equivalent" transfer matrix $P_4(s)$ for the standard problem of Fig. 8.9 is given by:

$$P_4(s) = \begin{bmatrix} [T(s) \;\; 0] & -I \\ [G_f(s) \;\; G_d(s)] & 0 \end{bmatrix} = \left[\begin{array}{cc|ccc} A_T & 0 & B_T & 0 & 0 \\ 0 & A & R_1 & E_1 & 0 \\ \hline C_T & 0 & D_T & 0 & -I \\ 0 & C & R_2 & E_2 & 0 \end{array}\right] \tag{8.110}$$

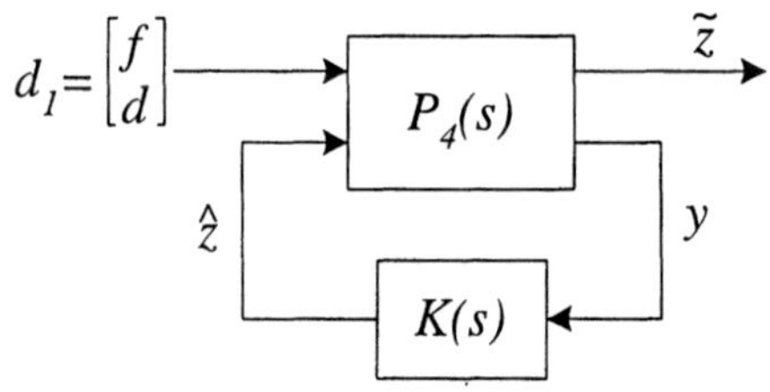

Figure 8.9. Standard problem formulation of filtered fault estimation with disturbance attenuation

where the transfer matrix $P_4(s)$ is defined as:

$$\begin{bmatrix} \tilde{z} \\ y \end{bmatrix} = P_4(s) \begin{bmatrix} d_1 \\ \hat{z} \end{bmatrix} \tag{8.111}$$

The sensitivity transfer matrix for this standard H_∞ formulation is given by:

$$G_{\tilde{z}f}(s) = LFT(P_4(s), K(s)) = [T(s) \;\; 0] - K(s)[G_f(s) \quad G_d(s)] \tag{8.112}$$

To estimate the filtered fault with the disturbance attenuation property within the standard H_∞ formulation, an optimal filter $K(s) \in RH_\infty$ should be found to satisfy the following condition:

$$||G_{\tilde{z}f}(s)||_\infty < \gamma \tag{8.113}$$

The signal $\hat{z}(t)$ in Figs.8.8 & 8.9 can be used as a residual signal as well as an estimate of the filtered fault.

8.3.4 *Robustness issues*

In Section 8.3.3, the estimation of the filtered fault with the disturbance attenuation property is studied. This filtered fault estimate can be used as a residual. Since the disturbance affect on this residual is minimized and the residual has been made close to the filtered fault, the robust FDI in terms of the disturbance effect minimization and fault effect maximization is achieved. However, the robustness is only achieved on the assumption that there are no modeling errors or the modeling errors have been approximately transformed into disturbances using techniques developed in Chapter 5. In reality, the modeling errors always exist and cannot totally be "transformed" into disturbances. This problem can be tackled using techniques exist in H_∞ control since we have formulated the FDI problem as a special H_∞ problem. To start with the investigation, let us ignore the control input and the system model is:

$$y(s) = (I + \Delta_d(s))G_d(s)d(s) + (I + \Delta_f(s))G_f(s)f(s) \tag{8.114}$$

If a residual generator $r(s) = H_y(s)y(s)$ is used to generate the residual, the residual response is:

$$\begin{aligned} r(s) &= H_y(s)G_d(s)d(s) + H_y(s)G_f(s)f(s) \\ &\quad + H_y(s)\Delta_d(s)G_d(s)d(s) + H_y(s)\Delta_f(s)G_f(s)f(s) \end{aligned} \tag{8.115}$$

To solve the problem of estimating faults for systems with disturbances and modeling errors, the standard H_∞ problem in Fig. 8.9 should be reformulated to incorporate the uncertainty block as shown in Fig. 8.10.

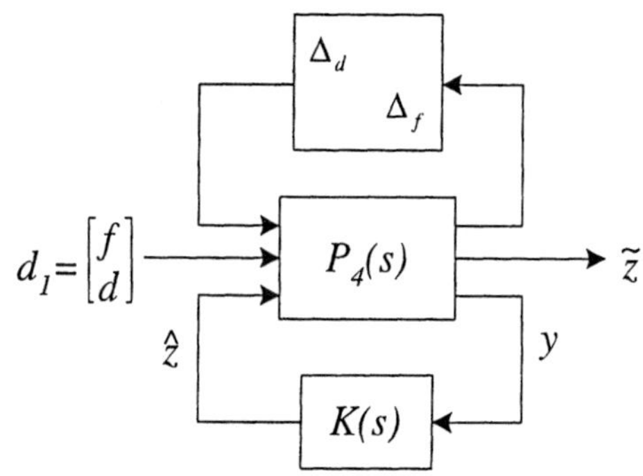

Figure 8.10. Standard problem formulation of robust fault estimation

As discussed in Section 8.3.1, the control input can only be ignored when there are no modeling errors in the transfer matrix between control input and the system output, i.e. $G_u(s)$. However, this is not always the case. A complete description of a system with all kinds of uncertainties is:

$$\begin{aligned} y(s) &= (I + \Delta_u(s))G_u(s)u(s) \\ &\quad + (I + \Delta_d(s))G_d(s)d(s) + (I + \Delta_f(s))G_f(s)f(s) \end{aligned} \tag{8.116}$$

If a residual generator $r(s) = H_y(s)y(s) + H_u(s)$ is used to generate the residual, the residual response is:

$$\begin{aligned} r(s) &= H_y(s)G_d(s)d(s) + H_y(s)G_f(s)f(s) \\ &\quad + H_y(s)\Delta_d(s)G_d(s)d(s) + H_y(s)\Delta_f(s)G_f(s)f(s) \\ &\quad + H_y(s)\Delta_u(s)G_u(s)u(s) \end{aligned} \tag{8.117}$$

It is not easy to incorporate the uncertainty $\Delta_u(s)$ in the standard problem formulation in Fig. 8.10. The only way is to transform the modeling error $\Delta_u(s)$ into an equivalent disturbance. This solution is feasible but it does not fully utilize the potential in H_∞ design. One way to solve this problem is the integrated design, i.e. to design controller and residual generator (fault estimator) simultaneously. The integrated design problem can be formulated according to Fig. 8.11 where y_e is the signal used to evaluate the control performance, $K_o(s)$ is the fault estimator and $K_c(s)$ is the controller.

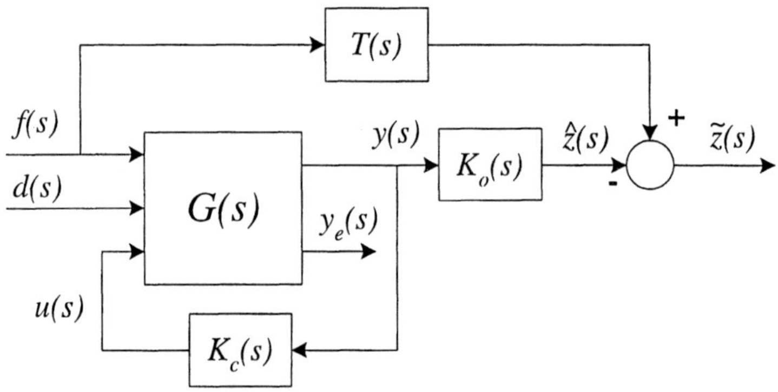

Figure 8.11. Formulation of integrated design

The integrated design problem depicted in Fig. 8.11 can be transformed into a standard H_∞ problem as given in Fig. 8.12.

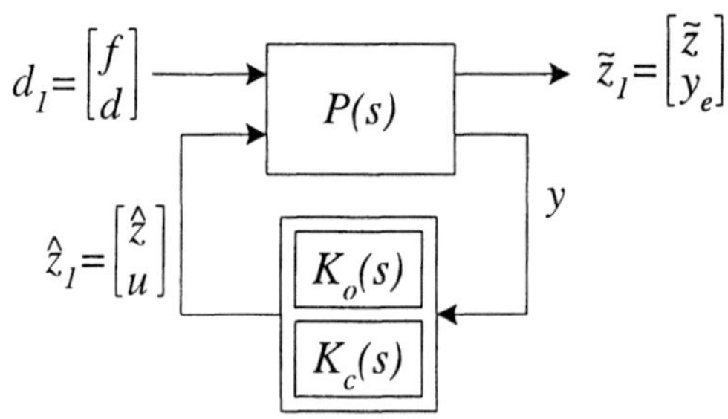

Figure 8.12. Standard problem formulation of integrated design

From Fig.8.12, it can be seen that there are two sub-blocks in the "controller" block. A similar scheme was investigated by Stoustrup et al. (1997) and they named it as the two-parameters integrated control structure. Stoustrup et al. (1997) also pointed out the four-parameters integrated control structure, studied in Nett et al. (1988) and Jacobson and Nett (1991), is just a special case of the two-parameters structure discussed here. The solutions for this integrated control structure for both nominal and robust cases are developed by Stoustrup *et al* (Kilsgaard et al., 1996; Stoustrup et al., 1997; Stoustrup and Grimble, 1997; Niemann and Stoustrup, 1997). With this standard H_∞ setting, all modeling error blocks can be considered and the idea is depicted in Fig. 8.13.

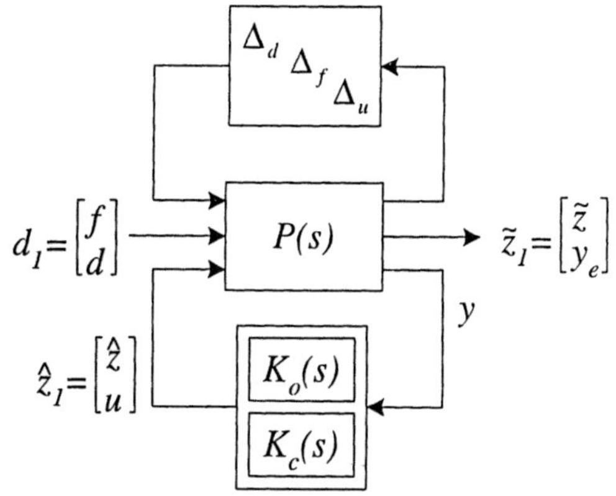

Figure 8.13. Standard problem formulation of integrated design with robustness consideration

8.4 LMI Approach for Robust Residual Generation

Two approaches for solving H_∞ optimization problems, namely the factorization approach and the algebraic Riccati equation approach, have been used to solve robust FDI problems in Sections 8.2 and 8.3. Now, let us consider to solve robust FDI using the method of linear matrix inequalities (LMI) (Boyd, Ghaoui, Feron and Balakrishnan, 1994; Gahinet et al., 1995).

As discussed before, the robustness against disturbance alone is not sufficient for ensuring good FDI performance. A suitable design of diagnostic observers should consider both robustness against disturbances and sensitivity to faults. Clearly, a trade-off has to be made sometimes in the design with respect to these two criteria. Using LMI formulation, a combined H_∞/H_- fault detection observer design was proposed by Hou and Patton (1996) and further developed in Patton and Hou (1997). The H_∞–norm is used to measure robustness against disturbances, while the H_-–norm is introduced as a measure for fault sensitivity. Based on the development of Patton and Hou (1997), the LMI approach for robust FDI is introduced here.

8.4.1 Problem formulation

Consider the linear time-invariant systems of the form

$$\left\{ \begin{array}{rcl} \dot{x}(t) & = & A\,x(t) + R_1\,f(t) + E_1\,d(t) \\ y(t) & = & C\,x(t) + R_2\,f(t) + E_2\,d(t) \end{array} \right. \tag{8.118}$$

where $f \in \mathbb{R}^g$ represents the faults, f can be component, actuator and sensor faults. It is sufficient to discuss the robust fault detection problem for (8.118) without the control input which is known.

Assume that the pair $\{C, A\}$ is observable and $\begin{bmatrix} C & E_2 \end{bmatrix}$ has full row rank. In addition, $\{A, R_1, C, R_2\}$ is assumed to have no transmission zeros, i.e. for any $s \in \mathbb{C}$, rank $\begin{bmatrix} -sI + A & R_1 \\ C & R_2 \end{bmatrix} = n + g$. Thus, detectability of the faults in (8.118) is guaranteed. Here, detectability of faults is understood in the sense of the effect that faults will have on the system output. A more precise concept of fault detectability can be found in Section 2.6.

A fault detection observer of full order has the form

$$\left\{ \begin{array}{rcl} \dot{\hat{x}}(t) & = & A\,\hat{x}(t) + K\,r(t) \\ r(t) & = & y(t) - C\,\hat{x}(t) \end{array} \right. \tag{8.119}$$

where r is the residual which is to be used as the fault detection signal.

The estimation error dynamic equation has the form

$$\left\{ \begin{array}{rcl} \dot{\tilde{x}} & = & (A - KC)\,\tilde{x} + (R_1 - KR_2)\,f + (E_1 - KE_2)\,d\,, \\ r & = & C\,\tilde{x} + R_2\,f + E_2\,d \end{array} \right. \tag{8.120}$$

The design of a robust fault detection observer, i.e. the determination of the observer gain matrix K, involves *three* objectives: (a) stabilizing the observer, (b) reducing the effects of disturbances on the residual and (c) increasing effects of faults on the residual. This can be formulated as a constrained H_∞ estimation problem as follows:

Definition 8.2 *Given two scalars $\beta > \gamma > 0$. Observer (8.119) is called an H_∞ fault detection observer if the following three conditions hold:*

1. $A - KC$ *is asymptotically stable*; *(8.121)*
2. $\|G_{rd}(j\omega)\|_\infty < \gamma$; *(8.122)*
3. $\|G_{rf}(j\omega)\|_- > \beta$ *(8.123)*

where

$$G_{rd}(s) := C(s\,I - A + KC)^{-1}(E_1 - KE_2) + E_2$$
$$G_{rf}(s) := C(s\,I - A + KC)^{-1}(R_1 - KR_2) + R_2\,.$$

The notation $\|G_{rf}(j\omega)\|_-$ is defined in the way:

$$\|G\|_- := \underline{\sigma}\,[G(j0)] \tag{8.124}$$

where $\underline{\sigma}(\cdot)$ for the smallest singular value.

The meaning of conditions (8.121) and (8.122) is well known. They correspond to the requirements for H_∞ estimation. Whilst condition (8.122) represents the worst-case criterion for the effect of disturbances on the residual r, condition (8.123) stands for the worst-case criterion for the sensitivity of r against f. (8.123) considers the case where the fault detection observer after the transient period reaches the stationary status. Clearly, these two criteria capture the most significant features of fault detection observers.

It should be pointed out that the norm $||G_{rf}(j\omega)||_-$ has been considered for the specific frequency $\omega = 0$, but not over the whole range $[0, \infty)$. Since the transfer matrix $G_{rf}(s)$ is strictly proper, $G_{rf}(j\infty) = 0$, which means that $||G_{rf}(j\omega)||_- = 0$ when $\omega \in [0, \infty)$. Hence, in a design of robust fault detection observers by using both robustness and sensitivity criteria, the norm $||G_{rf}(j\omega)||_-$ is valid over an interval $[\omega_1, \omega_2]$ for a finite number ω_2.

A combined criterion (8.122) with (8.123) such as in the way of $\dfrac{||G_{rf}||_-}{||G_{rd}||_\infty} > \alpha_2$ can, of course, be chosen and this combination also makes sense. However, this is not recommended. The reasons are as follows. Firstly, the problem formulation in Definition 8.2 has clear physical meaning. The criteria (8.122) and (8.123) provide directly quantitative measures for robustness and sensitivity of a fault detection observer. The value γ is very useful for threshold selection in detection decision-making. The ratio β/γ indicates how good a designed fault detection observer is and therefore can be used for evaluation of fault detection observers. Secondly, the present problem formulation enables the direct time-domain solution of the robust fault detection observer problem by using the recently developed LMI approach (Iwasaki and Skelton, 1994; Gahinet and Apkarian, 1994). Finally, as will be shown, the robust fault detection is equivalent to a constrained H_∞ estimation problem, the latter can be further reformulated as a standard problem of constrained optimization.

8.4.2 Analysis of sensitivity norm

In Section 8.4.1, the sensitivity measure has been defined as $\underline{\sigma}[G(j0)]$. It is however, not quite clear whether or not this definition makes sense. This section will provide a close look at the meaning of this sensitivity measure.

Because $G(j0) = G(j\omega)|_{\omega=0}$ is actually a constant matrix, for simplicity, denote $A = G(j0)$ and assume that A is a real matrix (as it is in real cases). In the following a more promising definition for $||A||_-$ will be given and $||A||_-$ will be shown to having the properties of a norm and hence being a norm in fact.

$||A||_-$ has been defined the smallest singular value of A, hence for an arbitrary A, two orthogonal matrices U and V ($UU^T = I$, $VV^T = I$) can be constructed such that $UAV^T = \Sigma$, the non-zero entries σ_i in the diagonal matrix Σ are the *singular values* of A. Thus, in fact, $\underline{\sigma}(A)$ has been defined as $||A||_-$. It is important here to give a more formal definition of $||A||_-$.

Definition 8.3 *A number denoted by $||A||_-$ is defined by*

$$||A||_- = \begin{cases} \min\limits_{Ax \neq 0} \frac{||Ax||}{||x||}, & A \neq 0 \\ 0, & A = 0 \end{cases} \tag{8.125}$$

Here, $||\cdot||$ is understood as the standard norm for vectors. It is useful to compare the above definition with the following standard matrix norm definition.

Definition 8.4 *(Strang, 1988): A number denoted by $||A||$ is called a norm of A and defined by*

$$||A|| = \max_{x \neq 0} \frac{||Ax||}{||x||} \tag{8.126}$$

If (8.125) had been defined by $||A||_- = \min_{x \neq 0} \frac{||Ax||}{||x||}$, then $||A||_- \equiv 0$ unless for all $x \neq 0$ it follows that $Ax \neq 0$, i.e. unless A has full column rank. In fact, if A has full column rank, $\min_{Ax \neq 0} \frac{||Ax||}{||x||} = \min_{x \neq 0} \frac{||Ax||}{||x||}$.

Apart from the less important point mentioned above, (8.125) and (8.126) show the nice nature of symmetry. Particularly, all these two definitions have clear physical interpretations: $||A||$ is a measure of the largest amount by which an arbitrary vector can be amplified by the multiplication Ax, whilst $||A||_-$ is a measure of the smallest amount of this. This is exactly what is needed in the robustness/sensitivity analysis of fault detection observers.

In view of $||A|| = \bar{\sigma}(A)$ (Strang, 1988), it is interesting to see if $||A||_- = \underline{\sigma}(A)$ as hoped. Since $||0||_- = \underline{\sigma}(0) = 0$, only the case $A \neq 0$ needs to be considered. Squaring both sides of (8.125) gives

$$\begin{aligned} ||A||_-^2 &= \min_{Ax \neq 0} \frac{||Ax||^2}{||x||^2} = \min_{Ax \neq 0} \frac{x^T A^T A x}{x^T x} \\ &= \frac{x^T \lambda_{\min}(A^T A) x}{x^T x} = \lambda_{\min}(A^T A) = \underline{\sigma}^2(A) \end{aligned} \tag{8.127}$$

due to the Rayleigh's principle (Strang, 1988). Here the notation $\lambda_{min}(\cdot)$ means the smallest eigenvalue of $(\cdot)$. So, this equation proves $||A||_- = \underline{\sigma}(A)$.

Recall that $UAV^T = \Sigma$. It is easy to see that the Moore-Penrose inverse of A (Campbell and Meyer, 1991) is $A^+ = V^T \Sigma^+ U$. This implies immediately that $\bar{\sigma}(A^+) = 1/\underline{\sigma}(A)$ which is significant because it establishes the interesting relations $||A||_- \, ||A^+|| = 1$ and $||A^+||_- \, ||A|| = 1$.

By now, as far as the fault diagnosis application is concerned, no further analysis may really be needed. This is because $||A||_-$ can be simply called a criterion which does make the precise sense as required in the robustness/sensitivity analysis of fault detection observers.

It is however, of theoretical interest in the following to see that $||A||_-$ as defined by (8.125) is really a *norm* although it is not an *induced norm* which requires a further condition on a norm and does not match the aim of defining $||A||_-$. The last point will be explained more precisely later.

Definition 8.5 *(Campbell and Meyer, 1991; Vidyasagar, 1993): A number denoted by, say, $||A||$ is called a norm of A if*

$$(i) \quad ||A|| = 0 \text{ iff } A = 0 \tag{8.128}$$

$$(ii) \quad ||\alpha A|| = |\alpha| \, ||A|| \,, \ \alpha - \textit{arbitrary constant} \tag{8.129}$$

$$(iii) \quad ||A + B|| \leq ||A|| + ||B|| \,. \tag{8.130}$$

A norm for which $||AB|| \leq ||A||\,||B||$ (assume A and B have compatible dimensions) is called an induced norm.

It is well known that $||A||$ is not only a norm but also an induced norm. In literature, the *induced norm* is sometimes also referred to as a *matrix norm.*

Firstly, it is clear that as $Ax \neq 0$ is required in the minimizing, $||A||_- \neq 0$ if $A \neq 0$, and trivially that $||A||_- = 0$ if $A = 0$. Because $||\alpha\, Ax||$ is a norm operation for a vector and thus $||\alpha\, Ax||=|\alpha|\,||Ax||$, so whether $A = 0$ or not, by Definition 8.4, it is evident that $||\alpha\, A||_- = |\alpha|\,||A||_-$. These verify that $||A||_-$ satisfies (i) and (ii).

By applying the Cauchy-Schwarz inequality (Strang, 1988) to the vector norm $||(A+B)x||$ gives $||Ax+Bx|| \leq ||Ax|| + ||Bx||$, and when dividing by $||x||$ and on further minimization, the inequality becomes $||A+B||_- \leq ||A||_- + ||B||_-$. Hence, $||A||_-$ is indeed a norm for A.

Finally, the following shows further that $||AB||_- \geq ||A||_-\,||B||_-$ and thus that the norm $||A||_-$ is *not* an induced norm. By the definition of $||A||_-$, $||ABx|| \geq ||A||_-\,||Bx||$ for all Bx. When divided by $||x||$ and further minimized, the inequality becomes $||AB||_- \geq ||A||_-\,||B||_-$.

Recall that by the definition of $||A||$ and the property $||AB|| \leq ||A||\,||B||$, $||ABx|| \leq ||AB||\,||x|| \leq ||A||\,||B||\,||x||$ which gives the upper bound of $||ABx||$ based merely on the information about $||A||$, $||B||$ and $||x||$. This inequality is clearly useful in the FDI robustness analysis.

In comparison with the above feature, by the definition of $||A||_-$ and the property $||AB||_- > ||A||_-\,||B||_-$, $||ABx|| \geq ||AB||_-\,||x|| \geq ||A||_-\,||B||_-\,||x||$. This gives the lower bound of $||ABx||$ based merely on the information about $||A||_-$, $||B||_-$ and $||x||$. The inequality is clearly desirable for the fault sensitivity analysis of fault detection observers.

The following is a concluding theorem which summarizes all the important points made in this section.

Theorem 8.7 *The number which is defined by*

$$||A||_- = \begin{cases} \min\limits_{Ax\neq 0} \frac{||Ax||}{||x||}, & A \neq 0 \\ 0, & A = 0 \end{cases} \tag{8.131}$$

has the following properties:

$$(i) \quad ||A||_- = 0 \text{ iff } A = 0 \tag{8.132}$$

$$(ii) \quad ||\alpha\, A||_- = |\alpha|\,||A||_-\,, \quad \alpha - \text{arbitrary constant} \tag{8.133}$$

$$(iii) \quad ||A+B||_- \leq ||A||_- + ||B||_- \tag{8.134}$$

$$(iv) \quad ||AB||_- \geq ||A||_-||B||_- \tag{8.135}$$

$$(v) \quad ||A||_- = \underline{\sigma}(A) \tag{8.136}$$

$$(vi) \quad ||A||_-\,||A^+|| = 1 \tag{8.137}$$

$$(vii) \quad ||A^+||_-\,||A|| = 1 \tag{8.138}$$

Having such nice properties, $||A||_-$ is clearly a norm and worthy of being pursued further to explore its potential in the robustness/sensitivity analysis of fault detection observers. Mathematically, the notion of matrix measures has been developed based on induced matrix norms $\|\cdot\|$. It may be of interest to explore the counterpart based on the norm $\|\cdot\|_-$ with the special property (iv).

Now it is clear that the norm $\|G(j0)\|_-$ makes good sense for the design of robust fault detection observers. In addition, the norm defined by the following equation

$$\|G(j\omega)\|_- = \underline{\sigma}(G(j\omega)), \qquad \omega \in [\omega_1, \omega_2] \tag{8.139}$$

also makes sense, where $[\omega_1, \omega_2]$ is a given frequency range. However, $\|G(j\omega)\|_-$ will not make sense without considering the frequency range constraint $\omega \in [\omega_1, \omega_2]$, since in that case $\|G(j\omega)\|_- \equiv 0$ due to $\lim_{\omega\to\infty} G(j\omega) = 0$ when $G(j\omega)$ is strictly proper. In the fault detection observer design problem, the corresponding transfer matrix from faults to outputs is indeed strictly proper.

8.4.3 LMI solution to H_∞ control

To introduce the LMI solution to H_∞ control, consider the linear time-invariant systems of the form

$$\left\{\begin{array}{rcl} \dot{x}(t) & = & A\,x(t) + B\,u(t) + E_1\,d(t) \\ y(t) & = & C\,x(t) + D\,u(t) + D\,d(t) \end{array}\right. \tag{8.140}$$

where $x \in \mathbb{R}^n$ is the state, $y \in \mathbb{R}^m$ stands for the output, $u \in \mathbb{R}^r$ represents the input, $d \in \mathbb{R}^q$ denotes for the disturbance. All real matrices in (8.140) are constant. For simplicity, without loss of generality, assume that the matrix $\begin{bmatrix} B \\ D \end{bmatrix}$ has full column rank.

It is notable that, in the measurement equation, the input u and disturbance d have the same coefficient matrix D. The reason for this special choice will be made clear in establishing duality between H_∞ control and estimation.

The problem of H_∞ control with full information of x is addressed as (Zhou and Khargonekar, 1988): finding a feedback control of the form $u = -F\,x$ such that the closed-loop system is asymptotically stable and the H_∞ norm of the transfer function from disturbances to outputs is less than a given positive number γ. That is to find a F such that $A_c := A - B\,F$ is asymptotically stable and $\|G_{yd}(s)\|_\infty < \gamma$ with

$$G_{yd}(s) := (C - DF)(s\,I - A_c)^{-1}E_1 + D \tag{8.141}$$

The following H_∞ control result is a simplified version of the more general one given by Iwasaki and Skelton (1994) and Gahinet and Apkarian (1994).

Theorem 8.8 *(Iwasaki and Skelton, 1994): Given a scalar $\gamma > 0$. There exists an H_∞ controller if and only if there exists $X = X^T > 0$ solving the LMI*

$$\begin{bmatrix} B \\ D \end{bmatrix}^{\perp} \begin{bmatrix} AX + XA^T + E_1E_1^T & XC^T + E_1D^T \\ CX + DE_1^T & DD^T - \gamma^2 I \end{bmatrix} \begin{bmatrix} B \\ D \end{bmatrix}^{\perp T} < 0 \tag{8.142}$$

When the LMI is solvable, $u = -B^T X x$ is an H_∞ controller.

Theorem 8.9 *(Iwasaki and Skelton, 1994) When the LMI (8.142) is solvable, all solutions of the H_∞ control are given by $u = F x$ with*

$$F = F_1 + F_2 M F_3 \tag{8.143}$$

where M is an arbitrary constant matrix with $M^T M < I$ and

$$\left\{ \begin{array}{rcl} F_1 & = & (C_1^T - Q_{12}Q_{22}^{-1}C_2^T)(C_2 Q_{22}^{-1}C_2^T)^{-1} \\ F_2 & = & (Q_{12}Q_{22}^{-1}Q_{12}^T - Q_{11} - F_1 F_3^{-2} F_1^T)^{1/2} \\ F_3 & = & (-C_2 Q_{22}^{-1} C_2^T)^{-1/2} \end{array} \right. \tag{8.144}$$

with

$$\begin{bmatrix} Q_{11} & Q_{12} \\ Q_{12}^T & Q_{22} \end{bmatrix} := \begin{bmatrix} \bar{B}^+ \\ \bar{B}^\perp \end{bmatrix} Q \begin{bmatrix} \bar{B}^+ \\ \bar{B}^\perp \end{bmatrix}^T$$
$$\begin{bmatrix} C_1 & C_2 \end{bmatrix} := \bar{C} \begin{bmatrix} \bar{B}^{+T} & \bar{B}^{\perp T} \end{bmatrix}$$

and

$$Q := \begin{bmatrix} AX + XA^T & XC^T & E_1 \\ CX & -\gamma I & D \\ E_1^T & D^T & -\gamma I \end{bmatrix},$$
$$\bar{B} := \begin{bmatrix} B \\ D \\ 0 \end{bmatrix}, \quad \bar{C} := \begin{bmatrix} -X & 0 & 0 \end{bmatrix}.$$

In the above theorem, A^T represents the transpose of the matrix A. A^+ denotes the Moore-Penrose inverse of A. Let $A \in \mathbb{R}^{m \times n}$ be not full row rank, $A^\perp$ represents an arbitrary matrix satisfying

$$A^\perp A = 0, \quad \text{and} \quad \operatorname{rank} A^\perp + \operatorname{rank} A = m.$$

Let $A = A^T \geq 0$, then $A^{1/2}$ represents the unique square root of A.

Remarks: The existence of inverses of the matrices involved in (8.144) is guaranteed by solvability of (8.142) (Iwasaki and Skelton, 1994).

8.4.4 *Duality and H_∞ estimation*

Here the duality of H_∞ estimation and H_∞ control problems is established in a straightforward way. Using the duality, the LMI solution to H_∞ estimation problem is then obtained.

Consider the linear time-invariant systems of the form

$$\left\{ \begin{array}{rcl} \dot{x}(t) & = & A\, x(t) + E_1\, d(t) \\ y(t) & = & C\, x(t) + E_2\, d(t) \end{array} \right. \tag{8.145}$$

As the control input does not affect to the estimation, it is sufficient to discuss the robust estimation problem for (8.145) ignoring the control input. Without loss of generality, assume that the pair $\{C, A\}$ is observable and $\begin{bmatrix} C & E_2 \end{bmatrix}$ has full row rank.

A full-order observer has the form

$$\left\{ \begin{array}{rcl} \dot{\hat{x}}(t) & = & A\,\hat{x}(t) + K\,r(t) \\ r(t) & = & y(t) - C\,\hat{x}(t) \end{array} \right. \tag{8.146}$$

The estimation error dynamic equation takes the form

$$\left\{ \begin{array}{rcl} \dot{\tilde{x}}(t) & = & (A - KC)\,\tilde{x}(t) + (E_1 - KE_2)\,d(t) \\ r(t) & = & C\,\tilde{x}(t) + E_2\,d(t) \end{array} \right. \tag{8.147}$$

where $\tilde{x}(t) := x(t) - \hat{x}(t)$.

The problem of H_∞ estimation may be addressed as follows: the determination of a standard observer in the form (8.146) such that the H_∞ norm of the transfer function from disturbances to the residual of the observer achieves a given small real number $\gamma > 0$.

Definition 8.6 *Given a real number $\gamma > 0$. Observer (8.146) is called an H_∞ observer if the following two conditions hold:*

$$1. \quad A_o \text{ is asymptotically stable}; \tag{8.148}$$

$$2. \quad \|G_{rd}(j\omega)\|_\infty < \gamma \tag{8.149}$$

where $G_{rd}(s) := C(s\,I - A_o)^{-1}(E_1 - KE_2) + E_2$ and $A_o := A - K\,C$.

Theorem 8.10 *(Patton and Hou, 1997): There exists the duality between the H_∞ control and estimation problems, i.e., by choosing the same performance bound γ in control and estimation problems, there are the exchangeable relations in these two problems:*

$$\{A, B, C, D, E_1, F\} \Longleftrightarrow \{A^T, C^T, E_1^T, E_2^T, C^T, K^T\}.$$

Proof: The duality is obvious because

$$\|G_{yd}(j\omega)\|_\infty = \|G_{rd}^T(j\omega)\|_\infty \;\; ; \quad \|G_{rd}^T(j\omega)\|_\infty = \{A^T, C^T, E_1^T, E_2^T, C^T, K^T\}$$

◊ **QED**

Remark: The duality is known for the case when $E_2 = 0$ was proved by Shaked (1990). The theorem above takes into account the general case and its proof is almost trivial. Taking the advantage of the duality, the LMI solution to the H_∞ estimation problem on the strength of Theorem 8.8 can be immediately obtained.

Theorem 8.11 *(Patton and Hou, 1997): Given a scalar $\gamma > 0$. There exists an H_∞ observer of the form (8.146) if and only if there exists $Y = Y^T > 0$ solving the LMI*

$$\begin{bmatrix} C^T \\ E_2^T \end{bmatrix}^{\perp} \begin{bmatrix} A^TY+YA+C^TC & YE_1+C^TE_2 \\ E_1^TY + E_2^TC & E_2^TE_2-\gamma^2 I \end{bmatrix} \begin{bmatrix} C^T \\ E_2^T \end{bmatrix}^{\perp T} < 0 \qquad (8.150)$$

When the LMI (8.150) is solvable, the H_∞ observer gain matrix is given by $K = YC^T$.

Theorem 8.12 *(Patton and Hou, 1997): When the LMI (8.150) is solvable, all solutions of the H_∞ estimation are given by*

$$K = K_1 + K_2NK_3 \qquad (8.151)$$

where N is an arbitrary constant matrix satisfying $N^TN < I$ and

$$\begin{cases} K_1 = (C_2P_{22}^{-1}C_2^T)^{-1}(C_1 - C_2P_{22}P_{12}^T) \\ K_2 = (-C_2P_{22}^{-1}C_2^T)^{-1/2} \\ K_3 = (P_{12}P_{22}^{-1}P_{12}^T - P_{11} - K_1^TK_2^{-2}K_1^T)^{1/2} \end{cases} \qquad (8.152)$$

with

$$\begin{bmatrix} P_{11} & P_{12} \\ P_{12}^T & P_{22} \end{bmatrix} := \begin{bmatrix} \tilde{B}^+ \\ \tilde{B}^\perp \end{bmatrix} P \begin{bmatrix} \tilde{B}^+ \\ \tilde{B}^\perp \end{bmatrix}^T,$$

$$\begin{bmatrix} C_1 & C_2 \end{bmatrix} := \tilde{C} \begin{bmatrix} \tilde{B}^{+T} & \tilde{B}^{\perp T} \end{bmatrix}$$

and

$$P := \begin{bmatrix} A^TY + YA & YE_1 & C^T \\ E_1^TY & -\gamma I & E_2^T \\ C & E_2 & -\gamma I \end{bmatrix},$$

$$\tilde{B} := \begin{bmatrix} C^T \\ H^T \\ 0 \end{bmatrix}, \quad \tilde{C} := \begin{bmatrix} -Y & 0 & 0 \end{bmatrix}.$$

8.4.5 Robust fault detection observer design

Due to a duality between H_∞ control and H_∞ estimation (Shaked, 1990; Patton and Hou, 1997), a solution to the robust fault detection observer problem is given in the following theorem in a constrained LMI setting.

Theorem 8.13 *Given two scalars $\beta > \gamma > 0$. There exists an H_∞ fault detection observer of the form (8.119 with the criteria (8.121-8.123) if and only if there exists $Y = Y^T > 0$ solving the LMI*

$$\begin{bmatrix} C^T \\ E_2^T \end{bmatrix}^{\perp} \begin{bmatrix} A^TY+YA+C^TC & YE_1+C^TE_2 \\ E_1^TY + E_2^TC & E_2^TE_2-\gamma^2 I \end{bmatrix} \begin{bmatrix} C^T \\ E_2^T \end{bmatrix}^{\perp T} < 0 \qquad (8.153)$$

with the quadratic matrix inequality constraint

$$\left[C(YC^TC-A)^{-1}(R_1-YC^TR_2)+R_2\right]^T$$
$$\left[C(YC^TC-A)^{-1}(R_1-YC^TR_2)+R_2\right] > \beta^2 I\,. \qquad (8.154)$$

The observer gain is given by $K = YC^T$.

Two difficulties may exist in solving the above problem directly. Firstly, as no special assumption has been made in the H_∞ observer design, it is possible that the matrix $(YC^TC - A)$ on the left hand side of (8.154) is singular, which means that (8.154) has no solution in such a case. Secondly, even if the above problem has a solution, in general, it still needs a numerical method to obtain the solution.

Suppose that, instead of (8.154), the following sensitivity criterion

$$\text{tr}\left[C(KC-A)^{-1}(R_1-KR_2)+R_2\right]^T$$
$$\left[C(KC-A)^{-1}(R_1-KR_2)+R_2\right] > \beta^2 \qquad (8.155)$$

is required. This criterion means that the worst-case of the total effect of the faults on the stationary residual is considered.

By substituting the LMI solution (8.151) into (8.155), the H_∞ fault detection observer design can be reformulated as a constrained optimization problem:

$$\min_N \ \text{tr}\left[C(KC-A)^{-1}(R_1-KR_2)+R_2\right]^T$$
$$\left[C(KC-A)^{-1}(R_1-KR_2)+R_2\right]$$
$$\text{s.t.} \qquad N^TN < I \qquad (8.156)$$

where K is an explicit function of N, given by (8.151). This a special type of constrained non-linear optimization, i.e. quadratic criterion with quadratic constraints. Several well-known effective non-linear programming algorithms can be used to solve this problem.

The optimization (8.156) may lead to a solution which does not satisfy (8.155). In such a case, the LMI (8.153) should be re-solved with some increased γ. Thus, several attempts may be needed in solving the robust fault detection observer design.

When a robust fault observer is obtained by using the design suggested above, quantitative assessment of robustness against disturbances and sensitivity to faults can easily be achieved according to two bound parameters γ and β.

8.4.6 Discussion on LMI robust FDI approach

It is important to take into account fault sensitivity in the design of robust fault detection observers. A measure for fault sensitivity has been investigated in this section. It has been shown that this sensitivity measure is a well-defined norm of matrices. Some properties of the norm have also be derived.

By using criteria for both robustness against disturbances and sensitivity to faults, the design of robust fault detection observers can be reformulated as the problem of H_∞ estimation with constraints. Furthermore, the LMI formulation and solutions of H_∞ estimation reduce the problem of designing robust fault detection observers to a particular constrained optimization problem. The optimization problem can be solved by using some well-known numerical algorithms.

The designed observer considers both robustness against disturbances and sensitivity to faults. The design procedure offers two bounds for the robustness and sensitivity, which can be used for threshold selection in the decision-making stage of fault diagnosis. Furthermore, this is important for the evaluation of fault diagnosis observers.

Unlike the H_∞–norm which can be evaluated for rational transfer matrices over the frequency $[0, \infty)$, the H_-–norm of a rational transfer matrix can merely expressed over a finite frequency interval $[\omega_1, \omega_2]$. For computational advantages, the robust fault detection observer is designed with a fault sensitivity H_-–norm evaluated at the frequency $\omega = 0$. This corresponds to the case that the fault detection observer reaches its steady-state, a situation where an observer becomes really useful for providing estimation/detection information.

Although the H_-–norm has been shown suitable for evaluation of fault sensitivity, introduction of this sensitivity measure in the robust fault detection observer design has increased the complexity of the design compared with the straightforward nature of the H_∞ observer design. Two research topics may therefore need further consideration. The first topic is the development of better design strategies for the fault detection observers with the combined robustness and sensitivity criteria. The second topic is to seek for another sensitivity measure which may supersede the properties of the H_-–norm whilst also leading to tractable designs.

8.5 Summary

This chapter discussed the frequency domain robust FDI problems. Three H_∞ frequency domain design approaches have been introduced. The emphasis is on the formulation of robust FDI problems. Once the problem is formulated, the solution can be readily found using techniques existing in robust and H_∞ control.

The first approach is based on the use of factorization. This approach developed before the other two approaches, has explicit frequency domain interpretation which is to enhance residual quality with a post-filter (or dynamic weighting matrix). The available information about faults and disturbances can easily be embedded into the design. One advantage is that it can be used to solve the threshold selection problem as well. However, there are some disadvantages which prevent the wide acceptance of this approach. The first one is that there is no solution for dealing with modeling errors although it can be tackled in the residual evaluation stage or approximately treated as disturbances. The second and most serious one is the lack of design software.

The second approach is the standard H_∞ problem approach. A robust FDI problem can be formulated into a standard H_∞ problem with the aid of the linear fraction transformation, then the algebraic Riccati equation-based solution can be found. Both robust fault detection and fault estimation problems have been formulated in this chapter. One advantage of this approach is the simplicity because its close association with H_∞ filter problems. The most important advantage is that it provides a framework to deal with modeling errors. The modeling uncertainties can easily be incorporated into the standard H_∞ formulation and then robust solutions can be found using the techniques in robust control such as μ synthesis. The great potential of this approach, especially in the integrated design, is waiting to be further exploited.

The factorization approach designs the optimal post-filter which is applied to the primary residual. The primary residual is generated by a full order observer and the design of this observer gain has not been considered by the factorization. The standard H_∞ approach, on the other hand, designs the full order observer gain matrix. The residual in this approach is the output estimation, which is the primary residual treated by the factorization. It is possible to combine these two approaches together in order to achieve the maximum possible FDI performance. The standard H_∞ approach is used to design a full order observer whose residual is further enhanced by a post-filter designed by the factorization approach.

The third approach introduced is based on the use of linear matrix inequalities. Use this approach, the residual robustness against disturbance is formulated as a convex optimization problem which can be solved by LMI tools. The residual sensitivity to faults is treated as a constraint condition for the optimization problem. The main advantage of this approach is the simultaneous consideration of fault sensitivity and disturbance robustness. However, the problem of dealing with modeling uncertainty within the LMI framework is still waiting to be seen.

9 FAULT DIAGNOSIS OF NON-LINEAR DYNAMIC SYSTEMS

9.1 Introduction

The majority of model-based fault diagnosis methods are based on linear system models. For non-linear systems, the fault diagnosis problem has been traditionally approached in two steps. Firstly, the model is linearized at an operating point, and then robust techniques are applied to generate residual signals which are insensitive to model parameter variations within a small neighborhood of the operating point. The robustness issue is tackled using techniques developed for linear system models. The strategy only works well when the linearization does not cause a large mismatch between linear model and non-linear behavior, the residual has been designed to be robust enough to tolerate small model perturbations around the operating point, and the system closely operates around the operating point specified. However, for systems with high nonlinearity and a wide dynamic operating range, the linearized approach fails to give satisfactory results. A linearized model is an approximate description of the non-linear system dynamics around the operating point. However, when the system operating range becomes wider, the linearized model is no longer able to represent the system dynamics. One solution is to use a large number of linearized models corresponding to a range of operating points. However, this would involve a large number of FDI systems corresponding to all operating points. This is not very practical for real-time application.

It is necessary to develop fault diagnosis methods which tackle non-linear dynamic system models directly. There have been some attempts to use non-linear observers to solve non-linear system FDI problems. The first reported work was carried out by Hengy and Frank (1986) where an identity non-linear observer was used. This method is further considered by Frank (1987) and Adjallah, Maquin and Ragot (1994). However, a method to design the gain matrix for ensuring the observer stability was not developed. The unknown input observer approach was extended to include non-linear terms by Frank *et al* (Wünnenberg, 1990; Frank and Seliger, 1991; Seliger and Frank, 1991a; Frank, Ding and Wochnik, 1991; Seliger and Frank, 1999). An adaptive observer-based FDI approach for time-varying non-linear systems was proposed by Ding and Frank (1993). Frank (1994b) and Krishnaswami and Rizzoni (1994b) surveyed pre-1994 developments of non-linear system FDI methods from different perspectives. From a purely mathematical point of view, the precise fault detection and isolation of a non-linear dynamic system is a formidable one. It becomes substantially more difficult when uncertainty also presents in the system. Therefore it is necessary to restricting the class of non-linear systems in the study of FDI problems. A class of non-linear systems which has been studied extensively is the system with bilinear dynamics. Important studies on bilinear systems FDI can be found in Yu et al. (1994b), Yu and Shields (1994), Yu et al. (1996), Yu and Shields (1996), Bennett, Patton, Daley and Newton (1996a), Shields (1997), Yu and Shields (1997), Yang and Saif (1997) and Mechmeche and Nowakowski (1997). The main idea is to treat non-linear terms as disturbances and decouple their effects from the residual using an unknown input observer. This idea has been demonstrated by a practical example in Section 3.3.3. The sliding mode observer (Slotine, Hedrick and Misawa, 1987; Edwards and Spurgeon, 1994) has also been applied to non-linear system FDI problems (Edwards, Spurgeon, Patton and Klotzek, 1997). There have also been some studies on extending the parity relations approach to non-linear systems (Krishnaswami and Rizzoni, 1994a; Krishnaswami et al., 1995; Krishnaswami and Rizzoni, 1997; Guernez, Cassar and Staroswiecki, 1997). Unlike linear systems, there is no direct link between parity relation and observer-based FDI approaches. Owe to new developments of non-linear observers (Raghavan and Hedrick, 1994; Xia and Zeitz, 1997), the FDI problem for more general non-linear systems has been investigated (Schreier, Ragot, Patton and Frank, 1997; Hammouri, Kinnaert and El Yaagoubi, 1998a; Hammouri, Kinnaert and El Yaagoubi, 1998b). Recent developments on non-linear observer-based FDI can be found in the survey paper by García and Frank (1997) and a selection of observer-based approaches is introduced in Section 9.2.

The analytical models, which the non-linear observer approaches based on, are not easy to obtain in practice. Sometimes, the system cannot be modeled by explicit mathematical models. Without a model, the observer-based FDI is impossible. To overcome this problem, it is desirable to find a "universal" approximate model which can be used to represent any non-linear system approximately. Moreover, there should be a mechanism which can automatically identify this universal model. The neural network is exactly such a powerful tool of handling non-linear problems. One of the most important advantages of neural networks is their ability to implement non-linear transformations for functional approximation problems, i.e., neural networks are capable of forming an arbitrarily close approximation to any continuous non-linear mapping (function), given suitable weighting factors and a network architecture (Narendra and Parthasarthy, 1990; Hunt, Sbarbaro, Zbikowski and Gawthrop, 1992; Narendra, 1996). There are many systematic ways to establish neural network models based on the powerful learning ability of neural networks. Neural networks have been widely used in many engineering domain and the fault diagnosis application field is also no exception. There have been a large number of publications on neural networks-based FDI, e.g. Watanabe et al. (1989), Naidu et al. (1990), Himmelblau et al. (1991), Sorsa et al. (1991), Willis et al. (1991), Sorsa and Koivo (1993), Kavuri and Venkatasubramanian (1994), Leonard and Kramer (1993), Napolitano et al. (1995), Napolitano et al. (1996), Maki and Loparo (1997), Patton, Chen and Siew (1994), Köppen-Seliger and Frank (1995), Patton and Chen (1996a) and Frank and Köppen-Seliger (1997a). These publications are cited here because each of them either represents an interesting aspect of the FDI problem or has close link with model-based approaches.

The neural network, as an optimal approximate tool for handling non-linear problems, can be used to overcome difficulties in conventional techniques for dealing with nonlinearity. It is the authors' opinion that there is little to be gained by applying neural networks to linear time-invariant systems. Neural networks are properly aimed at processes that are ill-defined, complex, non-linear and stochastic. Neural networks have many advantages and can be used in a number of ways to tackle fault diagnosis problems for non-linear dynamic systems.

In the use of neural networks for fault diagnosis, there are two major problems accompanying the majority early publications. The first problem is that most studies only deal with steady-state processes. To achieve on-line fault diagnosis in the presence of transient behaviors, the system dynamics have to be considered. The second problem is that the neural network is only used as a *fault classifier* and other advantages and potential of neural networks have not been fully exploited. In these applications, neural networks are merely used to examine the possibility of a fault or abnormal features in the system outputs and gives a fault classification signal to declare whether or not the system is faulty. It may be valid to use only system outputs to diagnose faults for some static systems, however this is not the case for diagnosing faults in dynamic systems because the change in system inputs can also affect certain features of the system outputs. A diagnosis method which only utilizes output information could give incorrect information about faults in the system when the system input has been changed. This is especially true for non-linear systems. It must be pointed out that this problem has already been solved in model-based FDI by using the residual generation concept in which both the input and output of the monitored system are used to generate a fault indicator - the residual. The input effect can be decoupled from the residual and hence the residual only carries fault information. Fault diagnosis based on this properly designed residual can give reliable diagnostic information.

Recently, the residual generation and evaluation concepts have been combined with neural networks to form a powerful FDI tool for non-linear dynamic systems. Patton, Chen and Siew (1994) studied dynamic rather than static processes and proposed ways of utilizing neural networks for residual generation as well as evaluation. In their study, the powerful features of neural networks in modeling and classification have been used. The ideas have been further developed in Köppen-Seliger and Frank (1995), Patton and Chen (1996a) and Frank and Köppen-Seliger (1997a). A discussion of the ideas used in these studies is presented in Section 9.3.

To overcome the problem of precision and accuracy in FDI, important approaches based on fuzzy-logic have also been developed (Dexter, 1995; Frank and Kiupel, 1993; Isermann and Ulieru, 1993; Schneider and Frank, 1994). However, the fuzzy-logic approach to FDI is not, on its own, efficient for detecting incipient (i.e. small, difficult to detect and slowly developing) faults. On their own, fuzzy-logic approaches are limited to relatively simple systems as a consequence of the large and unmanageable rule base that would otherwise result. Moreover, such schemes are difficult to relate to classical FDI techniques which provide a powerful tool for design and analysis and are not limited so severely by system complexity. There is great benefit to be gained by combining fuzzy logic with model-based FDI concepts (Benkhedda and Patton, 1996; Frank and Kiupel, 1993; Isermann and Ulieru, 1993; Schneider and Frank, 1994; Schneider and Frank, 1996; Patton et al., 1998). However, the majority of studies on fuzzy-logic based FDI use only the interpretation and reasoning capabilities of the fuzzy logic. Takagi and Sugeno (1985) proved that the fuzzy logic can be used to form the fuzzy model which is very powerful in modeling non-linear dynamic systems. This modeling ability has recently been used in the design of fuzzy observers for non-linear dynamic systems FDI (Patton et al., 1998). A detailed study on the use of fuzzy observers for non-linear dynamic systems FDI is introduced in Section 9.4.

Recently, there have been some studies in combining neural networks with fuzzy logic to form the so-called neuro-fuzzy approach for non-linear dynamic systems FDI (Benkhedda and Patton, 1996; Zhang and Morris, 1996; Ayoubi and Isermann, 1997; Pfeufer and Ayoubi, 1997; Frank and Köppen-Seliger, 1997b; Ballé et al., 1998). With a combined approach, the advantages in both methods can be fully exploited. A neuro-fuzzy FDI approach is introduced in Section 9.5 with an application example.

9.2 Linear and Non-linear Observer-based Approaches

In practice many non-linear systems cannot be represented by linear models, in particular when they are not operating at a fixed operation point. This is the normal case in FDI, because at the occurrence of a fault the process diverges from its operating point. Therefore, a non-linear model should be used.

$$\left\{\begin{array}{lcl} \dot{x}(t) & = & g(x(t), u(t), f(t), d(t)) \\ y(t) & = & h(x(t), u(t), f(t), d(t)) \end{array}\right. \tag{9.1}$$

where $x(t)$ in state vector, $y(t)$ is output vector, $u(t)$ is input vector, $f(t)$ is fault vector, $d(t)$ is disturbance vector and $g(\cdot,\cdot,\cdot,\cdot)$ and $h(\cdot,\cdot,\cdot,\cdot)$ denote non-linear mappings (functions). The FDI problem is to generate a residual vector $r(t)$ using the following observer-like structure:

$$\left\{\begin{array}{lcl} \dot{\xi}(t) & = & g_r(\xi(t), u(t), y(t)) \\ r(t) & = & h_r(\xi(t), u(t), y(t)) \end{array}\right. \tag{9.2}$$

The residual should satisfy the following conditions:

$$\|r(t)\| \left\{\begin{array}{lll} \approx 0 & \text{when} & f(t) = 0 \\ \gg 0 & \text{when} & f(t) \neq 0 \end{array}\right.$$

The main challenge is to design non-linear mappings $g_r(\cdot,\cdot,\cdot)$ and $h_r(\cdot,\cdot,\cdot)$.

(1) Non-linear identity observer approach

This approach was first proposed by Hengy and Frank (1986) for the detection and isolation of component faults. Further design considerations were described by Frank (1987) and Adjallah et al. (1994). Similar to the linear case, this approach is based on an identity observer for the monitored system. Consider a non-linear system model given as

$$\left\{\begin{array}{lcl} \dot{x}(t) & = & g(x(t), u(t)) + R_1 f(t) \\ y(t) & = & h(x(t), u(t)) + R_2 f(t) \end{array}\right. \tag{9.3}$$

a non-linear identity observer for this system can be designed as

$$\left\{\begin{array}{lcl} \dot{\hat{x}}(t) & = & g(\hat{x}(t), u(t)) + K(\hat{x}(t), u(t))[y(t) - \hat{y}(t)] \\ \hat{y}(t) & = & h(\hat{x}(t), u(t)) \\ r(t) & = & y(t) - \hat{y}(t) \end{array}\right. \tag{9.4}$$

The residual $r(t)$ and the state estimation error $e(t) = x(t) - \hat{x}(t)$ are governed by:

$$\left\{\begin{array}{lcl} \dot{e}(t) & = & F(t)e(t) + O_1(e^2(t), t) + R_1 f(t) - K(\hat{x}(t), u(t))R_2 f(t) \\ r(t) & = & H(t)e(t) + O_2(e^2(t), t) + R_2 f(t) \end{array}\right. \tag{9.5}$$

where $O_1(e^2(t), t)$ and $O_2(e^2(t), t)$ represent the second- and higher-order terms with respect to $e(t)$ and

$$\left\{\begin{array}{lcl} F(t) & = & \frac{\partial g(\hat{x}(t), u(t))}{\partial \hat{x}(t)} - K(\hat{x}(t), u(t))H(t) \\ H(t) & = & \frac{\partial h(\hat{x}(t), u(t))}{\partial \hat{x}(t)} \end{array}\right. \tag{9.6}$$

It can be seen that the residual only affected by the fault vector $f(t)$ if the state estimation error $e(t)$ converges asymptotically to zero. The remaining problem is to design a (generally time-varying) matrix $K(\hat{x}(t), u(t))$ such that $e(t) = 0$ is an asymptotically stable equilibrium point of (9.5). In many practical situations even a constant matrix K will be sufficient for this purpose (Frank, 1987).

It should be pointed out, however, that no generally applicable algorithm is known that provides a solution for this stabilization problem. Since the problem is non-linear, complex numerical and computational difficulties may arise. Therefore this approach is not really practical.

(2) Thau observer approach

Thau (1973) developed an observer for a special class of non-linear systems. This observer has been applied to the fault detection and isolation of non-linear dynamic systems (Krishnaswami and Rizzoni, 1994b; Schreier et al., 1997) and uses the following non-linear system model:

$$\left\{ \begin{array}{lcl} \dot{x}(t) & = & Ax(t) + Bu(t) + R_1 f(t) + g(x(t), u(t)) \\ y(t) & = & Cx(t) + R_2 f(t) \end{array} \right. \tag{9.7}$$

It can be seen that the system has linear and non-linear terms. The system model satisfies the following two conditions:

C1: The pair (C, A) is observable.

C2: The non-linear function $g(x(t), u(t))$ is continuously differentiable and locally Lipschitz with constant ρ, i.e.

$$\|g(x_1, u) - g(x_2, u)\| \leq \rho \|x_1 - x_2\|$$

When these two conditions are satisfied, a stable observer for the system (9.7) can be constructed as:

$$\left\{ \begin{array}{lcl} \dot{\hat{x}}(t) & = & A\hat{x}(t) + Bu(t) + g(\hat{x}(t), u(t)) + K(y(t) - \hat{y}(t)) \\ \hat{y}(t) & = & C\hat{x}(t) \end{array} \right. \tag{9.8}$$

where K is the observer gain matrix which is

$$K = P_\theta^{-1} C^T. \tag{9.9}$$

The matrix P_θ is the solution to the Lyapunov equation (Schreier et al., 1997):

$$A^T P_\theta + P_\theta A - C^T C + \theta P_\theta = 0 \tag{9.10}$$

where θ is a positive parameter which is chosen such that (9.10) has a positive definite solution.

A detailed treatment of the Thau observer and its application can be found in Schreier et al. (1997). This has been combined with the fault detection filter concept for generating directional residual vectors for non-linear systems (Garg and Hedrick, 1995; García and Frank, 1997). Recently, Shields (1997) developed an observer with Thau-like structure for non-linear descriptor systems.

(3) Non-linear unknown input observer approach

The idea of the unknown input FDI observer which discussed in Chapter 3 was first extended to a certain class of non-linear systems by Wünnenberg (1990). This class of non-linear systems is modeled by:

$$\left\{ \begin{array}{rcl} \dot{x}(t) & = & Ax(t) + B(y(t), u(t)) + E_1 d(t) + R_1 f(t) \\ y(t) & = & Cx(t) + E_2 d(t) + R_2 f(t) \end{array} \right. \tag{9.11}$$

where $f(t)$ is the fault vector and $d(t)$ is unknown input vector. Note that the non-linear term $B(y(t), u(t))$ depends only on y and u which are directly available. It is therefore possible to compensate completely the nonlinearity by reproducing it using an observer of the form:

$$\left\{ \begin{array}{rcl} \dot{\xi}(t) & = & F\xi(t) + J(y(t), u(t)) + Gy(t) \\ r(t) & = & L_1 \xi(t) + L_2 y(t) \end{array} \right. \tag{9.12}$$

The conditions which are to be met by the observer matrices in order to provide robustness to unknown inputs and sensitivity to faults can be stated as follows (Wünnenberg, 1990; Frank, 1994b; Frank and Ding, 1997):

$$\left\{ \begin{array}{c} TA - FT = GC \quad ; \quad F \text{ stable} \\ J(y, u) = TB(y, u) \\ GE_2 - TE_1 = 0 \\ L_2 E_2 = 0 \\ L_1 T + L_2 C = 0 \\ rank(GR_2 - TR_1) = rank(R_1) \\ rank(L_2 R_2) = rank(R_2) \end{array} \right. \tag{9.13}$$

If these requirements can be fulfilled, the dynamics of the residual $(r(t))$ and the estimation error $(e(t) = \xi(t) - Tx(t))$ are governed by:

$$\left\{ \begin{array}{rcl} \dot{e}(t) & = & Fe(t) + GR_2 f(t) - TR_1 f(t) \\ r(t) & = & L_1 e(t) + L_2 R_2 f(t) \end{array} \right. \tag{9.14}$$

The drawback of this extension to the linear unknown input observer theory to a class of non-linear systems is that the class of systems described by the model (9.11) is rather limited. Many physical non-linear systems cannot be modeled in this way. Therefore, a given physical model must be transformed into the required form by a suitable non-linear state transformation. The existence conditions for such a transformation are very restrictive. Consequently, the class of models that are actually transformable is rather small (Frank, 1994b; Frank and Ding, 1997). But even if the existence conditions can be satisfied, finding the transformation will be hampered by the necessity to solve the higher order partial differential equations.

An alternative approach to FDI for non-linear systems was developed by Seliger and Frank (Frank and Seliger, 1991; Seliger and Frank, 1991a; Seliger and Frank, 1991b; Seliger and Frank, 1993; Seliger and Frank, 1999). Requiring weaker existence conditions, this approach extends the class of transformable systems to the following more general model:

$$\left\{ \begin{array}{rcl} \dot{x}(t) & = & A(x) + B(x)u + E(x)d + R(x)f \\ y(t) & = & Cx(t) \end{array} \right. \tag{9.15}$$

The design task is to find a non-linear transformation $\xi = T(x)$, separating the disturbed from the undisturbed part of the model. This separation can be achieved if, and only if

$$\frac{\partial T(x)}{\partial x} E(x) = 0 \tag{9.16}$$

This relation constitutes a system of 1*st*-order linear partial differential equations which are to be solved simultaneously by $\xi = T(x)$. Suppose that solutions $\xi = T(x)$ of (9.16) and a relation $x = \Psi(\xi, y^*)$ exist, then the model can be rewritten as (Frank and Seliger, 1991; Seliger and Frank, 1991a; Seliger and Frank, 1991b; Seliger and Frank, 1993; Frank, 1994b; Frank and Ding, 1997):

$$\dot{\xi} = \frac{\partial T(x)}{\partial x} (A(x) + B(x)u + R(x)f)\Big|_{x=\Psi(\xi,y^*)} \tag{9.17}$$

where the output transformation $y^* = C^*(y)$ denotes a subset of the set of available measurements $y = C(x)$ which is subject to the condition

$$dim(y^*) < dim(y) \tag{9.18}$$

Suppose, furthermore, that a relation

$$Q(T(x), C(x)) = 0 \tag{9.19}$$

exists. Then a non-linear observer can be set up in order to estimate the undisturbed portion ξ of the state x. The resulting observer is of the form:

$$\dot{\hat{\xi}} = \frac{\partial T(\hat{x})}{\partial \hat{x}} \left(A(\hat{x}) + B(\hat{x})u + K(\hat{\xi}, y, u)Q(\hat{\xi}, y) \right)\Big|_{x=\Psi(\hat{\xi},y^*)} \tag{9.20}$$

where the design freedom provided by the feedback matrix $K(\hat{\xi}, y, u)$ can be used to stabilize the differential equation governing the dynamics of the estimation error $e = \hat{\xi} - \xi$. This is by nature a non-linear unknown input observer, also often referred to as a disturbance decoupled unknown input observer (Frank and Seliger, 1991; Seliger and Frank, 1991a; Seliger and Frank, 1991b). The relations $Q(\hat{\xi}, y)$ can be used conveniently as a residual:

$$r = Q(\hat{\xi}, y) = Q(\xi + e, y) \tag{9.21}$$

The estimation error e is governed by

$$\dot{e} = \rho(e, t) - \frac{\partial T(x)}{\partial x} R(x)f \tag{9.22}$$

where the system of the non-linear differential equations $\dot{e} = \rho(e,t)$ must be designed such that the equilibrium point $e = 0$ is at least locally asymptotically stable. The residual will then converge to zero in the fault-free case. On the other hand, all faults in vector f will be reflected in e, if

$$rank\left\{\frac{\partial T(x)}{\partial x}R(x)\right\} = rank\left\{R(x)\right\} \tag{9.23}$$

If, in addition to (9.16) and (9.23), the conditions

$$\left\{\begin{array}{lcl} \frac{\partial T(x)}{\partial x}A(x) & = & FT(x) + \Phi_0(C(x)) \\ \frac{\partial T(x)}{\partial x}B(x) & = & \Phi_1(C(x)) \end{array}\right. \tag{9.24}$$

are satisfied, a residual generator with stable linear error dynamics can be designed (Seliger and Frank, 1991a). In the above equation, F is a stable constant matrix, $\Phi_0(C(x))$ and $\Phi_1(C(x))$ are suitable output transformations.

The residual generator can be then implemented as follows:

$$\left\{\begin{array}{lcl} \dot{\hat{\xi}} & = & = F\hat{\xi} + \Phi_0(y) + \Phi_1(y)u \\ r & = & Q(\hat{\xi}, y) \end{array}\right. \tag{9.25}$$

In this case, the estimation error and the residual evolve according to

$$\left\{\begin{array}{lcl} \dot{e} & = & Fe - \frac{\partial T(x)}{\partial x}R(x)f \\ r & - & Q(T(x) + e, C(x)) \end{array}\right. \tag{9.26}$$

It is important to note that (9.12) and (9.25) are stable on the complete disturbance de-coupling domain (Frank, 1994b; Frank and Ding, 1997). No such statement can be made regarding the observer described by Eq. (9.20).

(4) Bilinear observer approach

As a special class of non-linear systems, bilinear models are often used to represented a variety of industrial systems including chemical processes, hydraulic drive systems, gas-burning furnace systems and heat exchange systems (Yu et al., 1996). Recently, FDI problems for bilinear systems have been studied extensively (Yu et al., 1994b; Yu and Shields, 1994; Yu et al., 1996; Yu and Shields, 1996; Bennett et al., 1996a; Shields, 1997; Yu and Shields, 1997; Yang and Saif, 1997; Mechmeche and Nowakowski, 1997). All studies are based on the use of bilinear observers. There are two main approaches in designing bilinear observers, the first approach originated by Funahashi (1976) uses the Lyapunov method. The second approach is based on the use of techniques developed for linear unknown input observers. The basic idea is to treat non-linear terms the same way as unknown inputs and decouple their effects from the state estimation error. Although the system is bilinear, the observer used to estimate the state is actually linear. This idea has been demonstrated in Section 3.3.3. This approach was first studied by Hara and Furuta (1976) for bilinear systems without unknown inputs and later extended by Hać (1992) for bilinear systems with unknown inputs. Apart from the study by Bennett et al. (1996a) which uses the Lyapunov method, most other studies on bilinear systems FDI are based on the extensions of the unknown input observer method developed by Hać (1992). The most recent development in designing bilinear observers can be found in Hou and Pugh (1997).

Consider the bilinear system of the form

$$\begin{cases} \dot{x}(t) = Ax(t) + \sum_{i=1}^{r} B_i u_i(t) x(t) + E_1 d(t) \\ y(t) = Cx(t) + E_2 d(t) \end{cases} \tag{9.27}$$

where x, y and d are the state, output and unknown disturbance vectors, respectively. u_i, $i = 1, \cdots, r$, are entries of the known input vector u. A, B_i, C, E_1 and E_2 are real matrices of appropriate dimensions. Without loss of generality, it is assumed that the matrix $[C \quad E_2]$ has full row rank.

Definition 9.1 *A bilinear system of the form*

$$\begin{cases} \dot{\zeta}(t) = F\zeta(t) + Gy(t) + \sum_{i=1}^{r} M_i u_i(t)\zeta(t) + \sum_{i=1}^{r} L_i u_i(t) y(t) \\ \hat{x}(t) = H\zeta(t) + Ny(t) \end{cases} \tag{9.28}$$

is known as a state observer for the system (9.27) if $\hat{x}(t)$ with an arbitrary $\zeta(0)$ converges to x for arbitrary $x(0)$, $u(t)$, $y(t)$ and $d(t)$.

According to Hać (1992) and Hou and Pugh (1997), the necessary and sufficient conditions for (9.28) to be a state observer of the system (9.27) are:

$$\begin{cases} F \text{ has stable eigenvalues} \\ TA - FT = GC \\ HT + NC = I \\ NE_2 = 0 \\ TE_1 = GE_2 \\ L_i E_2 = 0 \quad i = 1, \cdots, r \\ TB_i = L_i C \quad i = 1, \cdots, r \\ M_i = 0 \quad i = 1, \cdots, r \end{cases} \tag{9.29}$$

The conditions in (9.29) are general conditions. However, they are not easy to verify and they also do not provide any assistance for the observer design. Hou and Pugh (1997) developed a straightforward method for bilinear observer design with conditions which are easy to verify. This method is introduced here.

Theorem 9.1 *(Hou and Pugh, 1997): For the bilinear system (9.27), a state observer in the form of*

$$\begin{cases} \dot{\zeta}(t) = F\zeta(t) + Gy(t) + \sum_{i=1}^{r} L_i u_i(t) y(t) \\ \hat{x}(t) = H\zeta(t) + Ny(t) \end{cases} \tag{9.30}$$

exists if and only if

$$rank\begin{bmatrix} E_2 & CE_1 & C\tilde{E} \\ 0 & E_2 & 0 \end{bmatrix} = rank(E_2) + rank\begin{bmatrix} E_1 & \tilde{E} \\ E_2 & 0 \end{bmatrix} \tag{9.31}$$

$$rank\begin{bmatrix} -sI + A & E_1 & \tilde{E} \\ C & E_2 & 0 \end{bmatrix} = dim(x) + rank\begin{bmatrix} E_1 & \tilde{E} \\ E_2 & 0 \end{bmatrix} \quad \forall s \in \mathbb{C},\ Re(s) \geq 0 \tag{9.32}$$

where

$$\begin{aligned} \tilde{E} &= [\tilde{B} \ \ 0](I - [C \ \ E_2]^+[C \ \ E_2]) \\ &= \tilde{B}[I - C^T(CC^T + E_2E_2^T)^{-1}C - C^T(CC^T + E_2E_2^T)^{-1}E_2] && (9.33) \\ \tilde{B} &= [B_1 \ B_2 \ \cdots \ B_r] && (9.34) \\ \tilde{L} &= [L_1 \ L_2 \ \cdots \ L_r] && (9.35) \end{aligned}$$

Note that $(\cdot)^+$ denotes a pseudo-inverse.

To design a state observer for the bilinear system (9.27), Hou and Pugh (1997) suggested to transform the original system model into an equivalent quasi-linear system representation. The design principle is given in the following Theorem.

Theorem 9.2 *(Hou and Pugh, 1997): There exists a state observer in the form of (9.30) for the bilinear system (9.27) if and only if there exists a state observer for the equivalent quasi-linear system*

$$\begin{cases} \dot{\xi}(t) = \bar{A}\xi(t) + \sum_{i=1}^{r} \bar{B}_i u_i(t)y(t) + \bar{B}_2 y(t) \\ \eta(t) = \bar{C}\xi(t) \end{cases} \tag{9.36}$$

The relation between x and ξ is given by

$$x(t) = \xi(t) + \bar{B}(I - E_2E_2^+)y(t) \tag{9.37}$$

and the new measurement η is given by

$$\eta(t) = (I - \bar{C}\bar{B})(I - E_2E_2^+)y(t) \tag{9.38}$$

The matrices in (9.36)–(9.38) are defined as

$$\begin{cases} \bar{E} = [E_1 \ \ \tilde{E}] \\ \hat{E} = [E_2 \ \ 0] \\ \bar{C} = (I - E_2E_2^+)C \\ \bar{B} = \bar{E}(I - \hat{E}^+\hat{E})[\bar{C}\bar{E}(I - \hat{E}^+\hat{E})]^+ \\ \bar{A} = (I - \bar{B}\bar{C})(A - \bar{E}\hat{E}^+C) \\ \bar{B}_2 = (I - \bar{B}\bar{C})\bar{E}\hat{E}^+ + \bar{A}\bar{B}(I - E_2E_2^+) \\ \bar{B}_i = (I - \bar{B}\bar{C})B_iC^T(CC^T + E_2E_2^T)^{-1} \ ; \quad i = 1, \cdots, r \end{cases} \tag{9.39}$$

The condition for a state observer for the equivalent system (9.36) to exist is that the pair $\{\bar{C}, \bar{A}\}$ is detectable which is guaranteed by the existence of state observers for the bilinear system (9.27) (Hou and Pugh, 1997). Therefore, a bilinear observer for the system (9.27) can be easily designed by using the equivalent system (9.36) together with the relation (9.37).

To facilitate further bilinear observer applications, the following corollary provides the bilinear observer design method for systems with the linear known input term in the state and output equations.

Corollary 9.1 *(Hou and Pugh, 1997): There exists a state observer in the form (9.30) for the bilinear system*

$$\begin{cases} \dot{x}(t) = Ax(t) + Bu(t) + \sum_{i=1}^{r} B_i u_i(t)x(t) + E_1 d(t) \\ y(t) = Cx(t) + Du(t) + E_2 d(t) \end{cases} \tag{9.40}$$

if and only if there exists a state observer for the equivalent quasi-linear system

$$\begin{cases} \dot{\xi}(t) = \bar{A}\xi(t) + \sum_{i=1}^{r} \bar{B}_i u_i(t)(y(t) - Du(t) + \bar{B}_1 u(t) + \bar{B}_2 y(t) \\ \eta(t) = \bar{C}\xi(t) + \bar{D}u(t) \end{cases} \tag{9.41}$$

The relation between x and ξ is given by

$$x(t) = \xi(t) + \bar{B}(I - E_2 E_2^+)y(t) - \bar{B}(I - E_2 E_2^+)Du(t) \tag{9.42}$$

and the new measurement η is given by

$$\eta(t) = (I - \bar{C}\bar{B})(I - E_2 E_2^+)y(t) \tag{9.43}$$

In addition to (9.39), the matrices used in (9.41)–(9.43) are defined as

$$\begin{cases} \bar{D} = (I - \bar{C}\bar{B})(I - E_2 E_2^+)D \\ \bar{B}_1 = (I - \bar{B}\bar{C})(B - \bar{E}\hat{E}^+ D) - \bar{A}\bar{B}(I - E_2 E_2^+)D \end{cases} \tag{9.44}$$

Examples which illustrate the design of bilinear observers can be seen in Hou and Pugh (1997).

9.3 Neural Networks in Fault Diagnosis of Non-linear Dynamic Systems

The application of neural networks in non-linear dynamic systems FDI is discussed in this section. Problems such as network structure selection and training algorithms are not discussed here as the neural network is considered a well-established mathematical tool. This section discusses how neural networks can be used in FDI and the advantages over conventional model-based techniques. An approach developed by Patton, Chen and Siew (1994) for detecting and isolating faults in non-linear dynamic processes using neural networks is presented. A neural network is used to model a multi-input and multi-output non-linear dynamic system. After training, the neural network can give very accurate estimation of the system output. Using the residual generation concept developed in model-based fault diagnosis, the weighted difference between actual and estimated outputs is used as a residual to detect faults. When the magnitude of this residual exceeds a pre-defined threshold, it is likely the system is faulty. In order to locate faults in the system (fault isolation) reliably, a secondary neural network is used to examine features in the residual. A particular feature would correspond to a specific fault location. Based on feature extraction and classification principles, the second neural network can locate (or isolate) faults reliably. It is worth mentioning that unlike most of the previous work in which neural networks are used for pattern recognition, the approach presented here uses neural networks in all stages of FDI.

9.3.1 The application of neural networks to non-linear dynamic systems FDI

9.3.1.1 Neural networks as models of non-linear dynamic systems. The linearized model is not a very suitable representation for systems with a high nonlinearity and wide ranging dynamics. For such systems, it is more efficient to use a non-linear model for FDI. One of the most important advantages of neural networks is their ability to implement non-linear transformations for functional approximation problems, i.e., neural networks are capable of forming an arbitrarily close approximation to any continuous non-linear mapping (function), given suitable weighting factors and a network architecture (Narendra and Parthasarthy, 1990; Hunt et al., 1992; Narendra, 1996). A neural network is a massively parallel, interconnected network of elementary units called neurons. Inputs to each neuron are combined and the neuron produces an output. The multi-layer feed-forward percepton and recurrent networks are two main classes of networks. From a system theoretic point of view, the feed-forward network is only a *static non-linear mapping* between inputs and outputs. Without modification, the feed-forward network cannot be used to represent dynamic systems. The recurrent network can be used to model non-linear dynamic systems, as the dynamic feedback can be introduced in the network (Pham and Oh, 1992). Due to the involvement of feedback, the network can be unstable. This is perhaps the most important issue that has to be addressed before the full potential of the recurrent network can be applied in practice. Narendra and Parthasarthy (1990) studied the ways of representing non-linear dynamic systems by feed-forward networks. They used various combinations of linear dynamic systems with feed-forward networks. In this form, the non-linear dynamic system is separated into subsystems which fall into two catalogues: the linear dynamic system (represented by a transfer matrix) and the non-linear static system (represented by a feed-forward network).

The simplest approach in representing non-linear dynamic systems is to use a combination of a feed-forward network with some time delay units. Assume that a non-linear dynamic systems is described as:

$$y(k) = F(y(k-1), \cdots, y(k-n), u(k), \cdots, u(k-n)) \tag{9.45}$$

where $u(k) \in \mathbb{R}^r$ is the input vector and $y(k) \in \mathbb{R}^m$ is the output vector, $F(\cdot, \cdots, \cdot)$ represents a general non-linear function. A feed-forward network with weight matrix W can now be used to represent this static non-linear function, with output:

$$\hat{y}(k) = NN(W, y(k-1), \cdots, y(k-n), u(k), \cdots, u(k-n)) \tag{9.46}$$

where n is the system order. In this way, a non-linear dynamic system can be modeled by a feed-forward network combining with some time delay units. This model illustrated in Fig.9.1 is sometimes called the *one step prediction* model.

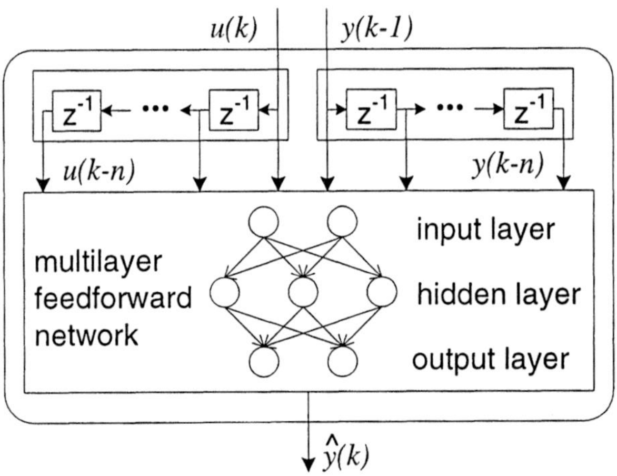

Figure 9.1. The neural network model of a non-linear dynamic system

9.3.1.2 Neural networks as modeling tools. For complex and uncertain systems, the derivation of analytical models from physical relationships can become very complicated and time consuming. Sometimes, it is very difficult or even impossible to derive analytical equations for some systems, such as hydraulic and aerodynamic systems. A "cheap" and universal tool for automatically building system models is desirable. The neural network can be used as such a tool as it can automatically extract the system features from historical training data using the learning algorithm, whilst requiring little or no *a priori* knowledge about the system. This provides a great flexibility for modeling complex non-linear systems. Moreover, the learning can be carried out on-line. This makes it easy to model time varying systems. The advantage of the neural networks over conventional identification methods is that it can handle all linear and non-linear systems universally, although it is not very efficient in representing linear systems. The network structure is normally chosen by experience together with a trial-and-error procedure. The network weighting matrix W will be determined during the training. The criterion is to minimize the modeling error $||y(k) - \hat{y}(k)||$. The back propagation algorithm is the most commonly used training methods (Narendra and Parthasarthy, 1990).

9.3.1.3 Neural networks as classifiers. After the residual has been generated, a decision-making mechanism should be used to determine fault occurrence and location. Traditionally, decision-making is implemented via a thresholding logic using either fixed or adaptive thresholds or statistical testing methods. The main task in decision-making is to classify the residuals into a number of distinguishable patterns corresponding to different faulty situations. Hence decision-making can be based on the pattern recognition principle. Pattern recognition implies initiating certain actions based on the observation of input data. The input representing a pattern is known as the measurement or feature vector. The function performed by a pattern recognition system is the mapping of the input feature vector into one of the various decision classes. In fault diagnosis, these decision classes are the different types (or locations) of faults occurring in the system. One of the advantages of neural networks is their ability to partition the pattern space for classification problems. Hence, a neural network can be used as a classifier (or pattern recognizer) to partition residual patterns and activate alarm signals. It can therefore detect and isolate the faults accordingly. In the training of neural networks to classify faults, output node values of 0.1 and 0.9 are used to indicate fault-free and faulty cases, respectively. In the application to fault diagnosis, output values above 0.5 indicate a fault. If fault patterns are known to occur for specific faults, this information could be stored in the neural network by choosing the training set of the neural network to co-ordinate with known faults.

9.3.1.4 Neural networks used to handle both analytical (quantitative) and heuristic (qualitative) information. It has been pointed in Section 2.14 that the fault diagnosis performance can be greatly enhanced by integrating together quantitative and qualitative information. There are many research studies in fault diagnosis using either quantitative or qualitative information. However, the theory and mathematic tools used in these two areas are so diversified that very few investigators consider to integrate them. This diversity in the use of terms and mathematical tools makes the information integration very difficult. It would be ideal to have an universal mathematical tool which can handle both quantitative and qualitative information. It is no doubt that neural networks can handle quantitative information. However, qualitative information is normally expressed as either Boolean or fuzzy logic. It is possible to implement any Boolean logic relation using neural networks (Morgan and Scofield, 1991). Moreover, fuzzy logic can also be handled by neural networks through the use of the fuzzy neural network concept (Hayashi, Buckley and Czogala, 1992). All these capabilities make the neural network an ideal and promising tool in the information integration. Moreover, a trained neural network can be used to evaluate the reliability of information provided by either quantitative or qualitative methods and decide which weights should used in the information fusion. The idea of integrating quantitative and qualitative diagnostic methods via neural networks is depicted in Fig.9.2.

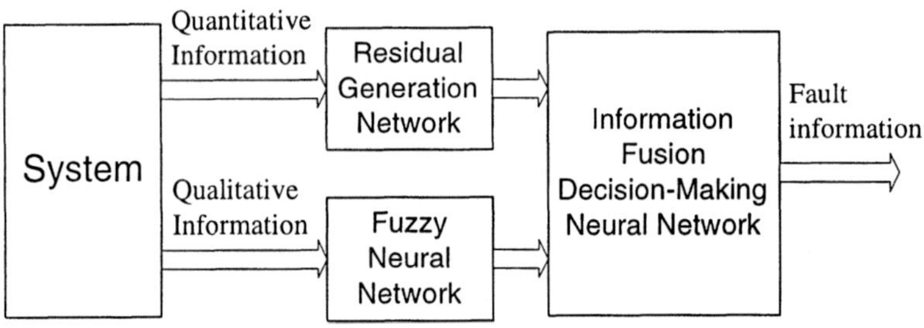

Figure 9.2. A conceptual structure of the integrated fault diagnosis approach

9.3.2 A fault diagnosis scheme based on neural networks

The neural networks-based FDI scheme developed by Patton, Chen and Siew (1994) is illustrated in Fig.9.3. This scheme comprises two stages: *residual generation* and *decision-making*. The residual generation scheme described here is based on the comparison of actual and anticipated system responses. The anticipated system response is generated by a neural network-based prediction model which illustrate in Fig.9.1. The difference between actual and predicated outputs gives rise a residual vector which is:

$$r(k) = y(k) - \hat{y}(k)$$

where

$$\hat{y}(k) = NN(W, y(k-1), \cdots, y(k-n), u(k), \cdots, u(k-n)) \tag{9.47}$$

where $NN(\cdot, \cdots, \cdot)$ denotes a neural network-based non-linear functional mapping. The residuals generated in this way should be independent of the system operating state under nominal plant operating conditions. In the absence of faults, the residual is only due to unmodeled noise and disturbance. When a fault occurs in the system, the residual deviates from zero in characteristic ways. In the second stage of fault diagnosis, a neural network-based classifier is used to partition the residual vector to patterns corresponding to different system faulty situations. The neural network is trained to recognize complex features in residuals and then gives detection and isolation information. To diagnose faults reliably, the classification network has to be trained using data from all possible fault situations. A neural network trained only in the fault-free situation cannot expected to perform well for faulty situations.

Concerning the approach introduced here, it may be said that the neural network is used as an alternative to the traditional state estimator such as a

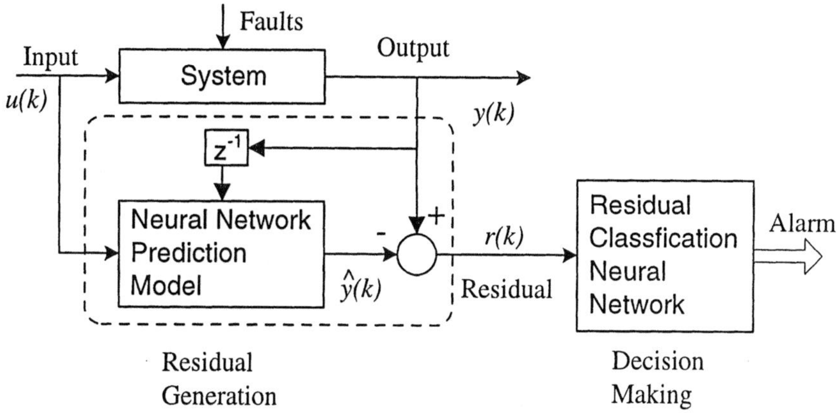

Figure 9.3. Fault detection and isolation scheme using two neural networks

Luenbeger observer or a Kalman filter. An important feature of a neural network used for output prediction is that it will learn the system dynamics during a training session made over several training cycles, with training data coming from either previous simulation or actual on-line data. After each training cycle, the neural network will know more about the system dynamic behavior. One of the important features of a neural network is its ability to learn the dynamic behavior of a non-linear system automatically, as long as the neural network architecture has at least 3 (input, output and one hidden) layers. Neural network-based approaches have several potential advantages over traditional estimation methods, to name a few: powerful non-linear mapping properties, noise tolerance, self-learning and parallel processing capabilities.

The two-stage approach is somewhat complicated because two neural networks need to be trained. The output prediction is unnecessary if we only need to diagnose faults, although it is very important for system reconfiguration. To increase efficiency in computation, two stages can be combined together as one, i.e, two networks are combined as single network. The FDI scheme using only one neural network is illustrated in Fig. 9.4. This scheme uses one neural network whose inputs are system input u and output y. The output of this network is the alarm signal which declares the occurrence and location of faults.

9.3.3 An application of the neural networks-based fault diagnosis in a laboratory system

A laboratory 3 tanks system is used here to demonstrate the effectiveness of the neural network-based fault diagnosis scheme. The system is widely accepted as a laboratory demonstrator and has industrial parallels including the cooling water circuits of chemical distillation columns and chemical reactors in general, and feed water systems in power stations. The 3-tanks system, although simple

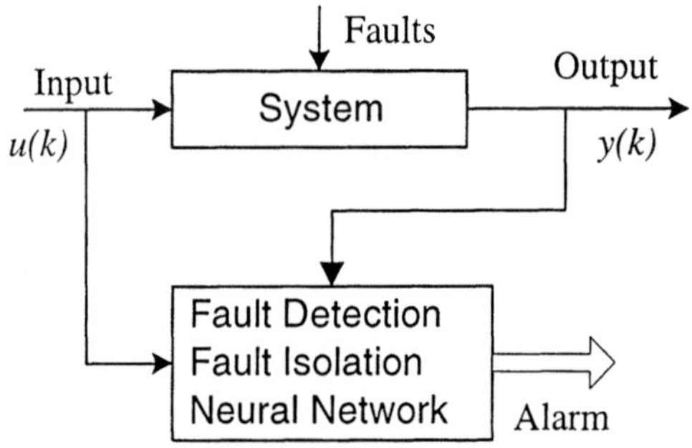

Figure 9.4. Fault detection and isolation scheme using two neural networks

in operation, enables generic concepts to be developed and, very importantly, to be tested physically. The simplicity of the system does not obscure the fundamentals and basic ideas which are being studied.

9.3.3.1 Description of the three tanks system. The 3 tanks system shown in Fig.9.5 is a non-linear controlled system consisting of three tanks of circular cross-section which are connected to each other through connecting pipes of circular cross-section. There are two pumps controlling the incoming flows $Q_1(t)$ and $Q_2(t)$ and driven by an electronic power device with appropriate control strategy implemented on microcomputer. All three water levels $h_1(t)$, $h_2(t)$ and $h_3(t)$ are measured via piezoresistive pressure sensors. For the simulation of plugging in the interconnections between the tanks and for the simulation of leaks in the tanks, various valves are installed. The numerical values of the physical parameters of the system can be found in Wünnenberg (1990).

The dynamic model of the system is derived using the incoming and outgoing mass flow under consideration of Torricellies law and is described by:

$$\begin{aligned}
A\frac{dh_1}{dt} &= -a_1 s_{13}\text{sgn}(h_1-h_3)\sqrt{2g|h_1-h_3|}+Q_1+Q_{f1} \\
A\frac{dh_2}{dt} &= a_3 s_{23}\text{sgn}(h_3-h_2)\sqrt{2g|h_3-h_2|}-a_2 s_o\sqrt{2gh_2}+Q_2+Q_{f2} \\
A\frac{dh_3}{dt} &= a_1 s_{13}\text{sgn}(h_1-h_3)\sqrt{2g|h_1-h_3|} \\
&\quad -a_3 s_{23}\text{sgn}(h_3-h_2)\sqrt{2g|h_3-h_2|}+Q_{f3}
\end{aligned}$$

where $Q_{fi}(t)$ $(i=1,2,3)$ denote the additional mass flows into the tanks cause by leaks or plugging in the various tanks or pipes, the incoming mass flows $Q_1(t)$ and $Q_2(t)$ are control inputs while the three state variables are the heights of the liquid in the tank. To define the state $x(t)$, input $u(t)$ and fault $f(t)$ vectors

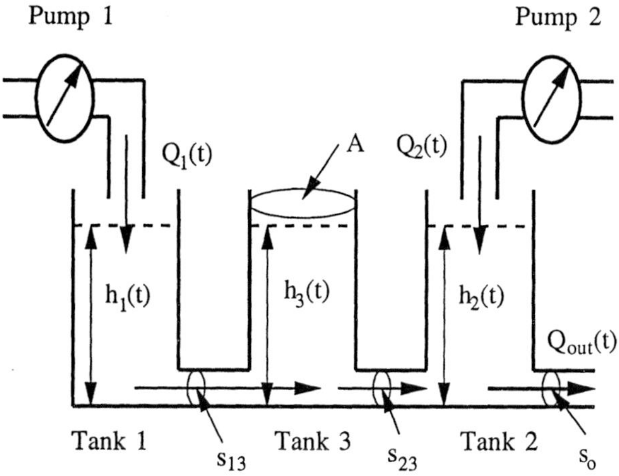

Figure 9.5. Three tanks system

as:

$$x(t) = \begin{bmatrix} h_1 \\ h_2 \\ h_3 \end{bmatrix} \quad u(t) = \begin{bmatrix} Q_1 \\ Q_2 \end{bmatrix} \quad f(t) = \begin{bmatrix} Q_{f1} \\ Q_{f2} \\ Q_{f3} \end{bmatrix}$$

The state space equation of the fault-free system is thus:

$$\frac{dx(t)}{dt} = F_1(x(t), u(t))$$

The discrete form of this model is:

$$x(k) = F_2(x(k-1), u(k-1))$$

where $F_1(.,.)$ and $F_2(.,.)$ denote two non-linear functions. In this particular application, all state variables are measurable and hence the system can be modeled as a first-order state space equation. As discussed before, a non-linear mapping can be approximated by a multi-layer feed-forward neural network. Hence, a prediction of the state can be obtained as:

$$\hat{x}(k) = NN(x(k-1), u(k-1))$$

where NN denotes a neural network which input is $\{x(k-1), u(k-1)\}$ and the output is $\hat{x}(t)$.

9.3.3.2 Simulated fault detection and isolation results. The first stage of fault diagnosis is to generate a residual signal which is the difference between the actual and predicted system state vectors. In the simulation, a $5-2-3$ neural network is trained to predict the state of the plant using the control variables as well as a set of past state variables. This network then runs in parallel with the system and the difference between the two state vectors is used as a residual vector:

$$r(k) = x(k) - \hat{x}(k) = x(k) - NN_{\{5-2-3\}}(x(k-1), u(k-1))$$

where NN denotes a multi-layer feed-forward network and the subscript $5-2-3$ denotes the network structure. After the residual has been generated, the second neural network is trained to classify the residual and subsequently detect and isolate faults. The output of the second neural network is the alarm (or indicating) signal $e = [e_1, e_2, e_3]$ and e_i indicating the leak (fault) in the *ith* tank. In our study, a $3-8-3$ neural network is used as the fault classifier to generate alarms.

$$e(k) = NN_{\{3-8-3\}}(r(k))$$

The desired training data for this neural network is assigned as in the Table 9.1. The trained neural network will then be used to evaluate the residuals.

Table 9.1. Training data for fault classifier

fault	e_1	e_2	e_3
tank 1	0.9	0.1	0.1
tank 2	0.1	0.9	0.1
tank 3	0.1	0.1	0.9

Ideally, the *ith* alarm e_i will be 0.9 when a fault occur in the *ith* tank, otherwise it will be 0.1. However, the realistic situation will be different due to noise and parameter perturbation and other uncertain factors. Therefore, a threshold (0.5 in this study) is used in the classification.

$$\begin{cases} e_i < 0.5 & \text{no fault in the } ith \text{ tank} \\ e_i > 0.5 & \text{fault in the } ith \text{ tank} \end{cases}$$

In this study, a variety of incipient and abrupt faults have been considered. The typical shapes of fault signals are show in Fig.9.6.

Fig.9.7 shows that an incipient fault occurs in tank 1 at 15 seconds and the network detects the fault and isolates it correctly. Fig.9.8 shows the response of the classifier to an abrupt fault which occurred in tank 3 at 10 seconds and disappeared after 25 seconds.

As discussed in Section 9.3.2, two FDI stages can be combined into one using one neural network. This one stage approach is also assessed here. The input-output relation of the combined neural network in this approach is as follows:

$$e(k) = NN_{\{8-6-3\}}(x(k), x(k-1), u(k-1))$$

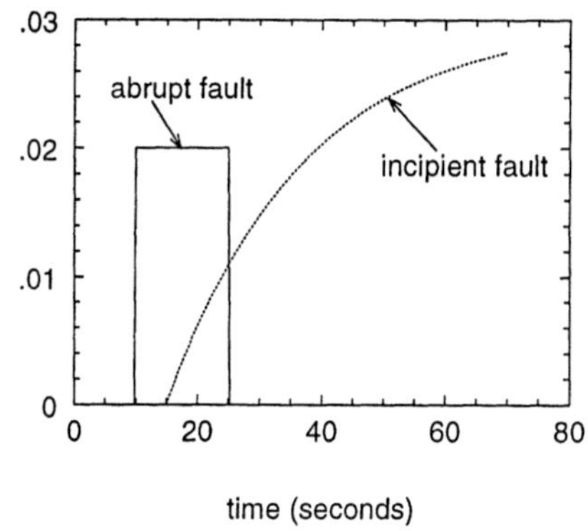

Figure 9.6. Typical shapes of fault signals

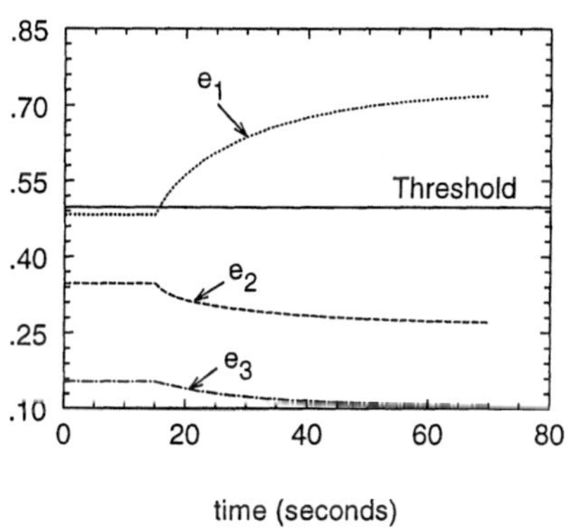

Figure 9.7. Alarm signals when an incipient fault occurred in tank 1

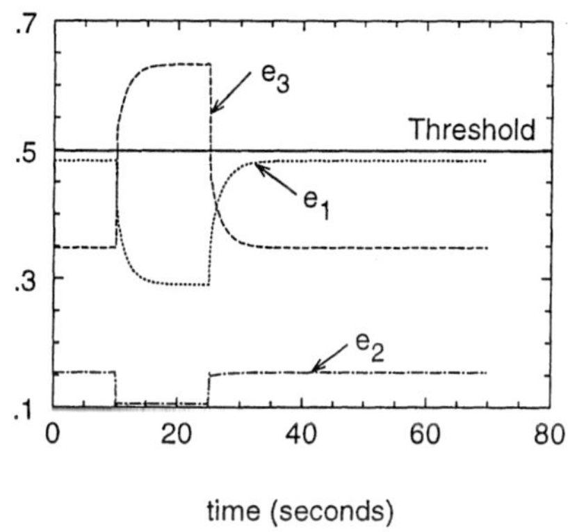

Figure 9.8. Alarm signals when an abrupt fault occurred in tank 3

This $8-6-3$ neural network is trained as the alarm signal generator to diagnose faults. Fig.9.9 shows a single fault occurring in tank 2 at 30 seconds and the network detects and isolates the fault correctly. Fig.9.10 shows two faults occurring in two tanks, a fault in tank 1 at 35 seconds and another one in tank 3 at 40 seconds. The network detects and isolates the two faults correctly.

Remarks: Neural networks present an alternative approach to the automated diagnosis of faults. By learning from the historical fault information, knowledge is stored in the neural network which provides it with associative memory. Neural networks are able to filter out noise and classify faults, and hence provide an automated, highly sensitive and economical diagnosis of faults even in the presence of noise. The advantages of neural networks in FDI of non-linear dynamic systems have been discussed and the ways of using these advantages are illustrated. The FDI scheme introduced is applied to a laboratory 3-tanks system. The simulation results show that, with two neural networks, leak, blockage and sensor faults can be correctly detected and isolated. The simulation results

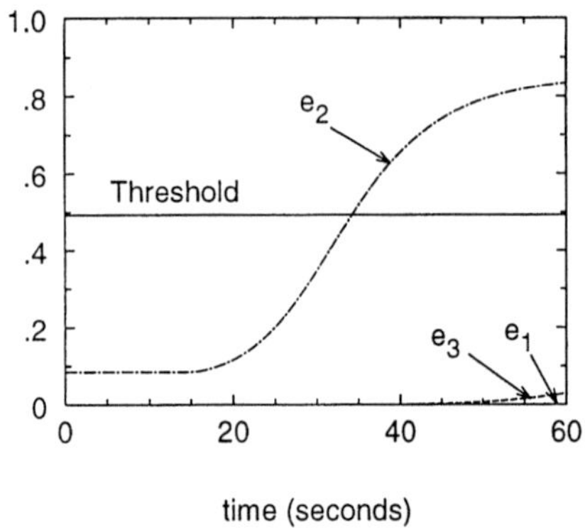

Figure 9.9. Alarm signals when an incipient fault occurred in tank 2

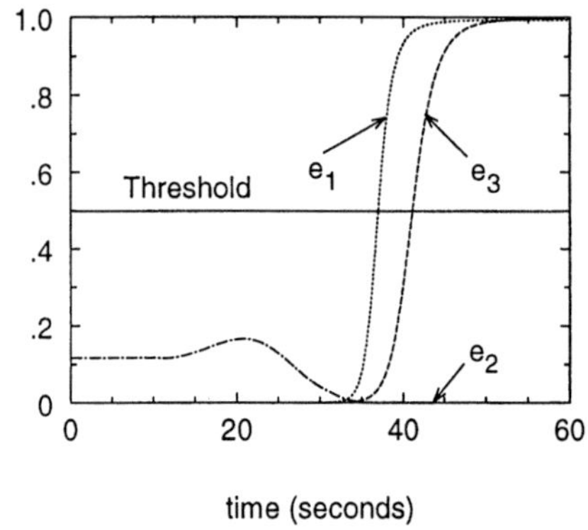

Figure 9.10. Alarm signals when two incipient faults occurred in tank 1 and tank 3

demonstrated that the neural network-based FDI scheme can diagnose faults in non-linear dynamic system reliably providing sufficient training. It has been proved that the scheme can be used to detect and isolate incipient faults. However, there is a certain limit in the detection of very small faults as it becomes impossible to train the neural network when the fault signal becomes comparable with noise and modeling errors. Further work will focus on examining the robustness of the scheme against noise and uncertainty. A good fault diagnosis scheme should perform well even when the operating conditions have been changed after training. One possible way forward is to train the network periodically.

9.4 Fuzzy Observers for Non-linear Dynamic Systems Fault Diagnosis

This section presents a novel fault detection and isolation scheme for non-linear dynamic systems. This scheme combines fuzzy logic with the model-based method to formulate the so-called fuzzy observers. The main idea is to use the Takagi-Sugeno fuzzy model (Takagi and Sugeno, 1985). Using this model, a non-linear dynamic system is described by a number of locally-linearized models. Under the fuzzy logic observer scheme, a number of local linear observers are designed and the state estimate is given by a fuzzy fusion of local observer outputs. The diagnostic signal - a residual is the difference between the estimated and real system outputs. Although all local observers are stable, the global observer is not necessarily stable (i.e. the state estimate may not be converge). In this section, the linear matrix inequality method is used to analyze the global stability of the fuzzy observer and some measures to achieve global stability are presented. In Section 9.4.1, the Takagi-Sugeno fuzzy model and stability analysis is discussed. The concept and structure of fuzzy observers is presented in Section 9.4.2. In Section 9.4.3, the fuzzy logic observer is used to detect and isolate faults in an induction motor of a rail traction system and the computer as well as test-rig simulation results demonstrate the effectiveness of the scheme.

9.4.1 Takagi-Sugeno fuzzy model and stability analysis

9.4.1.1 Takagi-Sugeno fuzzy model. A Takagi-Sugeno (T-S) fuzzy model is a simple way to describe a non-linear dynamic system using locally linearized linear models (Takagi and Sugeno, 1985). According to the T-S model, a non-linear dynamic system can be linearized around a number of operating points. Each linear model represents the local system behavior around the operating point. The global system behavior is described by a fuzzy fusion of all linear model outputs. The model is described by fuzzy IF-THEN rules which represent local linear relations of the non-linear system. At an operating point, the rule to describe the system is as follows.

Rule i: $(i = 1, 2, \ldots, N)$

$$\text{If } w(t) \text{ is } M_i \quad \text{Then} \quad \begin{cases} \dot{x}(t) = A_i x(t) + B_i u(t) \\ y(t) = C_i x(t) + D_i u(t) \end{cases} \tag{9.48}$$

where $x(t) \in \mathbb{R}^n$, $u(t) \in \mathbb{R}^r$, $y(t) \in \mathbb{R}^m$ and A_i, B_i, C_i and D_i are time invariant matrices of appropriate dimensions. The vector $w(t)$ is termed the *premise variable*, whereas M_i is a fuzzy set and N is the number of IF-THEN rules.

Given the input vector $u(t)$, the global state and output of the system are inferred as follows:

$$\begin{cases} \dot{x}(t) = \sum_{i=1}^{N} \mu_i(w(t)) \left[A_i x(t) + B_i u(t)\right] \\ y(t) = \sum_{i=1}^{N} \mu_i(w(t))[C_i x(t) + D_i u(t)] \end{cases} \tag{9.49}$$

$\mu_i(w(t))$ is the grade of membership of the premise variable, $w(t)$, or the tensor product of grade of memberships, if $w(t)$ is a vector. The membership grade functions $\mu_i(w(t))$ $(i = 1, \cdots, N)$ satisfy the following constraints

$$\begin{cases} \sum_{i=1}^{N} \mu_i(w(t)) = 1 \\ 0 \leq \mu_i(w(t)) \leq 1 \quad \forall i = 1, 2, ..., N \end{cases} \tag{9.50}$$

9.4.1.2 Stability analysis. To analyze the system stability, the unforced (zero-input) system model is considered:

$$\dot{x}(t) = \sum_{i=1}^{N} \mu_i(w(t)) A_i x(t) \tag{9.51}$$

According to Tanaka and Sugeno (1992), the stability of a system modeled by Eq.(9.51) can be verified using the following theorem.

Theorem 9.3 *(Tanaka and Sugeno, 1992): The system described by Eq.(9.51) is asymptotically stable if there exists a common positive matrix P such that the following inequality holds*

$$A_i^T P + PA_i < 0 \quad \textit{for all} \;\; i = 1, \cdots, N \tag{9.52}$$

This theorem, which gives a sufficient condition for ensuring stability, is an extension of the second Lyapunov theorem. The matrix inequality in Eq. (5) can be solved using the Linear Matrix Inequality (LMI) method (Boyd et al., 1994).

9.4.1.3 Eigenvalues assignment. Theorem 9.3 only guarantees the stability by restricting all eigenvalues of A_i $(i = 1, 2, \ldots, N)$ in the left hand side of the complex plane. To make sure the system is stable in the presence of uncertainty such as neglected high frequency dynamics etc, the eigenvalues are normally required to be assigned within a specific region in s-plane. The regional assignment is also required for achieving satisfactory time response and closed-loop damping. To discuss the assignment of eigenvalues in a specific region, the concept of an LMI region is introduced here.

Definition 9.2 (LMI Region) *(Chilali and Gahinet, 1996): A subset of the complex plane is called an LMI region if there exists a symmetric matrix $\Phi = [\phi_{ij}] \in \mathbb{R}^{k\times k}$ and a matrix $\Theta = [\theta_{ij}] \in \mathbb{R}^{k\times k}$, such that*

$$\Omega = \{z \in \Omega \;:\; f_\Omega(z) < 0\} \tag{9.53}$$

where: $f_\Omega(z) = \Phi + z\Theta + \bar{z}\Theta^T = [\phi_{ij} + \theta_{ij}z + \theta_{ji}\bar{z}]_{1<i,j<k}$ is called the characteristic function of the LMI region.

Theorem 9.4 *(Chilali and Gahinet, 1996): The matrix A has all its eigenvalues in Ω, if and only if there exists a symmetric matrix P such that*

$$M_\Omega(A, P) = \left[\phi_{ij}P + \theta_{ij}AP + \theta_{ji}PA^T\right]_{1<i,j<k} < 0 \tag{9.54}$$

This theorem provides a way of ensuring the eigenvalues of a matrix lie within a specific region. For example, if the eigenvalues of the matrix A are required to be assigned in a disc Ω_c, in the s-plane, of radius c_1 and center $(-c_0, 0)$, the LMI region characteristic function and associated matrices can be found as:

$$f_{\Omega_c}(z) = \begin{bmatrix} -c_1 & c_0 + z \\ c_0 + \bar{z} & -c_1 \end{bmatrix}; \quad \Phi = \begin{bmatrix} -c_1 & c_0 \\ c_0 & -c_1 \end{bmatrix}; \quad \Theta = \begin{bmatrix} 0 & 1 \\ 0 & 0 \end{bmatrix}$$

The condition for ensuring all eigenvalues of A within the disc Ω_c is thus:

$$\begin{bmatrix} -c_1 P & c_0 P + AP \\ c_0 P + PA^T & -c_1 P \end{bmatrix} < 0$$

Note that the classical Lyapunov stability theorem corresponds to the special case of Theorem 9.4 where $f_\Omega(z) = z + \bar{z}$. More complex convex regions can be formulated by the intersection of LMI regions as given by the following corollary (Chilali and Gahinet, 1996).

Corollary 9.2 *(Chilali and Gahinet, 1996): A matrix A has all its eigenvalues in the intersection of regions Ω_1 and Ω_2 ($\Omega_1 \cap \Omega_2$), if and only if there exists a positive definite matrix P such that*

$$M_{\Omega_1}(A, P) < 0 \quad \textit{and} \quad M_{\Omega_2}(A, P) < 0$$

Moreover if the characteristic functions of Ω_1 and Ω_2 are $f_{\Omega_1}(z)$ and $f_{\Omega_2}(z)$ respectively, the characteristic function of $\Omega_1 \cap \Omega_2$ is

$$f_{\Omega_1 \cap \Omega_2}(z) = diag\{f_{\Omega_1}(z),\ f_{\Omega_2}(z)\}$$

Theorem 9.4 only gives a sufficient condition for ensuring region eigenvalue constraint for a single matrix A. For the system described by T-S model (9.51), the global stability and satisfactory transient response and damping can be achieved by assigning eigenvalues of all linear sub-model matrices A_i ($i = 1, \cdots, N$) in the pre-defined region. This can be achieved if and only if there exists a symmetric matrix P such that

$$M_\Omega(A_l, P) = \left[\phi_{ij} P + \theta_{ij} A_l P + \theta_{ji} P A_l^T\right]_{1<i,j<k} < 0 \ \forall\, l \in \{1, \cdots N\} \quad (9.55)$$

9.4.2 Fuzzy observers and residual generation

For a non-linear dynamic system described by the T-S fuzzy model (Eqs. (9.48) & (9.49)), a fuzzy observer (Ma, Sun and He, 1998; Tanaka, Ikeda and Wang, 1998) can be designed to estimate the system state vector. For the fuzzy observer design, it is assumed that the fuzzy system model is locally observable, i.e., all (C_i, A_i) ($i = 1, \ldots, N$) pairs are observable. Using the same idea in T-S fuzzy model, a fuzzy observer utilizes a number of local linear time-invariant observers. Each local observer is associated with each fuzzy rule given below:

Rule i: $(i = 1, 2, \ldots, N)$

$$\text{IF } w(t) \text{ is } M_i \quad \text{Then} \begin{cases} \dot{\hat{x}}(t) = A_i \hat{x}(t) + B_i u(t) + K_i[y(t) - \hat{y}(t)] \\ \hat{y}(t) = C_i \hat{x}(t) + D_i u(t) \end{cases} \quad (9.56)$$

Using the idea of *parallel distributed compensation (PDC)* (Tanaka, Ikeda and Wang, 1996), the overall state estimation is a non-linear (fuzzy) combination of individual local observer outputs. The overall observer dynamics will then be a weighted sum of individual linear observers.

$$\left\{\begin{array}{lll}\dot{\hat{x}}(t) & = & \sum_{i=1}^{N}\mu_i(w)\left[A_i\hat{x}(t)+B_iu(t)+K_i(y(t)-\hat{y}(t))\right] \\ \hat{y}(t) & = & \sum_{i=1}^{N}\mu_i(w)[C_i\hat{x}(t)+D_iu(t)]\end{array}\right. \tag{9.57}$$

where the weights $\mu_i(w(t))$ $(i=1,\ldots,N)$ are the same as the weights used in T-S model (Eq.(2)). The implementation of this fuzzy observer can be carried out by a fuzzy fusion of N local linear observers. Suppose $\hat{x}_i \in \mathbb{R}^n$ and $\hat{y}_i \in \mathbb{R}^m$ are state and output estimates of the *ith* local observer, we can implement the local observers and the final state and output estimates as follows:

Local Observers:

$$i=1,\cdots,N \qquad \left\{\begin{array}{lll}\dot{\hat{x}}_i(t) & = & A_i\hat{x}(t)+B_iu(t)+K_i[y(t)-\hat{y}(t)] \\ \hat{y}_i(t) & = & C_i\hat{x}(t)+D_iu(t)\end{array}\right. \tag{9.58}$$

Global State and Output Estimates:

$$\left\{\begin{array}{lll}\hat{x}(t) & = & \sum_{i=1}^{N}\mu_i(w)\hat{x}_i(t) \\ \hat{y}(t) & = & \sum_{i=1}^{N}\mu_i(w)\hat{y}_i(t)\end{array}\right. \tag{9.59}$$

To analyze the convergence of the observer, the state estimation error ($e(t) = x(t)-\hat{x}(t)$) dynamics given by the following differential equation is examined.

$$\dot{e}(t)=\sum_{i=1}^{N}\sum_{j=1}^{N}\mu_i\mu_j(A_i-K_iC_j)e(t) \tag{9.60}$$

If the above error dynamic equation is stable, the state estimation will converge asymptotically to the real state. An observer with converging state estimation can be referred to as a stable observer. It can be proved that the stability of the above error dynamic equation can be verified by the following theorem.

Theorem 9.5 *(Ma et al., 1998; Tanaka et al., 1996): The fuzzy observer (Eq.(9.57)) is asymptotically stable if there exists a common positive definite matrix P such that for $i=1,\cdots,N$*

$$(A_i-K_iC_i)^TP+P(A_i-K_iC_i)<0 \tag{9.61}$$

and for $i<j\le N$

$$\left(\frac{A_i-K_iC_j+A_j-K_jC_i}{2}\right)^TP+P\left(\frac{A_i-K_iC_j+A_j-K_jC_i}{2}\right)<0 \tag{9.62}$$

The fuzzy observer given by Eq. (9.57) can be simplified if there are no uncertainty and non-linearity involved in the system output equation (i.e. $C_1 = C_2 = \cdots = C_N = C$) and there is no input term in the output equation. In this situation, the system output equation is:

$$y(t) = Cx(t) \tag{9.63}$$

This is very a common situation in practice because the output equation represents the relations between measurements and the system state variables. These relations are not normally affected by the system dynamics. The simplified fuzzy observer is given by:

$$\begin{cases} \dot{\hat{x}}(t) &= \sum_{i=1}^{N} \mu_i(w)[A_i\hat{x}(t) + B_iu(t) + K_i(y(t) - \hat{y}(t))] \\ \hat{y}(t) &= C\hat{x}(t) \end{cases} \tag{9.64}$$

This fuzzy observer can be implemented through the use of N local linear observers and a fuzzy fusion.

$$\begin{cases} \dot{\hat{x}}_i(t) &= A_i\hat{x}_i(t) + B_iu(t) + K_i[y(t) - \hat{y}(t)] \;\; ; \quad i = 1, \cdots, N \\ \hat{x}(t) &= \sum_{i=1}^{N} \mu_i(w)\hat{x}_i(t) \\ \hat{y}(t) &= C\hat{x}(t) \end{cases} \tag{9.65}$$

The schematic diagram of such a fuzzy observer is shown in Fig. 9.11 where it can be seen that a fuzzy inference engine is used to "select" the appropriate estimation, from those generated by the N parallel observers.

The transition between one observer to the other depends on the operating regime defined by the premise variable $w(t)$. The estimation error dynamics are given by the following differential equation:

$$\dot{e}(t) = \sum_{i=1}^{N} \mu_i(w(t))(A_i - K_iC)e(t) \tag{9.66}$$

The stability condition of the simplified fuzzy observer (Eq. (9.64)), given in the following corollary, is simpler than the original fuzzy observer (Eq. (9.60)).

Corollary 9.3 *The simplified fuzzy observer (described by Eq.(9.64)) is stable asymptotically if there exists a common positive definite matrix P such that:*

$$(A_i - K_iC)^TP + P(A_i - K_iC) < 0 \quad \textit{for } i = 1, \cdots, N \tag{9.67}$$

To ensure that the estimation error ($e(t)$) of the fuzzy observer has fast and well damped response, it is necessary to assign all local observer eigenvalues in a specific region in the s-plane. In the design of fuzzy observer given here, the eigenvalues of all local observer are assigned within a region $S(\alpha, \beta)$ shown in Fig. 9.12.

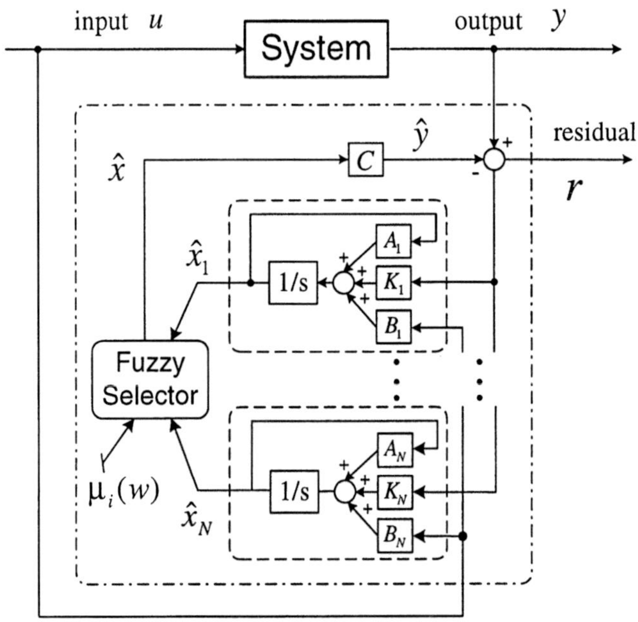

Figure 9.11. Fuzzy observer

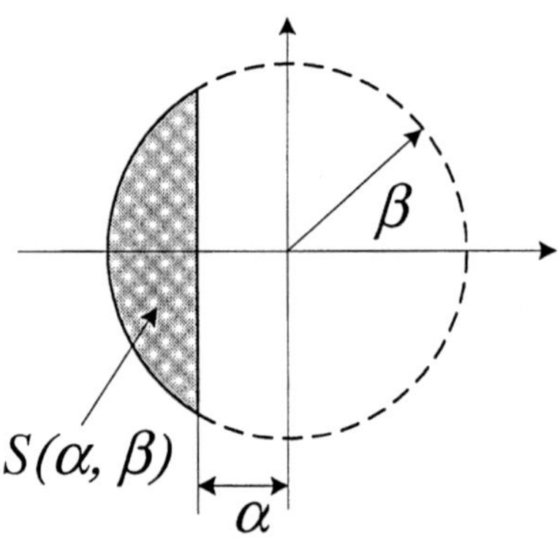

Figure 9.12. Eigenvalue region

According to the regional eigenvalue assignment technique discussed in Section 9.4.1.3, it can be shown that the fuzzy observer eigenvalue constraints can be verified using the following corollary.

Corollary 9.4 *All local observers in the fuzzy observer have their eigenvalues in the region $S(\alpha, \beta)$, if there exists a common positive definite matrix P such* *that:*

$$\begin{bmatrix} -\beta P & (A_i - K_iC)P \\ P(A_i - K_iC)^T & -\beta P \end{bmatrix} < 0 \quad \textit{for} \quad i = 1, \cdots, N \qquad (9.68)$$

and

$$(A_i - K_iC)^T P + P(A_i - K_iC) + 2\alpha P < 0 \quad \textit{for} \quad i = 1, \cdots, N \qquad (9.69)$$

These inequalities can be efficiently solved via numerical approach within the LMI framework (Boyd et al., 1994). If the eigenvalues of all local observer dynamic matrices $(A_i - K_iC)$ $(i = 1, \cdots, N)$ are within the region $S(\alpha, \beta)$, the observer error dynamics should have all its "poles" in the region $S(\alpha, \beta)$.

Once the state and/or output are estimated, the diagnostic signal - residual, can be generated by the comparison of measured and estimated outputs.

$$\text{Residual:} \quad r(t) = y(t) - \hat{y}(t) \qquad (9.70)$$

The faults can thus be diagnosed using a simple thresholding logic:

$$\|r(t)\| \begin{cases} \leq \text{Threshold} & \text{normal} \\ > \text{Threshold} & \text{faulty} \end{cases} \qquad (9.71)$$

9.4.3 Fault diagnosis of the induction motor in a railway traction system

9.4.3.1 System description. A rail traction control system illustrated by Fig. 9.13 is studied here. This system comprises a pulse width modulation (PWM) inverter which provides the necessary power to drive the machine. The induction motor is the most important part of this system and hence its reliability should be guaranteed. Since the torque and flux required by the controller cannot be measured, an estimator/observer should be used for control purposes. The possible faults in this system include incipient faults due to gain and bias changes, noise and inference pickup and intermittent faults such as the loss of measurement. The residual signal generated by the observer in this system is used as alarm signal for fault diagnosis.

The dynamics of an induction motor are somewhat complex because of the coupling effects between the stator and rotor phases. However, by using a power-invariant transform, the dynamics of a 3-phase induction motor could be represented by a 2-axis direct and quadrature (D-Q) model (Bose, 1986). The latter can then be modeled by the following bilinear differential equation (Bennett, Patton, Daley and Newton, 1996b; Bennett et al., 1996a; Bennett, 1998).

$$\begin{cases} \dot{x} = A(\omega)x + Bu = Ax + A_{non}\omega x + Bu \\ y = Cx \end{cases} \qquad (9.72)$$

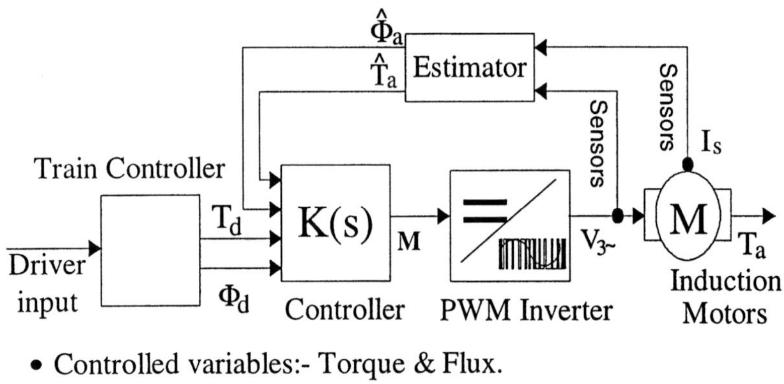

Figure 9.13. A rail traction control system

where $x = [I_{ds}\ I_{qs}\ I_{dr}\ I_{qr}]^T$, $y = [I_{ds}\ I_{dr}]^T$, $u = [V_{ds}\ V_{qs}]^T$, ω is the motor speed, I and V are the current and voltage respectively. This model has been derived based on three main assumptions, namely: ideal air gap flux distribution, linear magnetization characteristic and zero temperature coefficient in the windings. The model parameter matrices are given in Bennett et al. (1996a) and Bennett (1998):

$$A = \begin{bmatrix} -110.2424 & 0 & 200.9966 & 0 \\ 0 & -110.2424 & 0 & 200.9966 \\ 107.1489 & 0 & -206.7995 & 0 \\ 0 & 107.1489 & 0 & -206.7995 \end{bmatrix}$$

$$A_{non} = \begin{bmatrix} 0 & 17.0724 & 0 & 17.5653 \\ -17.0724 & 0 & -17.5653 & 0 \\ 0 & -17.5653 & 0 & -18.0724 \\ 17.5653 & 0 & 18.0724 & 0 \end{bmatrix}$$

$$B = \begin{bmatrix} 253.4307 & 0 \\ 0 & 253.4307 \\ -246.3194 & 0 \\ 0 & -246.3194 \end{bmatrix} \qquad C = \begin{bmatrix} 1 & 0 & 0 & 0 \\ 0 & 0 & 1 & 0 \end{bmatrix}$$

It can be seen that the induction motor is a bilinear system (special class of non-linear systems). This system is described by a T-S fuzzy representation here. The idea is to re-formulate input/output relations of the system into locally linear sub-models that depend on the angular speed, (premise variable in the sense of T-S model). The dynamics of the induction motor can then be described by the following T-S fuzzy rules:

$$\text{If } \omega(t) \text{ is } \omega_i \quad \text{Then } \dot{x}(t) = (A + A_{non}\omega_i)x(t) + Bu(t) \tag{9.73}$$

where $i = 1, 2, \cdots, N$ and $A + A_{non}\omega_i$ (A_i) is a time invariant matrix defined for each i ($i = 1, 2, \cdots, N$). The overall system dynamics and output are then given by

$$\begin{cases} \dot{x}(t) = \sum_{i=1}^{N} \mu_i(\omega_i) \{(A + A_{non}\omega_i)x(t) + B_i u(t)\} \\ y(t) = Cx(t) \end{cases} \tag{9.74}$$

For a T-S fuzzy model, the rule number is normally determined by the modeling accuracy required. A large rule number generally leads to higher accuracy. However, model complexity should also be considered. For the induction motor studied here, only 3 rules are used. Therefore, 3 models working at operating points: $\omega_1 = 0$, $\omega_2 = 32.8125$ and $\omega_3 = 86.2500$, are used. The universe of discourse is divided into three intervals defined by the linguistic variables *Small* ([0, 32.8125)), *Medium* ([32.8125, 86.25)) and *High* ([86.25, 176)). The membership grade functions are given in Fig. 9.14.

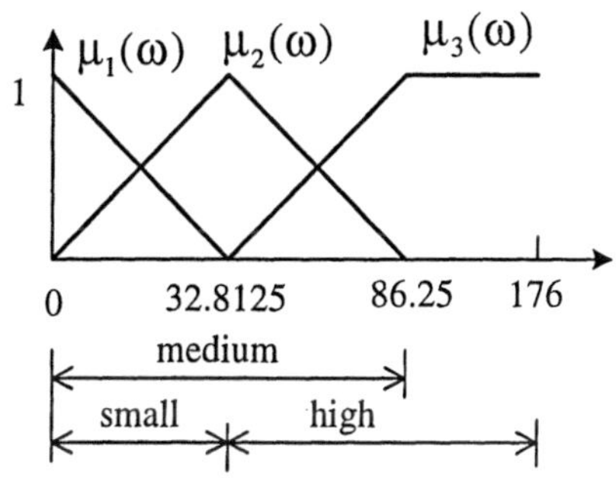

Figure 9.14. Membership grade functions

9.4.3.2 Fuzzy observer design. An fuzzy observer is designed to estimate the system output and generate the diagnostic residual signal. As three rules are involved in the system model, three local observers are required in the fuzzy observer. To produce fast diagnostic performance, the eigenvalues of all three local observers are assigned in the region $S(200, 653)$. The design involves an iterative procedure. Firstly, three sets of arbitrary eigenvalues are selected in the region $S(200, 653)$. Then, the pole-placement routine is used to find the gain matrices for all three local observers. After the observers are designed, Eq. (9.68) and (9.69) are solved using the LMI toolbox (Gahinet et al., 1995). If a positive definite matrix P exists, the fuzzy observer with required eigenvalues is found. Otherwise, another three sets of eigenvalues are chosen. This procedure is repeated until a positive definite matrix P is found. Since the region $S(200, 653)$ is very large, the probability of finding a satisfactory solution is very high.

To implement the observer, the output vector which comprises motor currents I_{ds} and I_{qs} is required. The motor currents are not measured in practice. There is a transform relationship between motor currents and three phase currents I_a, I_b, and I_c (Bennett et al., 1996a; Bennett, 1998). The latter three are measured by phase sensors.

$$\begin{bmatrix} I_a \\ I_b \\ I_c \end{bmatrix} = \begin{bmatrix} 0.8165 & 0 \\ -0.4082 & 0.7071 \\ -0.4082 & -0.7071 \end{bmatrix} \begin{bmatrix} I_{ds} \\ I_{qs} \end{bmatrix} \tag{9.75}$$

The motor currents can be obtained from any two of sensed three phase currents. Frequently, the first two phase current are used. Note that the same transform relation also exists between motor voltages and measured phase voltages.

The observer is first tested against an initial perturbation, and the resulting responses are shown in Fig. 9.15. It can be seen that the observer presents stable dynamics, although it exhibits a relatively large overshoot. However, this is tolerable since the transient time is very small (less than 1 sec).

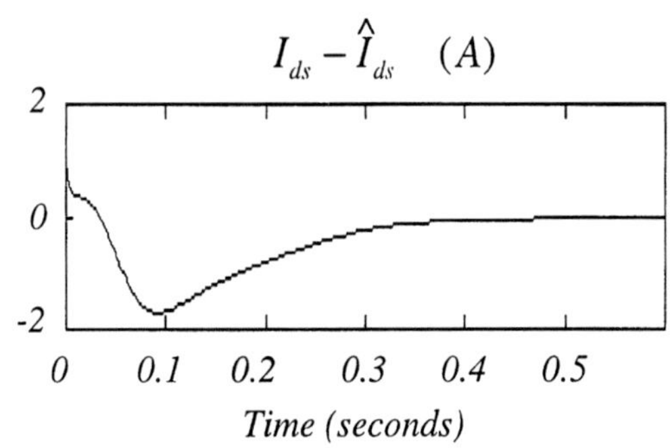

Figure 9.15. Output estimation error due to initial perturbations

9.4.3.3 Residual generation and fault detection and isolation. From the fuzzy observer, the state and output vectors are estimated. The diagnostic signal – the residual is normally generated by comparing the estimated and real system outputs. In this induction motor example and the fuzzy observer designed, the residual signal is given by:

$$r(t) = y(t) - \hat{y}(t) = \begin{bmatrix} I_{ds} - \hat{I}_{ds} \\ I_{qs} - \hat{I}_{qs} \end{bmatrix} \tag{9.76}$$

When a disconnection (fault) occurs in the sensor which measures the phase A current between times $t_1 = 1sec$ and $t_2 = 2sec$, the residual signal is shown in Fig. 9.16. The simulation demonstrates the fault can be easily detected using the residual generated by the fuzzy observer.

One observer is sufficient to detect faults, i.e. it can signal an alarm if a fault occurs in the system. To isolate faults, a number of observers should be designed based on the idea of generalized observer schemes (Patton et al., 1989; Frank, 1990). Since the output vector (consists of motor currents) can

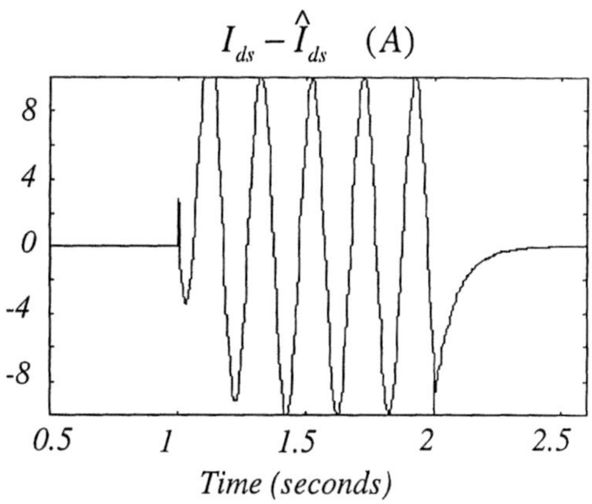

Figure 9.16. Residual response to fault

be determined using any two of phase currents. Three fuzzy observers can be designed for fault isolation purpose. Each observer is driven by two phase current sensors. The output vectors for these three observers are:

$$\text{Observer 1:} \quad \begin{bmatrix} I_{ds} \\ I_{qs} \end{bmatrix} = \begin{bmatrix} 1.2247 & 0 \\ 0.7070 & 1.4142 \end{bmatrix} \begin{bmatrix} I_a \\ I_b \end{bmatrix}$$

$$\text{Observer 2:} \quad \begin{bmatrix} I_{ds} \\ I_{qs} \end{bmatrix} = \begin{bmatrix} 1.2247 & 0 \\ -0.7070 & -1.4142 \end{bmatrix} \begin{bmatrix} I_a \\ I_c \end{bmatrix}$$

$$\text{Observer 3:} \quad \begin{bmatrix} I_{ds} \\ I_{qs} \end{bmatrix} = \begin{bmatrix} -1.2249 & -1.2249 \\ 0.7070 & -0.7071 \end{bmatrix} \begin{bmatrix} I_b \\ I_c \end{bmatrix}$$

The three observers fault isolation scheme is illustrated in Fig. 9.17. If one phase sensor malfunctions, residuals for two observers which use the failed sensor will exceed the threshold. The residual of the observer which uses two healthy sensors will remain under the threshold. Based on information given by three residuals, the failed sensor can be isolated. It is very interesting to point out that three fuzzy observers used in this scheme are identical because they are designed using the same system model. However, they are driven by different signals.

This fault detection and isolation scheme has been implemented in a test-rig. Two real-time simulations have been carried out. In the first simulation, a disconnection (fault) in phase A is injected between $t_1 = 1sec$ and $t_2 = 2sec$. During the second simulation, a disconnection (fault) in phase A is injected between $t_1 = 1.3sec$ and $t_2 = 2.3sec$. The residual norms from three observers are shown in Fig. 9.18. The simulation demonstrates that the fault can be detected easily using residual generated by the fuzzy observers.

9.4.3.4 Fault-tolerant control. To maintain the system stability and reliable operation under faulty conditions, some reconfiguration measures should

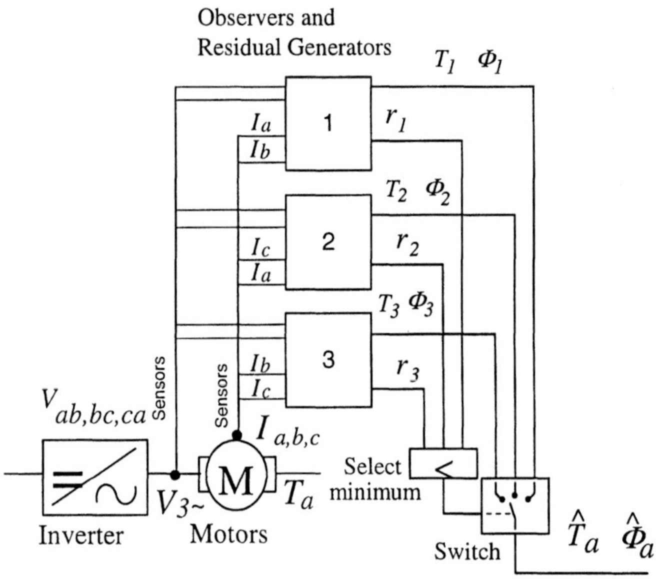

Figure 9.17. Fault isolation and fault-tolerant control scheme for the traction system

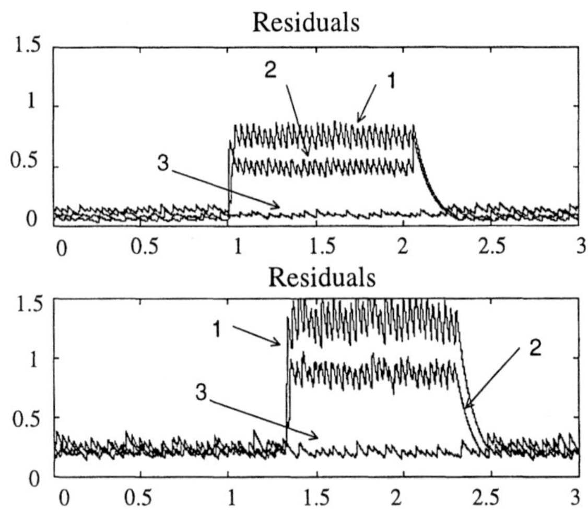

Figure 9.18. Fault diagnostic residuals from real-time simulations

be taken into account. In this study, a switch mechanism is used to switch between three estimates to provide the system with accurate torque and flux information. When a fault occurs in one phase sensor, the fuzzy observer which driven by two health phase sensors can provide reliable state variables which are used for estimating torque and flux. With healthy flux and torque estimates, the reliable system operation can be maintained. This is the underlying idea of the fault-tolerance using "inferred" measurements.

To assess the impact of faults on the system, intermittent faults are injected successively in each of the 3 phase sensors. With fault detection and reconfiguration operational, it is possible to detect, isolate and remove the effect of the three faults. This can be seen from the simulation shown in Fig. 9.19. The torque estimation error in this case remains close to zero except for some glitches due to the delay in detection and switching. It should be pointed out the fail-safe operation of the fault-tolerant control scheme is achieved without any additional hardware except some investment in computing power.

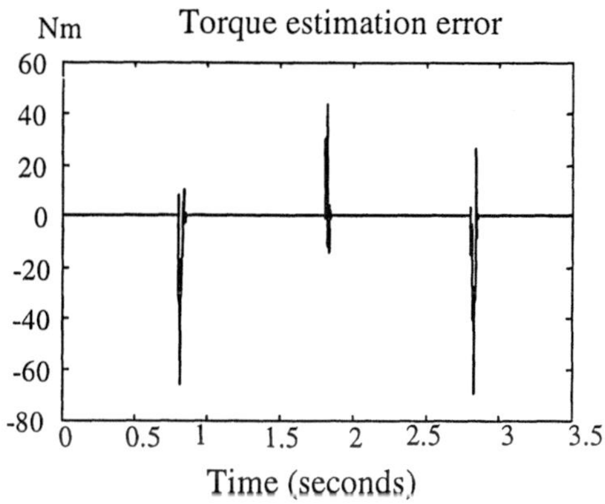

Figure 9.19. Torque error with reconfiguration action

If no action is taken to remedy sensor faults, the system performance will be degraded and this will damage the system. Fig. 9.20 shows how sensor faults cause the torque estimation to deviate from its zero value when no action is taken. As the estimated torque signal is needed for control action, the system may become unstable under this faulty condition.

Remarks: A new approach for generating diagnostic residual signals for non-linear systems is introduced here. In this approach, the residual generation is based on the use of fuzzy observers. The fuzzy observer used is based on the combination of the Takagi-Sugeno fuzzy model and the idea of parallel distributed compensators. To ensure the good residual response for fast and reliable detection, a technique has been introduced for assigning the fuzzy observer eigenvalues in a specific region. The stability conditions and/or eigenvalue constraints have been formulated and solved within a linear matrix inequality framework. The developed scheme has been successfully applied, in simulation, to detect and isolate faults in the induction motor of a rail traction system. It can be concluded that the fuzzy observer is an effective tool to generate residual signals for non-linear dynamic system fault diagnosis. The scope of application of this work extends to all non-linear systems with possible incipient faults.

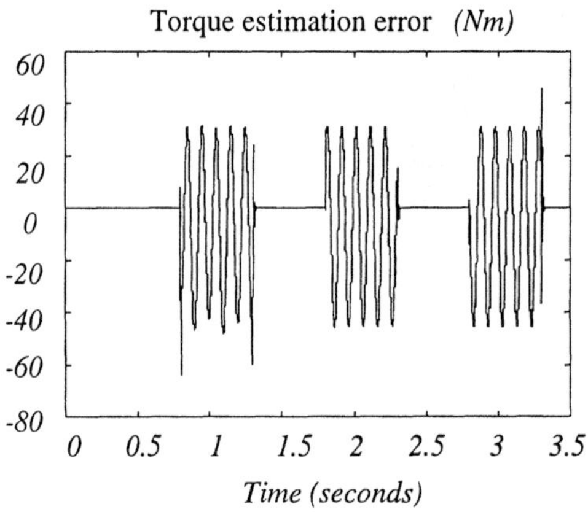

Figure 9.20. Torque error in the presence of phase sensor faults

9.5 A Neuro-Fuzzy Approach for Non-linear Dynamic Systems Fault Diagnosis

Fault diagnosis can be facilitated by using either quantitative and qualitative information of the system monitored. A novel neuro-fuzzy based approach which integrates quantitative and qualitative information for FDI is introduced in this section. Note that the material presented here is mainly taken from the paper by Benkhedda and Patton (1996). This approach combines, in a single framework, both numerical and symbolic knowledge about the process. The method is able to structure a quantitative model in a way that qualitative knowledge about the process can be included as well as extracted. The underlying concept is to structure a neural network, which can model non-linear systems efficiently, in a fuzzy-logic format. The network can therefore be trained more rapidly, and will also described explicitly the causes of faults. Expert-knowledge can also be included in the same framework. A B-Spline neural-network is used in the unified framework. This network which incorporates both quantitative and qualitative information is used for both residual generation and evaluation in fault diagnosis. In order to illustrate the method developed, the fault diagnosis of a laboratory two-tanks system is studied.

9.5.1 B-Spline neural networks and fuzzy logic interpretation

The B-Spline function network has been used as one of effective neuro-fuzzy methods for system modeling and control (Brown and Harris, 1994; Lane, Handelman and Gelfand, 1992). We take advantage of this effective tool to generate and evaluate residuals for the purpose of fault detection and isolation. A B-Spline neural-network illustrated in Fig. 9.21 involves input space, basis functions and weight factors.

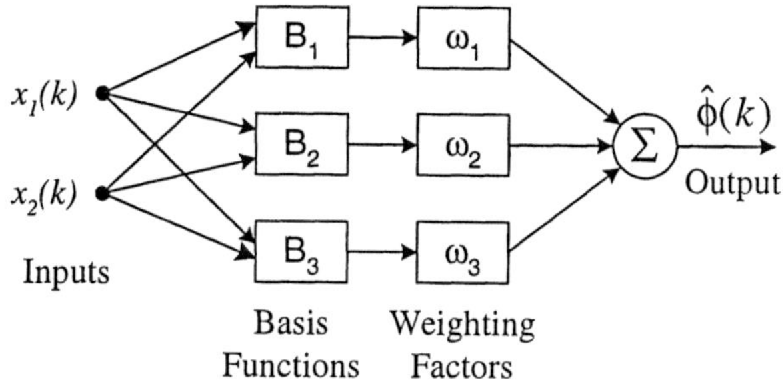

Figure 9.21. A B-Spine function network with two inputs, 3 basis functions and one output

From Fig. 9.21, it can be seen that all inputs of the network are applied to basis functions which are associative cells defined in the input space, and jointed at breakpoints, referred to as knots. These basis functions perform non-linear transformation of input into a bounded interval [0 1]. Their shape, size and overlap determine the modeling capabilities of the resulting network. The output is a weighted linear combination of basis function outputs. The weighting factors are adjusted by the network training. Given N partitions of the input within the interval $X = [x_{min}, x_{max}]$, the output of the one dimensional B-Spline network of the form

$$\hat{\phi} = \sum_{i=1}^{p} B_{n,j}\omega_i \tag{9.77}$$

can be constructed to approximate the target signal $\phi = F(x)$ using a linear combination of the B-Splines, $B_{n,j}(x)$, weighted by coefficients ω_i. The sequence of normalized B-Splines $B_{n,1}(x)$, $B_{n,2}(x)$, $\cdots$, $B_{n,p}(x)$ constructed on the knots λ_0, λ_1, $\cdots$, λ_N can be evaluated from the following recurrent relationship:

$$B_{n,j}(x) = \left(\frac{x - \lambda_{j-n}}{\lambda_{j-1} - \lambda_{j-n}}\right) B_{n-1,j-1}(x) + \left(\frac{\lambda_j - x}{\lambda_j - \lambda_{j-n+1}}\right) B_{n-1,j}(x) \tag{9.78}$$

where

$$B_{1,j}(x) = \begin{cases} 1 & \text{if } x \in I_j \\ 0 & \text{otherwise} \end{cases} \tag{9.79}$$

The B-Spline index j is associative with the region of local support $\lambda_{(j-n)} \leq x \leq \lambda_{(j)}$. The training of the network, then, consists of finding a set of weighting coefficients ω_i that minimizes the cost function

$$J = \frac{1}{N}\sum_{i=1}^{N}(\phi(t) - \hat{\phi}(t))^2 \tag{9.80}$$

where N denotes the number of training sets, whereas $\phi(t)$ and $\hat{\phi}(t)$ denote the target signal and the network's outputs respectively.

Several approaches can be used to train such a network, depending on whether the learning is done on-line or off-line. In our case, we assume that the designer has available to him a set of training data. Hence, and because the model is a linear combination of the set of basis functions, the weighted Moore-Penrose pseudo-inverse is used to find the optimal set of weighting terms.

For a Multiple-input-multiple-outputs (MIMO) system, multi-dimensional B-Spline models are used. They are constructed as tensor products of one-dimensional models. Their corresponding basis functions are formed by a direct multiplication of one-dimensional basis functions defined by Eqs.(9.78) & (9.79). For example, a p-dimensional basis function is defined as

$$B(x) = \prod_{i=1}^{p} B_p(x_p) \tag{9.81}$$

It is interesting to note that the output of the network, as given by Eq. (9.77), is very similar to that of a fuzzy associative memory network. Indeed, considering a fuzzy rule, R_{ij}, such as

$$R_{ij}: \quad \text{IF} \quad (x \text{ is } (\Omega_i) \quad \text{THEN} \quad (\phi \text{ is } \Psi_j) \qquad\qquad (c_{ij})$$

where Ω_i and Ψ_j denote respectively fuzzy sets in the input and output partition space, and c_{ij} the level of confidence in the rule R_{ij} being true, it can be shown that the output of a continuous fuzzy rule is given by

$$\phi(x) = \sum_{i=1}^{p} \mu_{\Phi_i}(x)\omega_i \tag{9.82}$$

where

$$\omega_i = \sum c_{ij} y_j^c \tag{9.83}$$

Brown and Harris (1994) showed that, given a set of optimal B-Spline networks weights ω_i, it is possible to find the equivalent fuzzy representation with the coefficients c_{ij} given by the relation

$$c_{ij} = \mu_{B_j}(\omega_i) \tag{9.84}$$

This relation, between B-Spline networks and fuzzy logic, is very important. It shows that, not only can a B-Spline network can be trained, from numerical data, but symbolic knowledge can be included (or extracted) from it too. This very important feature facilitates the integration of numerical and symbolic knowledge within a single framework.

9.5.2 Residual generation and fault detection using B-Spline networks

The main task of fault diagnosis is to generate diagnostic signals - residuals. The methods for generating residuals for non-linear dynamic systems via neural networks are discussed in Section 9.3. The underlying concept is to train the network to recognize the occurrence of a fault and to find the optimal function that maps the system inputs-outputs to a residual signal:

$$r(k) = F(\vec{u}(k), \vec{y}(k)) \tag{9.85}$$

where $\vec{u}(k) = [u(k), u(k-1), \cdots, u(k-n)]^T$ and $\vec{y}(k) = [y(k), y(k-1), \cdots, y(k-n)]^T$ are the input and output of the system over a time window. The input of the network includes past as well as current values of the measurements, to capture temporal information.

In this study, the B-Spline neural-networks is used to diagnose faults in non-linear systems and overcome the disadvantage of multi-layer perceptron networks. To be more specific, the measured inputs and outputs of the system are processed through an associative memory network, as opposed to a multi-layer perceptron network of in Section 9.3.

It is important to note that the aim of the fault detection observers, or residual generators, is not to estimate the state of the plant but rather to respond promptly to the occurrence of a fault. Hence, the residual generator should produce a value of 1 when a fault develops in the system, and 0 otherwise. In such an approach, it may be said that the network used is an alternative to the traditional fault detection observers.

An important feature of a neural-network is that it will learn during a training session made over several training cycles, with training data coming from different operating points. However, before the training is started, the order of the B-Spline network needs to be chosen. A network with a second order basis function, and 2 linear knots, is chosen here. As it can be seen from Fig. 9.22, such a configuration permits to divide the normalized input space into 3 linguistic variables Small, Medium and Large.

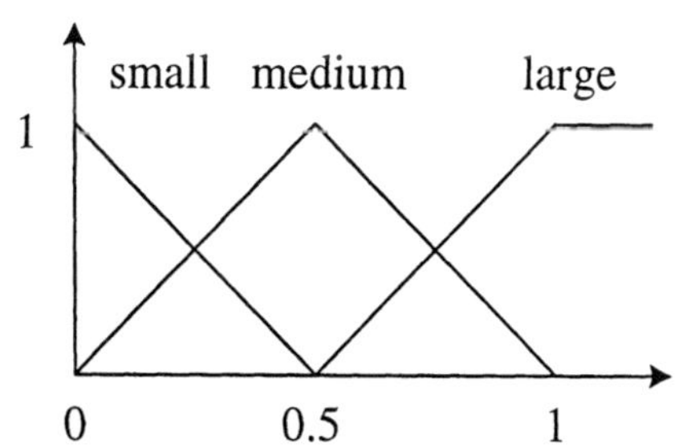

Figure 9.22. B-Spline functions defined over the normalized input space

Once the training is performed, a set of optimal weights ω_i $(i = 1, \cdots, p)$ are used to derive the corresponding fuzzy description of the residual generator. Moreover, since the basis functions can be interpreted linguistically, a qualitative model of the residual generator can be derived. This provides the operator with an explanation about the cause of the fault which is more understandable to him/her than using crisp neural-networks.

9.5.3 *Fault isolation via B-Spline function networks*

The B-Spline neural-network is used here to isolate faults. The locations of faults to be diagnosed are assumed to be known to the designer. Therefore, the designer can simulate the system with all possible faults to generate corresponding training data. The isolation network then has as many outputs as classes of behavior. Hence, for a system with 2 classes of faults, the output of the network will be a vector of dimension 3; this includes the models associated with the two faults as well as that corresponding to the *healthy* one.

To train the network, we need to decompose it into a set of $(M+1)$ multi-inputs-single-output (MIMO) sub-models, where M is the number of faulty classes, and find the set of optimal weighting factors with each sub-model. When the network is applied to a test point $(\vec{u}(k), \vec{y}(k))$, the network's output, $Flag$, is a real vector of dimension $(M+1)$. It can be seen from Fig. 9.23 that each component of that vector, $Flag\ i$ $(i = 1, 2, \cdots, M)$ is identified with a class of behavior, which can be either a $Fault\ i$, or the nominal model. When the system is operating at its nominal condition, all the network outputs are zero except the last one. However, when a specific fault develops in the system the corresponding output will deviate from zero, whereas the output flags become zero, confirming that the system is no longer healthy.

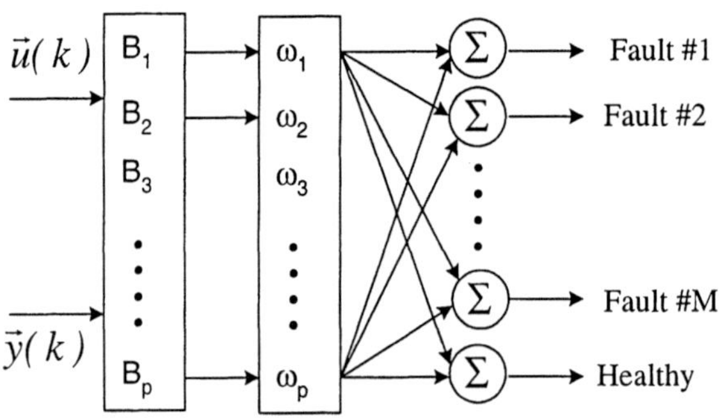

Figure 9.23. Network architecture used for fault isolation

9.5.4 *Fault diagnosis of a two-tanks system*

To evaluate the performance of the fault detection and isolation, a laboratory two-tanks system (see Fig. 9.24) is chosen. The system presents realistic challenges such as non-linearity and modeling uncertainty.

The two-tanks system consists of two interconnected tanks, connected to each other through connecting pipes of circular cross-section. Two pumps $P1$ and $P2$ control the incoming mass-flows $Q_1(t)$ and $Q_2(t)$. The two tanks are equipped with piezo-resistive pressure transducers for measuring the level of liquid, $h_1(t)$ and $h_2(t)$. For the dynamic model, the incoming mass flows $Q_1(t)$ and $Q_2(t)$ are defined as inputs, while the two measurements $h_1(t)$ and $h_2(t)$ are considered as outputs. The dynamic model is then derived using the incoming

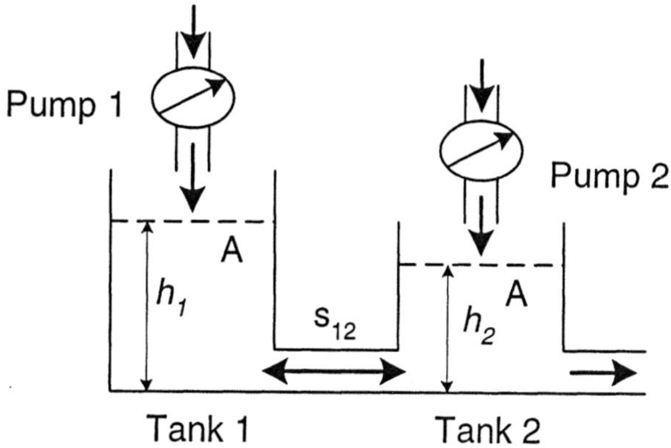

Figure 9.24. Two tanks system

and outgoing mass flows and is described by the following differential equations:

$$
\begin{aligned}
A\frac{dh_1(t)}{dt} &= Q_1(t) - a_1 s_{12}(h_1(t) - h_2(t))\sqrt{2g\,|h_1(t) - h_2(t)|} + Q_{f1}(t) \\
A\frac{dh_2(t)}{dt} &= Q_2(t) + a_1 s_{12} sgn(h_1(t) - h_2(t))\sqrt{2g\,|h_1(t) - h_2(t)|} + Q_{f2}(t)
\end{aligned}
$$

The $Q_{fi}(t)$ $(i = 1, 2)$ denote the additional mass flows into the tanks caused by leaks. The detailed description about this two tanks system can be found in Benkhedda and Patton (1996).

The fault diagnosis B-Spline neural-network is trained off-line, using a set of data from the simulation. After a set of optimal weighting terms i is found, the network is simulated, on-line to monitor the system. Faults are then introduced, and the performance of the fault diagnosis algorithm is assessed.

Fig. 9.25 shows the residual response to an impulse input of 25 seconds. A leak has also been included at time intervals indicated. It is interesting to note that, the fault signature is very clear in the residual response, $r(k)$, generated by the B-Spline network, as it shows a significant response each time a fault occurs.

As mentioned earlier, it is possible to extract symbolic knowledge from the residual generator. This is illustrated in following table where 2 rules are extracted from the B-Spline neural-network based residual generator. The universe of discourse of the residual generator output is divided into two overlapping intervals delimiting the linguistic variables *Healthy* and *Faulty.*

Once the residual generator has been trained, it is also possible to include some qualitative knowledge about the plant. Such a knowledge can either be obtained from the human expert, or generated from a qualitative model of the plant using qualitative reasoning physics. To illustrate this consider the case where the cross-pipe is jammed, and assume that the operator does not

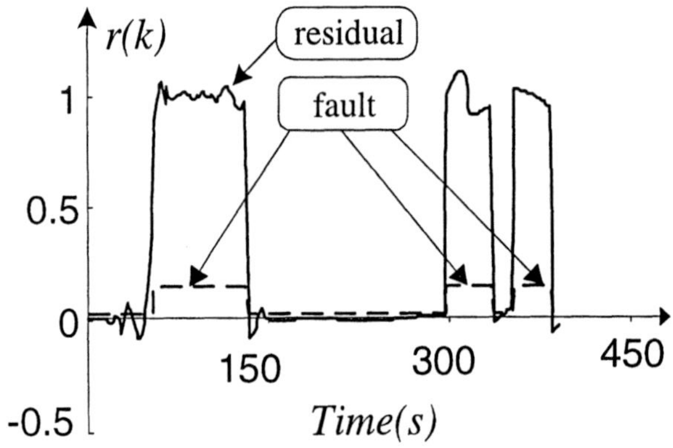

Figure 9.25. Residual response

Table 9.2. Qualitative model of the residual generator

IF $h_1(k)$, $h_1(k-1)$, $h_2(k)$ Small & $u(k)$ Large	Then Faulty	(1.00)
IF $h_1(k)$, $h_1(k-1)$, $h_2(k)$, $u(k)$ Small	Then Healthy	(0.998)

have any data associated with that behavior, except some heuristic rules gained through experience. These rules can be expressed as follows:

If $Q_1(k)$, $h_1(k)$, $h_1(k-1)$ High & $h_2(k)$ Small, the System Faulty ($c = 0.9$)

The aim is then to find the optimal weighting term w associated with the above fuzzy relation. For that purpose, fuzzy output membership functions are defined in the output universe of discourse. In this application it is assumed that the system falls in either a *Faulty* or a *Normal* behavior (although more behaviors could also be considered) and therefore only two functions are chosen. The weighting term, w, is then taken to be as the projection of the rule confidence index, c, (equal to 0.9 for this example) on the universe of discourse. For the particular example considered above, the corresponding weighting term was found to be $w = 0.9$.

A single mapping based on the use of a B-Spline neural-network is used here for the fault isolation purpose. The underlying concept is to train the network to classify faults directly from the measured inputs and outputs of the system. For that, a set of training data associated with different classes of faults needed to be generated. The network's output is, then, a vector, whose elements indicate the truth in a certain model being true. The following classes of behavior were identified in our application.

Fault 1: This fault is defined as being a leak in tank 1. A value of 1 in the signal, *Flag* 1, indicates that a leak in tank 1 has occurred.

Fault 2: Similarly to Fault 1, this class is associated with a leak in tank 2. A value of 1 in the signal *Flag* 2 indicates that the tank 2 is leaking.

Fault 3: he third fault considered in this work is where the pipe connecting the two tanks is jammed. The signal associated with that class of fault is *Flag* 3.

Healthy: The redundant signal, *Flag* 4, associated with the Healthy operating condition, indicates if a fault (any fault) has occurred in the system.

The network was trained to identify the above faults, and then tested, on-line, to detect and isolate faults. Figs. 9.26–9.28 show the responses of the different outputs to the three faults. It can be seen that, at each time, only the signal associated with the fault's class which deviates from the zero value.

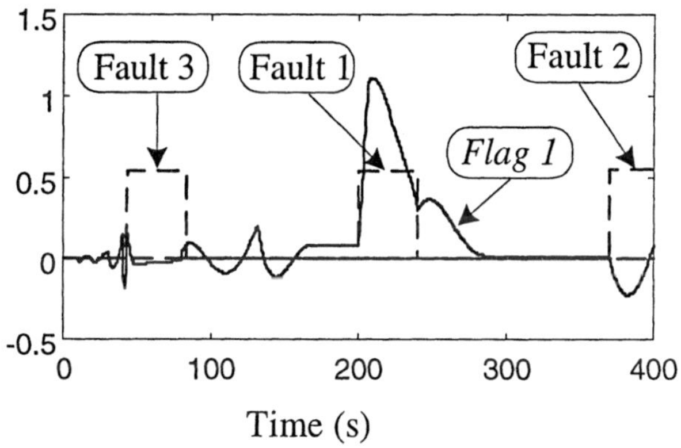

Figure 9.26. Flag 1 response to faults

After the training of the network has been completed, it is possible to find the causes of each fault. This is obtained by finding the equivalent fuzzy representation of the B-Spline network. The human operator can then modify such a knowledge, or even add any new information that has been missed during the training, a new fault for example.

Remarks: The neuro-fuzzy approach introduced here, has the advantage of a neural network but provides, also, information close to human common sense. This, indeed, allows the designer to find what the network has learnt, and check if the explanation provided by the network has a physical meaning; rules, and therefore weighting coefficients, can be altered if necessary. Moreover, because of the analogy between fuzzy-logic systems and B-Spline neural-networks, the designer can also include some heuristic (qualitative) knowledge about the either the plant or the fault. The combination of qualitative and quantitative

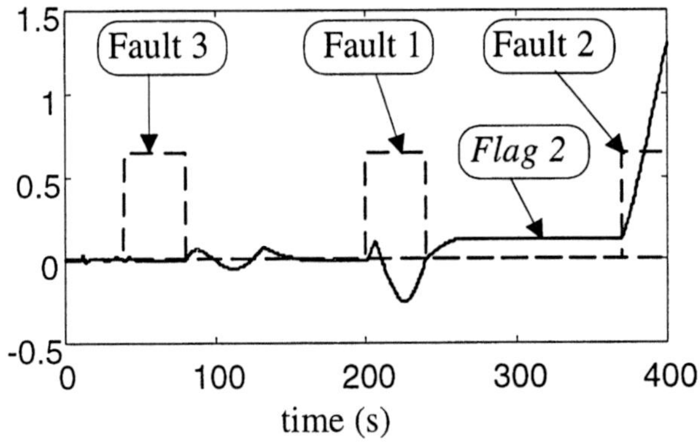

Figure 9.27. Flag 2 response to faults

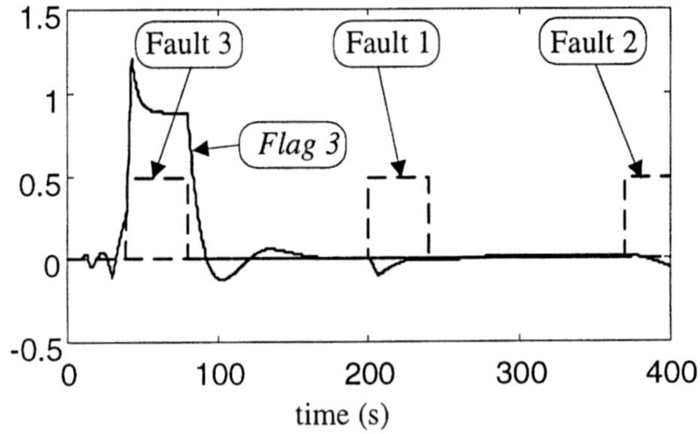

Figure 9.28. Flag 3 response to faults

information through the integrated framework proposed provides a powerful tool for reliable fault diagnosis. The scheme can be applied to a wide range of complex non-linear dynamic systems with uncertain factors.

9.6 Summary

The problems of fault detection and isolation for non-linear dynamic systems have been studied in this chapter. This chapter started with a brief review on non-linear systems FDI. Then the chapter preceded to the introduction of linear and non-linear observer approaches. Several observer-based approaches which were initially developed for linear systems can be extended to certain classes of non-linear systems, e.g. bilinear systems. Although the system is non-linear, the observer-based residual generator can be linear. The observer-based approaches are very important for non-linear dynamic systems FDI. However, they are only applicable to very limited class of non-linear systems with well-developed analytical models.

To tackle non-linear systems with ill-defined models, we need to develop techniques which do not critically depend on analytical models. The use of neural networks provides us with such a technique. The ways of using neural networks for non-linear systems FDI have been discussed. And then, a scheme of using neural networks for fault detection and isolation are introduced. This scheme utilizes both modeling and classification capabilities of neural networks and has been illustrated by a laboratory three tanks system example.

The Takagi-Sugeno fuzzy model provides a powerful way to model non-linear dynamic systems. With this model, the local linear model responses are mixed together using the fuzzy fusion mechanism. Even if we do not have an analytical model for a non-linear system, all local linear models can be identified. In this sense, the Takagi-Sugeno fuzzy model can be used to model very wide range of non-linear dynamic systems. With a fuzzy model, a fuzzy observer can be designed. The concept and design of fuzzy observers have been introduced in this chapter. Fuzzy observers have then been applied to fault detection, fault isolation and fault-tolerant control of an induction motor in a rail traction system.

With the combination of fuzzy logic and neural networks, the neuro-fuzzy approach can further enhance the fault detection and isolation performance. This chapter has, through a two tanks system, illustrated the effectiveness of a B-Spline function network-based neuro-fuzzy approach in detecting and isolating faults in non-linear dynamic systems.

Appendix A
Terminology in Model-based Fault Diagnosis

The following definitions are extract from the IFAC *SAFEPROCESS* Technical Committee's initiative of defining common terminology in the field of model-based fault diagnosis (Isermann and Ballé, 1997). However, these proposals are preliminary because the discussions are still going on.

States and Signals

Fault: An unpermitted deviation of at least one characteristic property or parameter of the system from the acceptable/usual/standard condition.

Failure: A permanent interruption of a system's ability to perform a required function under specified operating conditions.

Malfunction: An intermittent irregularity in the fulfilment of a system's desired function.

Error: A deviation between a measured or computed value (of an output variable) and the true, specified or theoretically correct value.

Disturbance: An unknown (and uncontrolled) input acting on a system.

Perturbation: An input acting on a system, which results in a temporary departure from the current state.

Residual: A fault indicator, based on a deviation between measurements and model-equation-based computations.

Symptom: A change of an observable quantity from normal behaviour.

Functions

Fault detection: Determination of the faults present in a system and the time of detection.

Fault isolation: Determination of the kind, location and time of detection of a fault. Follows fault detection.

Fault identification: Determination of the size and time-variant behaviour of a fault. Follows fault isolation.

Fault diagnosis: Determination of the kind, size, location and time of detection of a fault. Includes fault detection, isolation and identification.

Monitoring: A continuous real-time task of determining the conditions of a physical system, by recording information, recognising and indicating anomalies in the behaviour.

Supervision: Monitoring a physical system and taking appropriate actions to maintain the operation in the case of faults.

Protection: Means by which a potentially dangerous behaviour of the system is suppressed if possible, or means by which the consequences of a dangerous behaviour are avoided.

Models

Quantitative model: Use of static and dynamic relations among system variables and parameters in order to describe a system's behaviour in quantitative mathematical terms.

Qualitative model: Use of static and dynamic relations among system variables and parameters in order to describe a system's behaviour in qualitative terms such as causalities or if-then rules.

diagnostic model Diagnostic model: A set of static or dynamic relations which link specific input variables - the symptoms - to specific output variables - the faults.

analytical redundancy Analytical redundancy: Use of two or more (but not necessarily identical) ways to determine a variable, where one way uses a mathematical process model in analytical form.

System Properties

Reliability: Ability of a system to perform a required function under stated conditions, within a given scope, during a given period of time.

Safety: Ability of a system not to cause danger to persons or equipment or the environment.

Availability: Probability that a system or equipment will operate satisfactorily and effectively at any point of time.

Dependability: A form of availability that has the property of always being available when required. It is the degree to which a system is operable and capable of performing its required function at any randomly chosen time during its specified operating time, provided that the item is available at the start of that period.

Appendix B
Inverted Pendulum Example

The laboratory inverted pendulum system shown in Fig. B.1 has been used as a benchmark system to demonstrate fault diagnosis techniques and concepts due to its wide availability in the control laboratory.

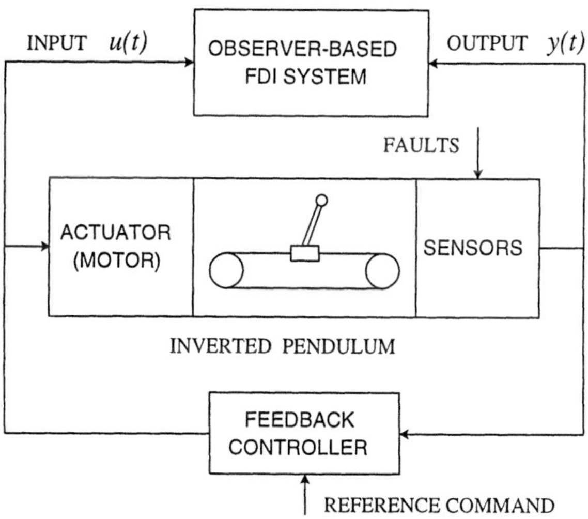

Figure B.1. The controlled inverted pendulum system

This is a non-linear system with some uncertain factors such as friction etc. A simplified linearized is used here to illustrate the fault detectability. The linearized state space model matrices are:

$$A = \begin{bmatrix} 0 & 0 & 1 & 0 \\ 0 & 0 & 0 & 1 \\ 0 & -1.93 & -1.99 & 0.009 \\ 0 & 36.9 & 6.26 & -0.174 \end{bmatrix}$$

$$B = [0 \quad 0 \quad -0.3205 \quad -1.009]^T$$

$$C = \begin{bmatrix} 1 & 0 & 0 & 0 \\ 0 & 1 & 0 & 0 \\ 0 & 0 & 1 & 0 \end{bmatrix} \qquad D = 0_{3\times 1}$$

where the state variables are: the cart position perturbation Δx, the pendulum angle ϕ, the cart velocity $\dot{x}$ and the pendulum angular velocity $\dot{\phi}$. The system is unstable and needs to be stabilized. Since the purpose of the example is to illustrate the fault detection capability, the system is simply stabilized via a state feedback controller.

A full-order observer with poles $\{-14, -20, -8 \pm 8i\}$ is used to estimate the output and the output estimation error is used as the residual signal. The steady-state gain between the residual and the faults is:

$$G_{rf}(0) = C(A - KC)^{-1}K + I$$

Where $K \in R^{4\times 3}$ is the observer gain matrix. Assume that:

$$K = [k_1 \;\; k_2 \;\; k_3] \qquad (k_i \in \mathbb{R}^4, i = 1, 2, 3)$$

$$A = [0 \;\; a_2 \;\; a_3 \;\; a_4] \qquad (a_i \in \mathbb{R}^4, i = 2, 3, 4)$$

$$(A - KC)^{-1} = [g_1, g_2, g_3, g_4]^T \qquad (g_i \in \mathbb{R}^4, i = 1, \cdots, 4)$$

Now, $G_{rf}(0) = C(A - KC)^{-1}K + I$ can be computed as:

$$G_{rf}(0) = \begin{bmatrix} 1 + g_1^T k_1 & g_1^T k_2 & g_1^T k_3 \\ g_2^T k_1 & 1 + g_2^T k_2 & g_2^T k_3 \\ g_3^T k_3 & g_3^T k_2 & 1 + g_3^T k_3 \end{bmatrix}$$

However

$$(A - KC)^{-1}(A - KC) = \begin{bmatrix} g_1^T \\ g_2^T \\ g_3^T \\ g_4^T \end{bmatrix} [-k_1 \;\; a_2 - k_2 \;\; a_3 k_3 \;\; a_4] = I_4$$

This leads to: $g_1^T k_1 = -1$, $g_2^T k_1 = 0$ and $g_3^T k_1 = 0$. Substituting these relations into $G_{rf}(0)$, we have:

$$G_{rf}(0) = \begin{bmatrix} 0 & * & * \\ 0 & * & * \\ 0 & * & * \end{bmatrix}$$

This proves that the strong detectability for faults in sensor 1 cannot be achieved no matter what observer gain matrix is used and what is the cart position, if the residual generator is based on a full-order observer.

Appendix C
Matrix Rank Decomposition

Proposition: Any $p \times q$ and rank r $(r \leq \min\{p, q\})$ matrix $E \in \mathbb{R}_r^{p \times q}$ can be decomposed as follows :

$$E = E_1 E_2$$

where $E_1 \in \mathbb{R}_r^{p \times r}$, $E_2 \in \mathbb{R}_r^{r \times q}$ and

$$rank(E_1) = rank(E_2) = r$$

Proof: According to the singular value decomposition (SVD) theorem, the matrix E can be decomposed as:

$$E = U \Sigma V^T$$

where $U \in \mathbb{R}^{p \times p}$ and $V \in \mathbb{R}^{q \times q}$ are orthogonal matrices and:

$$\Sigma = \begin{bmatrix} \Sigma_r^2 & 0_{r \times (q-r} \\ 0_{(p-r) \times r} & 0_{(p-r) \times (q-r)} \end{bmatrix} \in \mathbb{R}^{p \times q} \qquad \Sigma_r^2 = diag\{\sigma_1^2, \sigma_2^2, \cdots, \sigma_r^2\}$$

where $\sigma_1^2, \sigma_2^2, \cdots^2, \sigma_r^2$ are singular values of E.

The matrix E can be rewritten as:

$$E = U \begin{bmatrix} \Sigma_r \\ 0_{(p-r) \times r} \end{bmatrix} [\Sigma_r \quad 0_{r \times (q-r)}] V^T$$

Define:

$$E_1 = U \begin{bmatrix} \Sigma_r \\ 0_{(p-r) \times r} \end{bmatrix} = [u_1, u_2, \cdots, u_r] \Sigma \quad \in \mathbb{R}^{p \times r}$$

$$E_2 = [\Sigma_r \quad 0_{r \times (q-r)}] V^T = \Sigma_r [v_1, v_2, \cdots, v_r]^T \quad \in \mathbb{R}^{r \times q}$$

where $u_1, u_2, \cdots, u_r$ are first r columns of U and $v_1, v_2, \cdots, v_r$ are first r columns of V. It can be easily see that E_1 is a full column matrix and E_2 is a full row rank matrix.

◊ **QED.**

Appendix D
Proof of Lemma 3.2

The observability matrix of (C, A) is defined as:

$$W_0 = \begin{bmatrix} C \\ CA \\ \vdots \\ CA^{n-1} \end{bmatrix}$$

The observability matrix of (C_1, A) is also defined as:

$$W_{01} = \begin{bmatrix} C \\ CA \\ CA \\ CA^2 \\ \vdots \\ CA^{n-1} \\ CA^{n-1} \\ CA^n \end{bmatrix}$$

From Cayley-Hamilton theorem, one can seen that CA^n can be represented by a linear combination $C, CA, \cdots, CA^{n-1}$ and this leads to:

$$rank(W_0) = rank(W_{01}) = n_0$$

If we select n_0 linear independent rows vectors $p_1^T, \cdots, p_{n_0}^T$ from W_0 matrix, these row vectors are also the rows of the matrix W_{01}. These row vectors are now combined with another $n - n_0$ arbitrary independent row vectors

$p_{n_0+1}^T, \cdots, p_n^T$ to construct an non-singular matrix as:

$$P = \begin{bmatrix} p_1^T \\ \vdots \\ p_{n_0}^T \\ p_{n_0+1}^T \\ \vdots \\ p_n^T \end{bmatrix}$$

If one apply a transformation P to the system matrix pairs (C, A) and (C_1, A), the standard observability decompositions of (C, A) and (C_1, A) are formulated as (Chen, 1984):

$$PAP^{-1} = \begin{bmatrix} A_{11} & 0 \\ A_{12} & A_{22} \end{bmatrix} \qquad A_{11} \in \mathbb{R}^{n_0 \times n_0}$$

$$CP^{-1} = [C^* \quad 0] \qquad C^* \in \mathbb{R}^{m \times n_0}$$

$$C_1P^{-1} = [C_1^* \quad 0] \qquad C_1^* \in \mathbb{R}^{2m \times n_0}$$

where (C^*, A_{11}) and (C_1^*, A_{11}) are observable matrix pairs.

From the standardized observability decompositions shown above, it can be seen that (C, A) or (C_1, A) are detectable *iff* A_{22} is stable, i.e. the detectability of (C, A) is equivalent to the detectability of (C_1, A).

$\diamondsuit$ **QED.**

Appendix E
Low Rank Matrix Approximation

Eckart-Young Theorem (Eckart and Young, 1936; Tufts et al., 1982)**:** Let A be an $m \times n$ matrix of rank r which has complex elements. The singular value decomposition of the matrix A is:

$$A = U\Sigma V^* \quad ; \quad U \in \mathbb{C}^{m\times m} \quad , \quad V \in \mathbb{C}^{n\times n} \quad , \quad \Sigma \in \mathbb{C}^{m\times n}$$

The matrices U and V are unitary, and Σ is a rectangular diagonal matrix with real and nonnegative diagonal entries. These diagonal entries, called the singular values of A, are conventionally ordered in decreasing (or increasing) order.

Let S_p be the set of all $m \times n$ matrices of rank p $(< r)$. For all matrices B in S_p,

$$||A - \hat{A}|| \leq ||A - B||$$

where

$$\hat{A} = U\hat{\Sigma}V^*$$

and $\hat{\Sigma}$ is obtained from the matrix Σ by setting to zero all but p largest singular values. The matrix norm is the Frobenius norm. That is

$$||A - \hat{A}|| = \sqrt{trace[(A - B)^*(A - B)]}$$

Hence, in words, $\hat{A}$ is the best least squares approximation of lower rank p to the given matrix. That is to say that $\hat{A}$ is a low rank matrix approximation of A.

References

Adjallah, K., Maquin, D. and Ragot, J. (1994). Nonlinear observer-based fault detetcion, *Proc. of The Third IEEE Conf. on Control Applications*, Glasgow, Scotland, pp. 1115–1120.

Andry, A. N., Chung, J. C. and Shapiro, E. Y. (1984). Modalized observers, *IEEE Trans. Automat. Contr.* **AC-29**(7): 669–672.

Andry, A. N., Shapiro, E. Y. and Chung, J. C. (1983). Eigenstructure assignment for linear systems, *IEEE Trans. Aero. & Electron. Syst.* **AES-19**(5): 711–729.

Antsaklis, P. J. (1980). Maximal order reduction and supremal (A, B) invariant and controllability subspaces, *IEEE Trans. Automat. Contr.* **AC-25**(1): 44–49.

Appleby, B. D., Dowdle, J. R. and Vander Velde, W. (1991). Robust estimator design using μ synthesis, *Proc. of the 30th Conf. on Decision & Control*, Brighton, UK, pp. 640–644.

Arsan, M. I., Mouyon, P. and Magni, J. F. (1994). Fault diagnosis in the presence of parameter variations, *Preprints of the IFAC Sympo. on Fault Detection, Supervision and Safety for Technical Processes: SAFEPROCESS'94*, Espoo, Finland, pp. 93–98 (Vol.1).

Aström, K. J. (1991). Intelligent control, *Proc. First European Control Conf. ECC'91*, Grenoble, France, pp. 2328–2339. Plenary paper.

Ayoubi, M. and Isermann, R. (1997). Neuro-fuzzy systems for diagnosis, *Fuzzy Sets and Systems* **89**(3): 289–307.

Bakiotis, C., Raymond, J. and Rault, A. (1979). Parameter and discriminiant analysis for jet engine mechanical state diagnosis, *Proc. of The 1979 IEEE Conf. on Decision & Control*, Fort Lauderdale, USA.

Ballé, P., Fischer, M., Füssel, D., Nelles, O. and Isermann, R. (1998). Integrated control, diagnosis and reconfiguration of a heat exchanger, *IEEE Contr. Syst. Mag.* **18**(3): 52–63.

Basilevsky, A. (1983). *Applied Matrix Algebra in the Statistical Science*, North-Holland, New York.

Basseville, M. (1988). Detecting changes in signals and systems - a survey, *Automatica* **24**(3): 309–326.

Basseville, M. (1997). Information criteria for residual generation and fault detection and isolation, *Automatica* **33**(5): 783–803.

Basseville, M. and Benveniste, A. (eds) (1986). *Detection of Abrupt Changes in Signals and Dynamics Systems*, LNCIS 77, Springer-Verlag, Berlin.

Basseville, M. and Nikiforov, I. V. (1993). *Detection of Abrupt Changes: Theory and Application*, Information and System Science, Prentice Hall, New York.

Beard, R. V. (1971). *Failure Accommodation in Linear System Through Self Reorganization*, PhD thesis, Masschusetts Institute of Technology, Mass., USA.

Beattie, E. C., La Prad, R. F., McGlone, M. E., Rock, S. M. and Ahkter, M. M. (1981). Sensor failure detection for jet engines, *Technical Report NASA-CR-168190*, NASA.

Benkhedda, H. and Patton, R. J. (1996). B-spline network integrated qualitative and quantitative fault detection, *Proc. of the 13th IFAC World Congress*, San-Francisco, USA.

Bennett, S. M. (1998). *Fault-Tolerant Control Of A Rail Traction Induction Motor Drive*, PhD thesis, Univ. of Hull, Hull, UK.

Bennett, S. M., Patton, R. J., Daley, S. and Newton, D. A. (1996a). Model based intermittent fault tolerance in an induction motor drive, *Proc. of IMACS Multiconference: CESA'96*, Lille, France, pp. 678–683.

Bennett, S. M., Patton, R. J., Daley, S. and Newton, D. A. (1996b). Torque and flux estimation for a rail traction system in the preesence of intermittent sensor faults, *Proc. of the UKACC Int. Conf. on Contr.: CONTROL'96*, University of Exeter, UK, pp. 72–77 (Vol.I).

Berec, L. (1998). A multi-model method to fault detection and diagnosis: Bayesian solution. an introductory treatise, *Int. J. of Adaptive Contr. and Signal Processing* **12**(1): 81–92.

Betta, G., Dapuzzo, M. and Pietrosanto, A. (1995). A knowledge-based approach to instrument fault-detection and isolation, *IEEE Trans. On Instrumentation and Measurement* **44**(6): 1009–1016.

Bhattacharyya, S. P. (1978). Observer design for linear systems with unknown inputs, *IEEE Trans. Automat. Contr.* **AC-23**: 483–484.

Bogh, S. (1995). Multiple hypothesis-testing approach to FDI for the industrial actuator benchmark, *Contr. Eng. Practice* **3**(12): 1763–1768.

Bose, B. K. (1986). *Power Electronics and AC Drives*, Prentice Hall, London.

Boyd, S., Ghaoui, L., Feron, E. and Balakrishnan, V. (1994). *Linear Matrix Inequalities in System and Control Theory*, SIAM, Philadelphia.

Brown, M. and Harris, C. J. (1994). *Neurofuzzy Adaptive Modelling and Control*, Information and System Science, Prentice Hall, New York.

Brunet, J., Jaume, D., Labarrère, M., Rault, A. and Vergé, M. (1990). *Détection et Diagnostic de Panne - Approche par Modélisation*, Hermes Press, France. (In French).

Bundick, W. T. (1985). A preliminary evaluation of the generalized likelihood ratio for detecting and identifying control element failures in a transport aircraft, *Technical Report NASA-TM-87620*, NASA.

Bundick, W. T. (1991). Development of an adaptive failure-detection and identification for detecting aircraft control-element failures, *Technical Report NASA-TP-3051*, NASA.

Burrows, S. P. and Patton, R. J. (1991). Design of a low sensitivity, minimum norm and structurally constrained control law using eigenstructure assignment, *Optim. Contr. Appl. & Methods* **12**(3): 131–140.

Burrows, S. P. and Patton, R. J. (1992). Design of low-sensitivity modalized observers using eigenstructure assignment, *J. of Guidance, Contr. & Dynamics* **15**(3): 779–782.

Burrows, S. P., Patton, R. J. and Szymanski, J. E. (1989). Robust eigenstructure assignment with a control design package, *IEEE Contr. Syst. Mag.* **9**(4): 29–32.

Campbell, S. L. and Meyer, C. D. J. (1991). *Generalized Inverses of Linear Transformations*, Dover, New York.

Chang, C. T. and Chen, J. W. (1995). Implementation issues concerning the EKF-based fault-diagnosis techniques, *Chemical Engineering Science* **50**(18): 2861–2882.

Chang, I. C., Yu, C. C. and Liou, C. T. (1994). Model-based approach for fault-diagnosis. 1: Principles of deep model algorithm, *Industrial and Engineering Chemistry Research* **33**(6): 1542–1555.

Chang, S. and Hsu, P. L. (1993a). Fault detection observer design for linear systems with unknown inputs, *Proc. of 2nd European Control Conf.: ECC'93*, Groningen, Holland, pp. 1975–1980.

Chang, S. and Hsu, P. L. (1993b). State estimation using general structured observers for linear systems with unknown input, *Proc. of 2nd European Control Conf.: ECC'93*, Groningen, Holland, pp. 1794–1799.

Chang, S. K. and Hsu, P. L. (1995). A novel design for the unknown input fault-detection observer, *Control-Theory and Advanced Technology* **10**(4 Pt2): 1029–1051.

Chang, S. K., Hsu, P. L. and Lin, K. L. (1995). A parametric transfer-matrix approach to fault-identification filter design and threshold selection, *Int. J. Sys. Sci.* **26**(4): 741–754.

Chen, C. T. (1984). *Linear System Theory and Design*, Holt, Rinehart and Winston.

Chen, J. (1995). *Robust Residual Generation for Model-based Fault Diagnosis of Dynamic Systems*, PhD thesis, University of York, York, UK.

Chen, J. and Patton, R. J. (1994a). A re-examination of fault detectability and isolability in linear dynamic systems, *Preprints of the IFAC Sympo. on Fault Detection, Supervision and Safety for Technical Processes: SAFEPROCESS'94*, Espoo, Finland, pp. 590–596 (Vol.2).

Chen, J. and Patton, R. J. (1994b). Robust fault diagnosis of stochastic systems with unknown disturbances, *Proc. of the IEE Int. Conf.: Control' 94*, Peregrinus Press, Conf. Pub. No. 389, Warwick, UK, pp. 1340–1345.

Chen, J. and Patton, R. J. (1996). Optimal filtering and robust fault-diagnosis of stochastic-systems with unknown disturbances, *IEE Proc.-D: Contr. Theory & Appl.* **143**(1): 31–36.

Chen, J., Patton, R. J. and Liu, G. P. (1994a). Design of optimal residuals for detecting sensor faults using multi-object optimization and genetic algorithms, *Proc. of 1994 AIAA Guidance, Navigation and Control Conf.*, The Phoenician Scottsdale, AZ, USA, pp. 349–357. AIAA-94-3581-CP.

Chen, J., Patton, R. J. and Liu, G. P. (1994b). Detecting incipient sensor faults in flight control systems, *Proc. of The Third IEEE Conf. on Control Applications*, Glasgow, Scotland, pp. 871–876.

Chen, J., Patton, R. J. and Liu, G. P. (1996). Optimal residual design for fault-diagnosis using multiobjective optimization and genetic algorithms, *Int. J. Sys. Sci.* **27**(6): 567–576.

Chen, J., Patton, R. J. and Zhang, H. Y. (1993). A multi-criteria optimization approach to the design of robust fault detection algorithm, *Proc. of Int. Conf. on Fault Diagnosis: TOOLDIAG'93*, Toulouse, France.

Chen, J., Patton, R. J. and Zhang, H. Y. (1996). Design of unknown input observers and robust fault-detection filters, *Int. J. Contr.* **63**(1): 85–105.

Chen, J., Zhang, H. G. and Zhang, H. Y. (1990). A modified separated-bias estimation approach to the detection and estimation of failures in linear systems, *Preprints of 1990 XI IFAC World Congress*, Tallin.

Chen, J. and Zhang, H. Y. (1990). Parity vector approach for detecting failures in dynamic systems, *Int. J. Sys. Sci.* **21**(4): 765–770.

Chen, J. and Zhang, H. Y. (1991). Robust detection of faulty actuators via unknown input observers, *Int. J. Sys. Sci.* **22**(10): 1829–1839.

Chilali, M. and Gahinet, P. (1996). H_∞ design with pole-placement constraints - An LMI approach, *IEEE Trans. Automat. Contr.* **AC-41**(3): 358–367.

Choi, J. W. (1998). A simultaneous assignment methodology of right/left eigenstructures, *IEEE Trans. Aero. & Electron. Syst.* **34**(2): 625–634.

Choi, J. W., Lee, J. G., Kim, Y. and Kang, T. (1995). Design of an effective controller via disturbance accommodating left eigenstructure assignment, *J. of Guidance, Contr. & Dynamics* **18**(2): 347–354.

Chow, E. Y. and Willsky, A. S. (1980). Issues in the development of a general algorithm for reliable failure detection, *Proc. of the 19th Conf. on Decision & Control*, Albuquerque, NM.

Chow, E. Y. and Willsky, A. S. (1984). Analytical redundancy and the design of robust detection systems, *IEEE Trans. Automat. Contr.* **AC-29**(7): 603–614.

Chung, W. H. and Speyer, J. L. (1998). A game theoretic fault detection filter, *IEEE Trans. Automat. Contr.* **43**(2): 143–161.

Clark, R. N. (1978a). Instrument fault detection, *IEEE Trans. Aero. & Electron. Syst.* **AES-14**: 456–465.

Clark, R. N. (1978b). A simplified instrument failure detection scheme, *IEEE Trans. Aero. & Electron. Syst.* **AES-14**: 558–563.

Clark, R. N. (1979). The dedicated observer approach to instrument failure detection, *Proc. of The 18th IEEE Conf. on Decision & Control*, Fort Lauderdale, Fla., pp. 237–241.

Clark, R. N. (1989). State estimation schemes for instrument fault detection, *in* R. J. Patton, P. M. Frank and R. N. Clark (eds), *Fault Diagnosis in Dynamic Systems: Theory and Application*, Prentice Hall, chapter 2, pp. 21–45.

Clark, R. N., Fosth, D. C. and Walton, V. M. (1975). Detecting instrument malfunctions in control systems, *IEEE Trans. Aero. & Electron. Syst.* **AES-11**: 465–473.

Da, R. (1994). Failure detection of dynamical systems with the state χ^2 test, *J. of Guidance, Contr. & Dynamics* **17**(2): 271–277.

Da, R. and Lin, C. F. (1995). A new failure-detection approach and its application to GPS autonomous integrity monitoring, *IEEE Trans. Aero. & Electron. Syst.* **31**(1): 499–506.

Da, R. and Lin, C. F. (1996). Sensitivity analysis algorithm for the state chi-square test, *J. of Guidance, Contr. & Dynamics* **19**(1): 219–222.

Daley, S. and Wang, H. (1991). A parameteric design approach for observer based fault detection, *Proc. of 8'th Int. Conf. on Systems Engineering (ICSE'91)*, Coventry, UK, pp. 248–255.

Daley, S. and Wang, H. (1992). On the generation of an optimally robust residual signal for systems with structured model uncertainty, *Proc. of the 1992 American Control Conf.*, USA, pp. 2104–2108.

Dalton, T., Patton, R. J. and Chen, J. (1996). An application of eigenstructure assignment to robust resiudal design for fdi, *Proc. of the UKACC Int. Conf. on Contr.: CONTROL'96*, IEE, Univ. of Exter, UK, pp. 78–83.

Daly, K. C., Gai, E. and Harrison, J. V. (1979). Generalized likelihood test for FDI in redundancy sensor configurations, *J. of Guidance, Contr. & Dynamics* **2**(1): 9–17.

Darouach, M., Zasadzinski, M. and Keller, J. Y. (1992). State estimation for discrete systems with unknown inputs using state estimation of sigular systems, *Proc. of the 1992 American Control Conf.*, USA, pp. 3014–3015.

Darouach, M., Zasadzinski, M. and Xu, S. J. (1994). Full-order observers for linear systems with unknown inputs, *IEEE Trans. Automat. Contr.* **39**(3): 606–609.

Davis, L. (1991). *Handbook of Genetic Algorithms*, Van Nostrand Reinhold, New York.

Deckert, J. C., Desai, M. N., Deyst, J. J. and Willsky, A. S. (1977). F-8 DFBW sensor failure identification using analytic redundancy, *IEEE Trans. Automat. Contr.* **AC-22**(5): 795–803.

Deckert, J. C., Desai, M. N., Deyst, J. J. and Willsky, A. S. (1978). Reliable dual-redundant sensor failure detection and identiication for the NASA F-8 DFBW aircraft, *Technical Report NASA-CR-2944*, NASA.

Desai, M. and Ray, A. (1984). A fault detection and isolation methodology – theory and application, *Proc. 1984 Amer. Control Conf.*, pp. 262–270.

Dexter, A. L. (1995). Fuzzy model-based fault-diagnosis, *IEE Proc.-D: Contr. Theory & Appl.* **142**(6): 545–550.

Dexter, A. L. and Benouarets, M. (1997). Model-based fault diagnosis using fuzzy matching, *IEEE Trans. On Sys. Man and Cyber. Part A-Sys. & Humans* **27**(5): 673–682.

Ding, X. and Frank, P. M. (1989). Fault detection via optimally robust detection filters, *Proc. of 28th IEEE Conf. on Decision & Control*, Tampa, FL, pp. 1767–1772.

Ding, X. and Frank, P. M. (1990). Fault detection via factorization approach, *Syst. Contr. Lett.* **14**(5): 431–436.

Ding, X. and Frank, P. M. (1991). Frequency domain approach and threshold selector for robust model-based fault detection and isolation, *Peprint of IFAC/IMACS Symp. SAFEPROCESS'91*, Baden-Baden, pp. 307–312 (Vol.1).

Ding, X. and Frank, P. M. (1993). An adaptive observer-based fault detection scheme for nonlinear dynamic systems, *Preprints of The 12th IFAC World Congress*, Australia, pp. 307–310 (Vol.7).

Ding, X. and Guo, L. (1997). On observer based fault detection, *Proc. of the IFAC Sympo. on Fault Detection, Supervision and Safety for Technical Processes: SAFEPROCESS'97*, Pergamon 1998, Univ. of Hull, UK, pp. 115–124.

Ding, X. and Guo, L. (1998). An approach to time domain optimization of observer-based fault detection systems, *Int. J. Contr.* **69**(3): 419–442.

Ding, X., Guo, L. and Frank, P. M. (1993). A frequency domain approach to fault detection of uncertain dynamic systems, *Proc. of the 32nd IEEE Conf. on Decision and Contr.*, Texas, pp. 1722–1727.

Ding, X., Guo, L. and Frank, P. M. (1994). Parameterization of linear observers and its application to observer design, *IEEE Trans. Automat. Contr.* **AC-39**(8): 1648–1652.

Doraiswami, R. and Stevenson, M. (1996). A robust influence matrix approach to fault-diagnosis, *IEEE Trans. Contr. Sys. Techno.* **4**(1): 29–39.

Douglas, R. K. and Speyer, J. L. (1996). Robust fault-detection filter design, *J. of Guidance, Contr. & Dynamics* **19**(1): 214–218.

Doyle, J. C., Glover, K., Khargonekar, F. P. and Francis, B. A. (1989). State-space solutions to standard H_2 and H_∞ control problems, *IEEE Trans. Automat. Contr.* **AC-34**(8): 831–847.

Duan, G. R. (1993). Solution to matrix equation $AV + BW = VF$ and their application to eigenstructure assignment in linear systems, *IEEE Trans. Automat. Contr.* **AC-38**(2): 276–280.

Duan, G. R., Patton, R. J., Chen, J. and Chen, Z. (1997). A parameteric approach for robust fault detection in linear systems with unknown disturbances, *Proc. of the IFAC Sympo. on Fault Detection, Supervision and Safety for Technical Processes: SAFEPROCESS'97*, Pergamon 1998, Univ. of Hull, UK, pp. 307–312.

Duyar, A., Eldem, V., Merrill, W. C. and Guo, T. H. (1991). State-space representation of the open-loop dynamics of the space-shuttle main engine, *J. Dyn. Sys., Meas. & Contr. - Trans. of the ASME* **113**(4): 684–690.

Duyar, A., Eldem, V., Merrill, W. and Guo, T. H. (1994). Fault-detection and diagnosis in propulsion systems - a fault parameter-estimation approach, *J. of Guidance, Contr. & Dynamics* **17**(1): 104–108.

Duyar, A., Eldem, V. and Saravanan, N. (1990). A system identification approach for failure diagnosis and detection, *Proc. IEEE Int. Workshop on Intelligent Motion Control*, Istanbel, pp. 61–64.

Duyar, A. and Merrill, W. C. (1992). Fault-diagnosis for the space-shuttle main engine, *J. of Guidance, Contr. & Dynamics* **15**(2): 384–389.

Dvorak, D. L. (1992). *Monitoring and Diagnosis of Continuous Dynamic Systems Using Semi-Quantitative Simulation*, PhD thesis, The University of Texas at Austin, Austin, Texas 78712, USA.

Eckart, C. and Young, G. (1936). The approximation of one matrix by another of lower rank, *Psychometrika* **1**: 211–218.

Edelmayer, A., Bokor, J. and Keviczky, L. (1994). An H_∞ filtering approach to robust detection of failures in dynamic systems, *Proc. of the 33rd IEEE Conf. on Decision & Control*, Lake Buena Vista, USA, pp. 3037–3039.

Edelmayer, A., Bokor, J. and Keviczky, L. (1996). H_∞ detection filter design for linear systems: Comparison of two approaches, *Preprints of the 13th IFAC World Congress*, San-Francisco, USA, pp. 37–42.

Edelmayer, A., Bokor, J. and Keviczky, L. (1997a). Improving sensitivity of H_∞ detection filters linear systems, *Proc. of the IFAC Sympo.: SYSID'97 SICE*, pp. 1195–1200 (Vol.3).

Edelmayer, A., Bokor, J. and Keviczky, L. (1997b). A scaled L_2 optimisation approach for improving sensitivity of H_∞ detection filters for LTV systems, *in* C. Bányász (ed.), *Preprints of the 2nd IFAC Symp. on Robust Control Design: RECOND97*, Budapest, Hungary, pp. 543–548.

Edelmayer, A., Bokor, J., Szigeti, F. and Keviczky, L. (1997). Robust detection filter design in the presence of time-varying system perturbations, *Automatica* **33**(3): 471–475.

Edwards, C. and Spurgeon, S. K. (1994). On the development of discontinuous observers, *Int. J. Contr.* **59**(5): 1211–1229.

Edwards, C., Spurgeon, S. K., Patton, R. J. and Klotzek, P. (1997). Sliding mode observers for fault detection, *Proc. of the IFAC Sympo. on Fault Detection, Supervision and Safety for Technical Processes: SAFEPROCESS'97*, Pergamon 1998, Univ. of Hull, UK, pp. 507–512.

Eich, J. and Oehler, R. (1997). On the application of the generalised structured singular value to robust fdi system design, *Proc. of the IFAC Sympo. on Fault Detection, Supervision and Safety for Technical Processes: SAFEPROCESS'97*, Pergamon 1998, Univ. of Hull, UK, pp. 875–880.

Eide, P. and Maybeck, B. (1996). An MMAE failure-detection system for the F-16, *IEEE Trans. Aero. & Electron. Syst.* **32**(3): 1125–1136.

Emami-Naeini, A., Akhter, M. M. and Rock, M. M. (1986). Robust detection, isolation, and accommodation for sensor failures, *Technical Report NASA CR-174825*, NASA.

Emami-Naeini, A. E., Akhter, M. M. and Rock, S. M. (1988). Effect of model uncertainty on failure detection: the threshold selector, *IEEE Trans. Automat. Contr.* **AC-33**(2): 1106–1115.

Fahmy, M. M. and O'Reilly, J. (1982). On eigenstructure assignment in linear multivariable systems, *IEEE Trans. Automat. Contr.* **AC-27**(3): 690–693.

Fairman, F. W., Mahil, S. S. and Luk, L. (1984). Disturbance decoupled observer design via singular value decomposition, *IEEE Trans. Automat. Contr.* **AC-29**(1): 84–86.

Faitakis, Y. E. and Kantor, J. C. (1996). Residual generation and fault-detection for discrete-time-systems using an L_∞ technique, *Int. J. Contr.* **64**(1): 155–174.

Faitakis, Y. E., Thapliyal, S. and Kantor, J. C. (1994). Computing bounds for a simple fault detection scheme, *Proc. of the 1994 Amer. Contr. Conf.*, Maryland, pp. 2638–2642.

Fathi, Z., Ramirez, W. F. and Korbicz, J. (1993). Analytical and knowledge-based redundancy for fault-diagnosis in process plants, *AICHE J.* **39**(1): 42–56.

Favre, C. (1994). Fly-by-wire for commercial aircraft: the Airbus experience, *Int. J. Contr.* **59**(1): 139–157.

Francis, B. A. (1987). *A Course in H_∞ Control Theory*, Springer-Verlag, Berlin.

Frank, P. M. (1987). Fault diagnosis in dynamic system via state estimation - a survey, *in* Tzafestas, Singh and Schmidt (eds), *System Fault Diagnostics, Reliability & Related Knowledge-based Approaches*, D. Reidel Press, Dordrecht, pp. 35–98 (Vol. 1).

Frank, P. M. (1990). Fault diagnosis in dynamic system using analytical and knowledge based redundancy – a survey and some new results, *Automatica* **26**(3): 459–474.

Frank, P. M. (1991a). Enhancement of robustness in observer-based fault detection, *Preprints of IFAC/IMACS Sympo. SAFEPROCESS'91*, Baden-Baden, pp. 275–287 (Vol.1). "A modified version also published in *Int. J. Control*, Vol.59, No.4, 955-981, 1994".

Frank, P. M. (1991b). Fault diagnosis in dynamic systems using software redundancy, *European J. of Diagnosis and Safety in Automation (Revue européenne Diagnostic et sûreté de fonctionnement)* **1**(2): 113–143.

Frank, P. M. (1992a). Model-based fault diagnosis, *in* D. Atherton and P. Borne (eds), *Concise Encyclopedia of Simulation and Modelling*, Pergamon Press, pp. 262–269.

Frank, P. M. (1992b). Robust model-based fault detection in dynamic systems, *Preprints of IFAC Int. Sympo. "On-line fault detection and supervision in the chemical process industries"*, Delaware, USA, pp. 1–13.

Frank, P. M. (1993). Advances in observer-based fault diagnosis, *Proc. of Int. Conf. on Fault Diagnosis: TOOLDIAG'93*, Toulouse, pp. 817–836.

Frank, P. M. (1994a). Application of fuzzy logic process supervision and fault diagnosis, *Preprints of the IFAC Sympo. on Fault Detection, Supervision and Safety for Technical Processes: SAFEPROCESS'94*, Espoo, Finland, pp. 531–538 (Vol.2).

Frank, P. M. (1994b). Online fault-detection in uncertain nonlinear-systems using diagnostic observers - a survey, *Int. J. Sys. Sci.* **25**(12): 2129–2154.

Frank, P. M. (1996). Analytical and qualitative model-based fault diagnosis - a survey and some new results, *European J. of Contr.* **2**(1): 6–28.

Frank, P. M. and Ding, X. (1993). Frequency domain approach to minimizing detectable faults in FDI systems, *Appl. Math. & Comp. Sci.* **3**(3): 417–443.

Frank, P. M. and Ding, X. (1994). Frequency domain approach to optimally robust residual generation and evaluation for model-based fault diagnosis, *Automatica* **30**(4): 789–804.

Frank, P. M. and Ding, X. (1997). Survey of robust residual generation and evaluation methods in observer-based fault detection systems, *J. of Process Control* **7**(6): 403–424.

Frank, P. M., Ding, X. and Köppen, B. (1993). A frequency domain approach for fault detection at the inverted pendulum, *Proc. of Int. Conf. on Fault Diagnosis: TOOLDIAG'93*, Toulouse, pp. 987–994.

Frank, P. M., Ding, X. and Wochnik, J. (1991). Model based fault detection in diesel-hydraulically driven industrial trucks, *Proc. of 1991 Amer. Control Conf.*, Boston, pp. 1528–1533.

Frank, P. M. and Keller, L. (1981). Sensitivity discriminating observer design for instrument failure detection, *IEEE Trans. Aero. & Electron. Syst.* **AES-16**: 460–467.

Frank, P. M. and Kiupel, N. (1993). Fuzzy supervision and application to lean production, *Int. J. Sys. Sci.* **24**(10): 1935–1944.

Frank, P. M. and Köppen, B. (1993). Review of optimal solutions to the robustness problem in observer-based fault detection, *Proc. IMechE, Part J: J. of Syst. & Contr. Eng.* **207**: 105–112.

Frank, P. M. and Köppen-Seliger, B. (1997a). Fuzzy logic and neural network applications to fault diagnosis, *Int. J. of Approximate Reasoning* **16**(1): 67–88.

Frank, P. M. and Köppen-Seliger, B. (1997b). New developments using AI in fault diagnosis, *Eng. Appl. of AI* **10**(1): 3–14.

Frank, P. M. and Seliger, R. (1991). Fault detection and isolation in automatic processes, *in* C. Leondes (ed.), *Control and Dynamic Systems*, Vol. 49, Academic Press, pp. 241–287.

Frank, P. M. and Wünnenberg, J. (1987). Sensor fault detection via robust observers, *in* S. G. Tzafestas, M. G. Singh and G. Schmidt (eds), *System Fault Diagnostics, Reliability & Related Knowledge-based Approaches*, D. Reidel Press, Dordrecht, pp. 147–160 (Vol. 1).

Frank, P. M. and Wünnenberg, J. (1989). Robust fault diagnosis using unknown input schemes, *in* R. J. Patton, P. M. Frank and R. N. Clark (eds), *Fault Diagnosis in Dynamic Systems: Theory and Application*, Prentice Hall, chapter 3, pp. 47–98.

Frenzel, J. F. (1993). Genetic algorithms, *IEEE Potentials* **12**(3): 21–24.

Friedland, B. and Grabousky, S. (1982). Estimating sudden changes of biases in linear dynamic systems, *IEEE Trans. Automat. Contr.* **AC-27**(1): 237–240.

Funahashi, Y. (1976). Stable state estimator for bilinear systems, *Int. J. Contr.* **29**(2): 181–188.

Gahinet, P. and Apkarian, P. (1994). A linear matrix inequality approach to H_∞ control, *Int. J. Robust and Nonlinear Contr.* **4**(4): 421–448.

Gahinet, P., Nemirovski, A., Laub, A. and Chilali, M. (1995). *The LMI Control Toolbox*, The MathWorks Inc.

Gai, E. G., Adams, M. B. and Walker, B. K. (1976). Determination of failure thresholds in hybrid navigation, *IEEE Trans. Aero. & Electron. Syst.* **AES-12**(6): 744–755.

Gai, E. G., Adams, M. B., Walker, B. K. and Smestad, T. (1978). Correction to "Determination of failure thresholds in hybrid navigation", *IEEE Trans. Aero. & Electron. Syst.* **AES-14**(4): 696–697.

Gai, E. and Gurry, R. E. (1977). Failure detection by pilots during automatic landing: Models and experiments, *J. Aircraft* **14**(2): 135–141.

Gai, E., Harrison, J. V. and Daly, K. C. (1978). Failure detection and isolation performance of two redundancy sensor configurations, *Proc. of Position Location and Navigation Symposium (PLANS)*, San Diego, pp. 122–131.

Gantmacher, F. R. (1959). *The Theory of Matrices*, Vol. I & II, Chelsea Publishing Co.

Gao, Z. and Antsaklis, P. J. (1989). On the stable solution of the one- and two-sided model matching problem, *IEEE Trans. Automat. Contr.* **AC-34**(9): 978–982.

García, E. A. and Frank, P. M. (1997). Deterministic nonlinear observer-based approaches to fault diagnosis: a survey, *Contr. Eng. Practice* **5**(5): 663–670.

García, E. A., Köppen-Seliger, B. and Frank, P. M. (1995). A frequency-domain approach to residual generation for the industrial actuator benchmark, *Contr. Eng. Practice* **3**(12): 1747–1750.

Garg, V. and Hedrick, J. K. (1995). Fault detection filter for a class of nonlinear systems, *Proc. of the 1995 American Control Conf.*, Seattle, USA, pp. 1647–1651.

Ge, W. and Fang, C. Z. (1988). Detection of faulty components via robust observation, *Int. J. Contr.* **47**(2): 581–599.

Ge, W. and Fang, C. Z. (1989). Extended robust observation approach for failure isolation, *Int. J. Contr.* **49**(5): 1537–1553.

Geiger, G. (1982). Monitoring of an electrical driven pump using continuous-time parameter estimation methods, *Proc. the 6th IFAC Sympo. on Identification and Parameter Estimation*, Pergamon Press, Washington.

Gertler, J. (1988). Survey of model-based failure detection and isolation in complex plants, *IEEE Contr. Syst. Mag.* **8**(6): 3–11.

Gertler, J. (1991). Analytical redundancy methods in failure detection and isolation, *Preprints of IFAC/IMACS Sympo.: SAFEPROCESS'91*, Baden-Baden, pp. 9–21 (Vol.1). also published in a revised version in "*Control – Theory and Advanced Technology*, Vol. 9, No.1, 259-285, 1993".

Gertler, J. (1994). Modelling errors as unknown inputs, *Preprints of IFAC Sympo. on Fault Detection, Supervision and Safety for Technical Processes: SAFEPROCESS'94*, Espoo, Finland, pp. 266–271 (Vol.1).

Gertler, J. (1997). Fault detection and isolation using parity relations, *Contr. Eng. Practice* **5**(5): 653–661.

Gertler, J. (1998). *Fault Detection and Diagnosis in Engineering Systems*, Marcel Dekker, New York.

Gertler, J. and Costin, M. (1994). Model-based diagnosis of automotive engines, *Preprints of IFAC Sympo. on Fault Detection, Supervision and Safety for Technical Processes: SAFEPROCESS'94*, Espoo, Finland, pp. 421–430 (Vol.2).

Gertler, J. and DiPierro, G. (1997). On the relationship between parity relations and parameter estimation, *Proc. of the IFAC Sympo. on Fault Detection, Supervision and Safety for Technical Processes: SAFEPROCESS'97*, Pergamon 1998, Univ. of Hull, UK, pp. 453–458.

Gertler, J., Fang, X. W. and Luo, Q. (1990). Detection and diagnosis of plant failures; the orthogonal parity equation approach, *in* C. Leondes (ed.), *Control & Dynamics Systems*, Vol.37,Academic Press, pp. 157–216.

Gertler, J. and Kunwer, M. K. (1993). Optimal residual decoupling for robust fault diagnosis, *Proc. of Int. Conf. on Fault Diagnosis: TOOLDIAG'93*, Toulouse, pp. 436–452. also published in a revised version in "*Int. J. Contr.*, Vol.61, No.2, 395-421, 1995".

Gertler, J. and Luo, Q. (1989). Robust isolable models for failure diagnosis, *AIChE J.* **35**(11): 1856–1868.

Gertler, J., Luo, Q., Anderson, K. and Fang, X. W. (1990). Diagnosis of plant failures using orthogonal parity equations, *Proc. of the 11th IFAC World Congress*, Tallin.

Gertler, J. and Monajemy, R. (1995). Generating directional residuals with dynamic parity relations, *Automatica* **31**(4): 627–635.

Gertler, J. and Singer, D. (1990). A new structural framework for parity equation-based failure detection and isolation, *Automatica* **26**(2): 381–388.

Goldberg, D. E. (1989). *Genetic Algorithms in Search, Optimization, and Machine Learning*, Addison Wesley Publishing Company.

Golub, G. H. and Van Loan, G. F. (1989). *Matrix Computations*, John Hopkins Series in the Mathematical Sciences, The Johns Hopkins University Press, Baltimore and London.

Grainger, R. W., Holst, J., Isaksson, A. J. and Ninness, B. M. (1995). A parametric statistical approach to FDI for the industrial actuator benchmark, *Contr. Eng. Practice* **3**(12): 1757–1762.

Grimble, M. J. and Elsayed, A. (1990). Solution of the H_∞ optimal linear-filtering problem for discrete-time-systems, *IEEE Tran. On Acoustics Speech and Signal Proc.* **38**(7): 1092–1104.

Guan, Y. and Saif, M. (1991). A novel approach to the design of unknown input observers, *IEEE Trans. Automat. Contr.* **AC-36**(5): 632–635.

Guernez, C., Cassar, J. P. and Staroswiecki, M. (1997). Extension of parity space to nonlinear polynominal dynamic systems, *Proc. of the IFAC Sympo. on Fault Detection, Supervision and Safety for Technical Processes: SAFEPROCESS'97*, Pergamon 1998, Univ. of Hull, UK, pp. 857–862.

Hać, A. (1992). Design of disturbance decoupled observer for bilinear systems, *J. Dyn. Sys., Meas. & Contr. - Trans. of the ASME* **114**(4): 556–562.

Hammouri, H., Kinnaert, M. and El Yaagoubi, E. H. (1998a). Fault detection and isolation for state affine systems. To appear in *European J. of Contr.*

Hammouri, H., Kinnaert, M. and El Yaagoubi, E. H. (1998b). Observer-based approach to fault detection and isolation for nonlinear systems. To appear in *IEEE Trans. on Auto. Contr.*

Handelman, D. A. and Stengel, R. F. (1989). Combinating expert and analytical redundancy concepts for fault-tolerant flight control, *J. of Guidance, Contr. & Dynamics* **12**(1): 39–45.

Hara, A. and Furuta, K. (1976). Minimal order state observers for bilinear systems, *Int. J. Contr.* **24**(5): 705–718.

Hayashi, Y., Buckley, J. J. and Czogala, E. (1992). Systems engineering applications of fuzzy neural networks, *J. of Systems Eng.* **2**(4): 232–236.

Healey, A. J. (1998). Analytical redundancy and fuzzy inference in AUV fault detection and compensation, *Proc. of Oceanology 1998*, Brighton.

Hengy, D. and Frank, P. M. (1986). Component failure detection via nonlinear state observers, *Proc. of IFAC Workshop on Fault Detection and Safety in Chemical Plants*, Kyoto, Japan, pp. 153–157.

Himmelblau, D. M. (1978). *Fault Detection and Diagnosis in Chemical and Petrochemical Processes*, Chemical Engineering Monograph 8, Elsevier.

Himmelblau, D. M. (1986). Fault detection and diagnosis - today and tomorrow, *Proc. IFAC Workshop on Fault Detection and Safety in Chemical Plants*, Kyoto, Japan, pp. 95–105.

Himmelblau, D. M., Barker, R. W. and Suewatanakul, W. (1991). Fault classification with the aid of artificial neural networks, *Peprint of IFAC/IMACS Symp.: SAFEPROCESS'91*, Baden-Baden, pp. 369–373 (Vol.2).

Höfling, T. and Isermann, R. (1996). Fault-detection based on adaptive parity equations and single- parameter tracking, *Contr. Eng. Practice* **4**(10): 1361–1369.

Holland, J. H. (1975). *Adaption in Natural and Artificial Systems*, The University of Michigan Press, Ann Arbor.

Hou, M. and Müller, P. C. (1991). Design of robust observers for fault isolation, *Peprint of IFAC/IMACS Symp.: SAFEPROCESS'91*, Baden-Baden, pp. 295–300 (Vol.1).

Hou, M. and Müller, P. C. (1992). Design of observers for linear systems with unknown inputs, *IEEE Trans. Automat. Contr.* **AC-37**(6): 871–875.

Hou, M. and Müller, P. C. (1993). Unknown input decoupled Kalman filter for time-varying systems, *Proc. of 2nd European Control Conf.: ECC'93*, Groningen, Holland, pp. 2266–2270.

Hou, M. and Müller, P. C. (1994a). Disturbance decoupled observer design: A unified viewpoint, *IEEE Trans. Automat. Contr.* **AC-39**(6): 1338–1341.

Hou, M. and Müller, P. C. (1994b). Fault-detection and isolation observers, *Int. J. Contr.* **60**(5): 827–846.

Hou, M. and Patton, R. J. (1996). An LMI approach to H_-/H_∞ fault detection observers, *Proc. of the UKACC Int. Conf. on Contr.: CONTROL'96*, IEE, Univ. of Exter, UK, pp. 305–310.

Hou, M. and Pugh, A. C. (1997). Observing state in bilinear systems: a UIO approach, *Proc. of the IFAC Sympo. on Fault Detection, Supervision and Safety for Technical Processes: SAFEPROCESS'97*, Pergamon 1998, Univ. of Hull, UK, pp. 783–788.

Howell, J. (1994). Model-based fault-detection in information poor plants, *Automatica* **30**(6): 929–943.

Hung, Y. S. (1993). Model-matching approach to H_∞ filtering, *IEE Proc.-D: Contr. Theory & Appl.* **140**(2): 133–139.

Hunt, K. J., Sbarbaro, D., Zbikowski, R. and Gawthrop, P. J. (1992). Neural networks for control systems - a survey, *Automatica* **28**(6): 1083–1112.

Hwang, D. S., Chang, S. K. and Hsu, P. L. (1997). A practical design for a robust fault detection and isolation system, *Int. J. Sys. Sci.* **28**(3): 265–275.

Isaksson, A. J. (1993). An on-line threshold selector for failure detection, *Proc. of Int. Conf. on Fault Diagnosis: TOOLDIAG'93*, Toulouse, pp. 628–634.

Isermann, R. (1984). Process fault detection based on modelling and estimation methods: A survey, *Automatica* **20**(4): 387–404.

Isermann, R. (1987). Experiences with process fault detection via parameter estimation, *in* S. G. Tzafestas, M. G. Singh and G. Schmidt (eds), *System Fault Diagnostics, Reliability & Related Knowledge-based Approaches*, D. Reidel Press, Dordrecht, pp. 3–33.

Isermann, R. (1991a). Fault diagnosis of machine via parameter estimation and knowledge processing - tutorial paper, *Preprints of IFAC/IMACS Sympo.: SAFEPROCESS'91*, Baden-Baden, pp. 121–133 (Vol.1). "A modified version also published Also published in *Automatica*, Vol.29, No.4, 815-835".

Isermann, R. (1993). On the applicability of model-based fault detection for technical processes, *Preprints of The 12th IFAC World Congress*, Sydney, pp. 195–200 (Vol.9). Also published in "*Contr. Eng. Practice*, Vol.2, No.3, 439-450, 1994".

Isermann, R. (1994). Integration of fault detection and diagnosis methods, *Preprints of the IFAC Sympo. on Fault Detection, Supervision and Safety for Technical Processes: SAFEPROCESS'94*, Espoo, Finland, pp. 597–612 (Vol.2).

Isermann, R. (1997). Supervision, fault-detection and fault-diagnosis methods - an introduction, *Contr. Eng. Practice* **5**(5): 639–652.

Isermann, R. (1998). On fuzzy logic applications for automatic control, supervision, and fault diagnosis, *IEEE Trans. On Sys. Man and Cyber. Part A-Sys. & Humans* **28**(2): 221–235.

Isermann, R. and Ballé, P. (1997). Trends in the application of model-based fault detection and diagnosis of technical processes, *Contr. Eng. Practice* **5**(5): 709–719.

Isermann, R. (ed.) (1991b). *Preprints of IFAC/IMACS Symposium on Fault Detection, Supervision and Safety for Technical Processes – SAFEPROSS'91*, Baden-Baden, Germany.

Isermann, R. and Freyermuth, B. (1990). Process fault diagnosis based on process model knowledge, *Journal A* **31**(4): 58–65.

Isermann, R. and Freyermuth, B. (1991a). Process fault diagnosis based on process model knowledge – Part I: Principles for fault diagnosis with parameter estimation, *J. Dyn. Sys., Meas. & Contr. - Trans. of the ASME* **113**(4): 620–626.

Isermann, R. and Freyermuth, B. (1991b). Process fault diagnosis based on process model knowledge – Part II: Case-study experiments, *J. Dyn. Sys., Meas. & Contr. - Trans. of the ASME* **113**(4): 627–633.

Isermann, R. and Ulieru, M. (1993). Integrated fault detection and diagnosis, *Proc. of the 1993 IEEE Conference on Systems, Man & Cybernetics*, Le Touquet, France, pp. 743–748.

Iwasaki, T. and Skelton, R. E. (1994). All controllers for the general H_∞ control problem: LMI existence conditions and state space formulas, *Automatica* **30**(8): 1307–1317.

Jacobson, C. A. and Nett, C. N. (1991). An integrated approach to controls and diagnostics using the four parameter control, *IEEE Contr. Syst. Mag.* **11**(6): 22–29.

Jones, H. L. (1973). *Failure Detection in Linear Systems*, PhD thesis, Masschusetts Institute of Technology, Mass.,USA.

Jorgensen, R. B., Patton, R. J. and Chen, J. (1994). Fault detection and isolation using eigenstructure assignment, *Preprints of the IFAC Sympo. on Fault Detection, Supervision and Safety for Technical Processes: SAFEPROCESS'94*, Espoo, Finland, pp. 491–497 (Vol.2). also published in a revised version in "*Contr. Eng. Practice*, Vol. 3, No.12, 1751-1756, 1995".

Karcanias, N. and Kouvaritakis, B. (1979). The output zeroing problem and its relationship to the invariant zero structure, *Int. J. Contr.* **30**: 395–415.

Kautsky, J., Nichols, N. K. and Van Dooren, P. (1985). Robust pole assignment in linear state feedback, *Int. J. Contr.* **41**(5): 1129–1155.

Kavuri, S. N. and Venkatasubramanian, V. (1994). Neural network decomposition strategies for large-scale fault diagnosis, *Int. J. Contr.* **59**(3): 767–792.

Keller, J. Y., Nowakowski, S. and Darouach, M. (1992). State estimation and failure-detection in singular systems, *Control – Theory and Advanced Technology* **8**(4): 755–762.

Keller, J. Y., Summerer, L., Boutayeb, M. and Darouach, M. (1996). Generalized likelihood ratio approach for fault detection in linear dynamic stochastic systems with unknown inputs, *Int. J. Sys. Sci.* **27**(12): 1231–1241.

Keviczky, L., Bokor, J., Szigeti, F. and Edelmayer, A. (1993). Modelling time varying system perturbations: Application to robust change detection and identification, *Preprints of The 12th IFAC World Congress*, Australia, pp. 517–520 (Vol.7).

Kilsgaard, S., Rank, M. L., Niemann, H. and Stoustrup, J. (1996). Simultaneous design of controller and fault detector, *Proc. of the 35th IEEE Conf. on Decision and Contr.*, Kobe, Japan, pp. 628–629.

Kinnaert, M. (1993a). Design of redundancy relations for failure detection and isolation by constrained optimization, *Proc. of Int. Conf. on Fault Diagnosis: TOOLDIAG'93*, Toulouse, pp. 808–816.

Kinnaert, M. (1993b). Design of redundancy relations for failure detection and isolation by constrained optimization, *Preprints of The 12th IFAC World Congress*, Australia, pp. 227–230 (Vol.6).

Kinnaert, M. (1996). Design of redundancy relations for failure-detection and isolation by constrained optimization, *Int. J. Contr.* **63**(3): 609–622.

Kinnaert, M. and Peng, Y. B. (1995). Residual generator for sensor and actuator fault-detection and isolation - a frequency-domain approach, *Int. J. Contr.* **61**(6): 1423–1435.

Kobayashi, N. and Nakamizo, R. (1982). An observer design for linear systems with unknown inputs, *Int. J. Contr.* **35**: 605–619.

Köppen-Seliger, B. and Frank, P. M. (1995). Fault detection and isolation in technical processes with neural networks, *Proc. of the 34th Conf. on Decision & Control*, New Orleans, USA, pp. 2414–2419.

Korbicz, J., Fathi, Z. and Ramirez, W. F. (1993). State estimation schemes for fault-detection and diagnosis in dynamic-systems, *Int. J. Sys. Sci.* **24**(5): 985–1000.

Krishnaswami, V., Luh, G. C. and Rizzoni, G. (1995). Nonlinear parity equation based residual generation for diagnosis of automotive engine faults, *Contr. Eng. Practice* **3**(10): 1385–1392.

Krishnaswami, V. and Rizzoni, G. (1994a). Nonlinear parity equation residual generation for fault detection and isolation, *Preprints of the IFAC Sympo. on Fault Detection, Supervision and Safety for Technical Processes: SAFEPROCESS'94*, Espoo, Finland, pp. 317–322 (Vol.1).

Krishnaswami, V. and Rizzoni, G. (1994b). A survey of observer based residual generation for fdi, *Preprints of the IFAC Sympo. on Fault Detection, Supervision and Safety for Technical Processes: SAFEPROCESS'94*, Espoo, Finland, pp. 34–39 (Vol.1).

Krishnaswami, V. and Rizzoni, G. (1997). Robust residual generation for non-linear system fault detection and isolation, *Proc. of the IFAC Sympo. on Fault Detection, Supervision and Safety for Technical Processes: SAFE-PROCESS'97*, Pergamon 1998, Univ. of Hull, UK, pp. 163–168.

Kudva, P., Viswanadham, N. and Ramakrishna, A. (1980). Observers for linear system with unknown inputs, *IEEE Trans. Automat. Contr.* **AC-25**(2): 113–115.

Kurek, J. (1982). Observations of the state vector of linear multivariable systems with unknown inputs, *Int. J. Contr.* **36**(3): 511–515.

Labarrère, M. (ed.) (1993). *Proceedings of International Conference on Fault Diagnosis: TOOLDIAG'93*, Toulouse, France.

Labarrère, M. and Patton, R. J. (1993). Detection of sensor failures, *in* M. Pelegrin and W. M. Hollister (eds), *Concise Encyclopedia of Aeronautics & Space Systems*, Pergamon Press, pp. 101–110.

Lane, S. H., Handelman, D. A. and Gelfand, J. J. (1992). Theory and development of higher-order CMAC neural-networks, *IEEE Contr. Syst. Mag.* **12**(2): 23–30.

Leininger, G. G. (1981). Model degradation effects on sensor failure detection, *Proc. of the 1981 Joint Amer. Control Conf.*, Charlottesville, VA, pp. Paper FP-3A (Vol.3).

Leitch, R. (1993). Engineering diagnosis: match problems to solutions, *Proc. of Int. Conf. on Fault Diagnosis: TOOLDIAG'93*, Toulouse, pp. 837–844.

Leitch, R., Kraft, R. and Luntz, R. (1991). RESCU: a real-time knowledge based system for process control, *IEE Proc.-D: Contr. Theory & Appl.* **138**(3): 217–227.

Leitch, R. and Quek, C. (1992). Architecture for integrated process supervision, *IEE Proc.-D: Contr. Theory & Appl.* **139**(3): 317–327.

Leonard, J. A. and Kramer, M. A. (1993). Diagnosing dynamic faults using modular neural-nets, *IEEE Expert-Intelligent Systems & Their Applications* **8**(2): 44–53.

Leonhardt, S. and Ayoubi, M. (1997). Methods of fault diagnosis, *Contr. Eng. Practice* **5**(5): 683–692.

Leyval, L., Montmain, J. and Gentil, S. (1994). Qualitative-analysis for decision-making in supervision of industrial continuous-processes, *Mathematics and Computer in Simulation* **36**(2): 149–163.

Liu, B. F. and Si, J. N. (1997). Fault isolation filter design for linear time-invariant systems, *IEEE Trans. Automat. Contr.* **42**(5): 704–707.

Liu, G. P. and Patton, R. J. (1996). Robust control design using eigenstructure assignment and multiobjective optimization, *Int. J. Sys. Sci.* **27**(9): 871–879.

Liu, G. P. and Patton, R. J. (1998). *Eigenstructure Assignment for Control System Design*, John Wiley & Sons Ltd, Chichester.

Loparo, K. A., Buchner, M. R. and Vasudeva, K. S. (1991). Leak detection in an experimental heat exchanger process: A multiple model approach, *IEEE Trans. Automat. Contr.* **AC-36**(2): 167–177.

Lou, X., Willsky, A. S. and Verghese, G. C. (1986). Optimally robust redundancy relations for failure detection in uncertain systems, *Automatica* **22**(3): 333–344.

Lunze, J. (1991). A method for logic-based fault-diagnosis, *Preprints of IFAC/IMACS Sympo.: SAFEPROCESS'91*, Baden-Baden, pp. 45–52 (Vol.2).

Lunze, J. (1994). Qualitative modelling of linear dynamical systems with quantized state measurements, *Automatica* **30**(3): 417–431.

Lunze, J. and Schiller, F. (1992). Logic-based process diagnosis utilising the causal structure of dynamical systems, *Preprints of IFAC/IFIP/IMACS Int. Sympo. on Artificial Intelligence in Real-Time Control: AIRTC'92*, Delft, pp. 649–654.

Ma, X. J., Sun, Z. Q. and He, Y. Y. (1998). Analysis and design of fuzzy controller and fuzzy observer, *IEEE Trans. Fuzzy Sys.* **6**(1): 41–51.

Maciejowski, J. M. (1989). *Multivariable Feedback Design*, Addison-Wesley, Wokingham, U.K.

Magni, J. F. and Mouyon, P. (1991). A generalized approach to observers for fault diagnosis, *Proc. of the 30th IEEE Conf. on Decision & Control*, Brighton, UK, pp. 2236–2241. A modified version was also published in "*IEEE Trans. on Auto. Contr.*, Vol.39, No.2, 441-447, 1994".

Magni, J. F. and Mouyon, P. (1992). On residual generation by observer and parity space approaches, *Proc. of the 31st Conf. on Decision & Control*, Tucson, AZ, pp. 185–190.

Magni, J. F., Mouyon, P. and Arsan, M. I. (1993). Parameteric observation for fault diagnosis, *Preprints of The 12th IFAC World Congress*, Australia, pp. 287–290 (Vol.8).

Maki, Y. and Loparo, K. A. (1997). A neural-network approach to fault detection and diagnosis in industrial processes, *IEEE Trans. Contr. Sys. Techno.* **5**(6): 529–541.

Mangoubi, R., Appleby, B. D. and Farrell, J. R. (1992). Robust estimation in fault detection, *Proc. of the 31st Conf. on Decision & Control*, Tucson, AZ, USA, pp. 2317–2322.

Mangoubi, R., Appleby, B. D., Verghese, G. C. and Vander Velde, W. E. (1995). A robust failure detection and isolation algorithm, *Proc. of the 34th Conf. on Decision & Control*, New Orleans, USA, pp. 2377–2382.

Marquez, H. J. and Diduch, C. P. (1992). Sensitivity of failure detection using generalized observers, *Automatica* **28**(4): 837–840.

Martin, E. B., Morris, A. J. and Zhang, J. (1996). Process performance monitoring using multivariate statistical process-control, *IEE Proc.-D: Contr. Theory & Appl.* **143**(2): 132–144.

Martin, K. F. (1993). Review by discussion of condition monitoring and fault diagnosis in machine tools, *Int. J. of Machine Tools & Manufacture* **34**(4): 527–551.

Massoumnia, M. A. (1986a). *A Geometric Approach to Failure Detection and Identification in Linear Systems*, PhD thesis, Masschusctts Institute of Technology, Mass., USA.

Massoumnia, M. A. (1986b). A geometric approach to the synthesis of failure detection filters, *IEEE Trans. Automat. Contr.* **AC-31**(9): 839–846.

Massoumnia, M. A. and Vander Velde, W. E. (1988). Generating parity relations for detecting and identifying control system component failures, *J. of Guidance, Contr. & Dynamics* **11**(1): 60–65.

Mechmeche, C. and Nowakowski, S. (1997). Residual generator sythesis for bilinear systems with unknown inputs, *Proc. of the IFAC Sympo. on Fault Detection, Supervision and Safety for Technical Processes: SAFEPROCESS'97*, Pergamon 1998, Univ. of Hull, UK, pp. 765–770.

Medvedev, A. (1995). Fault-detection and isolation by a continuous parity space method, *Automatica* **31**(7): 1039–1044.

Medvedev, A. (1996). Fault-detection and isolation by functional continuous deadbeat observers, *Int. J. Contr.* **64**(3): 425–439.

Mehra, R. K. and Peschon, J. (1971). An innovations approach to fault detection and diagnosis in dynamic systems, *Automatica* **7**: 637–640.

Menke, T. E. and Maybeck, P. S. (1995). Sensor/actuator failure-detection in the vista F-16 by multiple model adaptive estimation, *IEEE Trans. Aero. & Electron. Syst.* **31**(4): 1218–1229.

Merrill, W. C. (1985). Sensor failure detection for jet engines using analytical redundancy, *J. of Guidance, Contr. & Dynamics* **8**(6): 673–682.

Merrill, W. C. (1990). Sensor failure detection for jet engines, *in* C. Leondes (ed.), *Control and Dynamics - Advances in Aerospace Systems Dynamics and Control Systems*, Vol.33, Academic Press, pp. 1–33.

Merrill, W. C., DeLaat, J. C. and Abdelwahab, M. (1991). Turbofan engine demonstration of sensor failure-detection, *J. of Guidance, Contr. & Dynamics* **14**(2): 337–349.

Merrill, W. C., DeLaat, J. C. and Bruton, W. M. (1988). Advanced detection, isolation, and accommodation of sensor failures – a real-time evaluation, *J. of Guidance, Contr. & Dynamics* **11**(6): 517–526. Also NASA TP-2740.

Merrill, W. C., DeLaat, J. C., Kroszkewicz, S. N. and Abdelwahab, M. (1987). Full-scale engine demonstration of an advanced sensor failure detection, isolation and accommodation algorithm - preliminary results, *AIAA Guidance, Navigation and Control Conference*, Monterey, Ca., USA. AIAA-87-2259.

Merrill, W. C., DeLaat, J. C., Kroszkewicz, S. N. and Abdelwahab, M. (1988). Advanced detection, isolation, and accommodation of sensor failures – engine demonstration results, *Technical Report NASA-TP-2836*, NASA.

Merrill, W. C. and Leininger, G. (1981). Identification and dual adaptive control of a turbojet engine, *Int. J. Contr.* **34**(3): 529–546.

Meserole, J. S. (1981). *Detection Filters for Fault-tolerant Control of Turbofan Engines*, PhD thesis, Masschusetts Institute of Technology, Mass., USA.

Miller, R. J. and Mukundan, R. (1982). On desinging reduced-order observers for linear time-invariant systems subject to unknown inputs, *Int. J. Contr.* **35**(1): 183–188.

Milne, R. (1987). Strategies for diagnosis, *IEEE Trans. on Sys., Man & Cybernetics* **SMC-17**(3): 333–339.

Mironovski, L. A. (1979). Functional diagnosis of linear dynamic systems, *Automn Remote Control* **40**: 1198–1205.

Mironovski, L. A. (1980). Functional diagnosis of dynamic system – a survey, *Automn Remote Control* **41**: 1122–1143.

Montgomery, R. C. and Caglayan, A. K. (1976). Failure accommodation in digital flight control systems by bayesian decision theory, *J. Aircraft* **13**(2): 69–75.

Moore, B. C. (1976). On the flexibility offered by state feedback in multivariable systems beyond closed-loop eigenvalue assignment, *IEEE Trans. Automat. Contr.* **AC-21**: 689–692.

Morgan, D. P. and Scofield, C. L. (1991). *Neural Networks and Speech Processing*, Kluwer Acadenic Publishers.

Morse, A. S. (1973). Structural invariants of linear multivariablr systems, *SIAM J. Contr. & Optimiz.* **11**(3): 446–465.

Mudge, S. K. and Patton, R. J. (1988). Analysis of the techniques of robust eigenstructure assignment with application to aircraft control, *IEE Proc.-D: Contr. Theory & Appl.* **135**(7): 275–281.

Naidu, S., Zafiriou, E. and McAvoy, T. J. (1990). Use of neural networks for failure detection in a control system, *IEEE Contr. Syst. Mag.* **10**: 49–55.

Napolitano, M. R., Casdorph, V., Neppach, C., Naylor, S., Innocenti, M. and Silvestri, G. (1996). Online learning neural architectures and cross-correlation analysis for actuator failure-detection and identification, *Int. J. Contr.* **63**(3): 433–455.

Napolitano, M. R., Neppach, C., Casdorph, V., Naylor, S., Innocenti, M. and Silvestri, G. (1995). Neural-network-based scheme for sensor failure-detection, identification, and accommodation, *J. of Guidance, Contr. & Dynamics* **18**(6): 1280–1286.

Narendra, K. S. (1996). Neural networks for control: Theory and practice, *Proc. IEEE* **84**(10): 1385–1406.

Narendra, K. S. and Parthasarthy, K. (1990). Identification and control of dynamical systems using neural networks, *IEEE Trans. on Neural Networks* **1**(1): 4–27.

Nett, C. N., Jacobson, C. A. and Balas, M. J. (1984). A connection between state space and doubly coprime fractional representations, *IEEE Trans. Automat. Contr.* **AC-29**(9): 831–832.

Nett, C. N., Jacobson, C. A. and Miller, A. T. (1988). An integrated approach to controls and diagnostics: The 4-parameter controller, *Proc. 1988 Amer. Contr. Conf.*, pp. 824–835.

Niemann, H. and Stoustrup, J. (1996). Filter design for failure detection and isolation in the presence of modelling errors and disturbances, *Proc. of the 35th IEEE Conf. on Decision and Contr.*, Kobe, Japan, pp. 1155–1160.

Niemann, H. and Stoustrup, J. (1997). Integration of control and fault detection: nominal and robust design, *Proc. of the IFAC Sympo. on Fault Detection, Supervision and Safety for Technical Processes: SAFEPROCESS'97*, Pergamon 1998, Univ. of Hull, UK, pp. 331–336.

Nikiforov, I., Varavva, V. and Kireichikov, V. (1993). Application of statistical fault-detection algorithms to navigation systems monitoring, *Automatica* **29**(5): 1275–1290.

Nikoukhah, R. (1994). Innovations generation in the presence of unknown inputs - application to robust failure-detection, *Automatica* **30**(12): 1851–1867.

Ninness, B. M. and Goodwin, G. C. (1991). Reliable model based failure detection via low order modelling, *Technical Report EE9117*, Univ. of Newcastle, Australia.

O'Reilly, J. (1983). *Observers for Linear Systems*, Academic Press.

Owens, T. J. (1988). Parametric state-feedback control with model-reduction, *Int. J. Contr.* **47**(5): 1299–1306.

Owens, T. J. and O'Reilly, J. (1989). Parametric state-feedback control for arbitrary eigenvalue assignment with minimum sensitivity, *IEE Proc.-D: Contr. Theory & Appl.* **136**(6): 307–313.

Palhares, R. M. and Peres, L. D. (1998). Optimal filtering schemes for linear discrete-time systems: a linear matrix inequality approach, *Int. J. Sys. Sci.* **29**(6): 587–593.

Park, J. H., Halevi, Y. and Rizzoni, G. (1994). A new interpretation of the fault-detection filter. 2: The optimal detection filter, *Int. J. Contr.* **60**(6): 1339–1351.

Park, J. H. and Rizzoni, G. (1994a). A new interpretation of the fault-detection filter. 1: Closed- form algorithm, *Int. J. Contr.* **60**(5): 767–787.

Park, J. and Rizzoni, G. (1993). A closed-form expression for the fault detection filter, *Proc. of The 32nd IEEE Conf. on Decision & Control*, Texas, USA, pp. 259–264.

Park, J., Rizzoni, G. and Ribbens, W. B. (1994). On the representation of sensor faults in fault-detection filters, *Automatica* **30**(11): 1793–1795.

Park, P. and Kailath, T. (1997). H_∞ filtering via convex optimization, *Int. J. Contr.* **66**(1): 15–22.

Park, Y. and Rizzoni, G. (1994b). An eigenstructure assignment algorithms for the design of fault detection filters, *IEEE Trans. Automat. Contr.* **39**(7): 1521–1524.

Park, Y. and Stein, J. L. (1988). Closed-loop, state and inputs observer for systems with unknown inputs, *Int. J. Contr.* **48**(3): 1121–1136.

Patton, R. J. (1988). Robust fault detection using eigenstructure assignment, *Proc. 12th IMACS World Congress on Scientific Computation*, Paris, pp. 431–434 (Vol.2).

Patton, R. J. (1989). The design of a sensor fault diagnosis system for a gas turbine engine - a feasibility study: Part I & Part II, *Technical report*, Lucas Aerospace Contract Report through the York Electronics Centre.

Patton, R. J. (1991). Fault detection and diagnosis in aerospace systems using analytical redundancy, *IEE Computing & Control Eng. J.* **2**(3): 127–136.

Patton, R. J. (1993). Robustness issues in fault tolerant control, *Proc. of Int. Conf. on Fault Diagnosis: TOOLDIAG'93*, Toulouse, France. Plenary Paper.

Patton, R. J. (1994). Robust model-base fault diagnosis: The state of the art, *Preprints of the IFAC Sympo. on Fault Detection, Supervision and Safety for Technical Processes: SAFEPROCESS'94*, Espoo, Finland, pp. 1–27 (Vol.1).

Patton, R. J. (1997a). Fault-tolerant control: the 1997 situation (survey), *Proc. of the IFAC Sympo. on Fault Detection, Supervision and Safety for Technical Processes: SAFEPROCESS'97*, Pergamon 1998, Univ. of Hull, UK, pp. 1029–1052.

Patton, R. J. (1997b). Robustness in model-based fault diagnosis: the 1995 situation, *A. Rev. Control* **21**: 103–123.

Patton, R. J. and Chen, J. (1990). The design of a robust fault diagnosis scheme for a engine sensor system, *Proc. of the IMACS Annals on Computing and Applied Mathematics $MIM - S^2$: 90*, Brussels, pp. IV.A.5-1 - IV.A.5.7. also published in *Mathematical and Intelligent models in systems simulation*, R. Hanus, P. Kool & S. Tzafestas (eds), 489-495, J.C. Baltzer AG Scientific Publishing Co., 1991.

Patton, R. J. and Chen, J. (1991a). Detection of faulty sensors in aero jet engine systems using robust model-based methods, *IEE Colloquium on "Condition monitoring for Fault Diagnosis"*. IEE Colloquium Digest No. 1991/156.

Patton, R. J. and Chen, J. (1991b). Optimal selection of unknown input distribution matrix in the design of robust observers for fault diagnosis, *Preprints of IFAC/IMACS Sympo.: SAFEPROCESS'91*, Baden-Baden, pp. 221–226 (Vol.1). also published in revised form in *Automatica*, Vol.29, No.4, 837-841.

Patton, R. J. and Chen, J. (1991c). A parity space approach to robust fault diagnosis using eigenstructure assignment, *Proc. First European Control Conf. ECC'91*, Grenoble, France, pp. 1419–1424.

Patton, R. J. and Chen, J. (1991d). A re-examination of the relationship between parity space and observer-based approaches in fault diagnosis, *European J. of Diagnosis and Safety in Automation (Revue européenne Diagnostic et sûreté de fonctionnement)* **1**(2): 183–200.

Patton, R. J. and Chen, J. (1991e). A review of parity space approaches to fault diagnosis, *Preprints of IFAC/IMACS Sympo.: SAFEPROCESS'91*, Baden-Baden, pp. 239–255 (Vol.1). Invited Survey Paper.

Patton, R. J. and Chen, J. (1991f). Robust fault detection of jet engine sensor systems using eigenstructure assignment, *Proc. of the 1991 AIAA Guidance, Navigation and Control Conf.*, New Orleans, pp. 1666–1975, AIAA–91–2797–CP. also published in revised form in *J. of Guidance, Control and Dynamics*, Vol.15, No.6, 1491-1497, 1992.

Patton, R. J. and Chen, J. (1991g). Robust fault detection using eigenstructure assignment: A tutorial consideration and some new results, *Proc. of the 30th IEEE Conf. on Decision & Control*, Brighton, UK, pp. 2242–2247.

Patton, R. J. and Chen, J. (1991h). A robust parity space approach to fault diagnosis based on optimal eigenstructure assignment, *Proc. of the IEE Int. Con.: Control' 91*, Peregrinus Press, IEE Conf. Pub. No. 332, Edinburgh, pp. 1056–1061.

Patton, R. J. and Chen, J. (1992a). A review of parity space approaches to fault diagnosis applicable to aerospace systems, *1992 AIAA Guidance, Navigation, and Control Conf.*, South Carolina, pp. AIAA–92–4538. also published in revised form in *J. of Guidance, Control and Dynamics*, Vol.17, No.2, 278-285, 1994.

Patton, R. J. and Chen, J. (1992b). Robustness in model-based fault diagnosis, *in* D. Atherton and P. Borne (eds), *Concise Encyclopedia of Simulation and Modelling*, Pergamon Press, pp. 379–392.

Patton, R. J. and Chen, J. (1992c). Robustness in quantitative model-based fault diagnosis, *IEE Colloquium on "Intelligent Fault Diagnosis - Part 2: Model-based Techniques"*. IEE Colloquium Digest No. 1992/048.

Patton, R. J. and Chen, J. (1992d). A robustness study of model-based fault detection for jet engine systems, *Proc. of the 1st IEEE Conf. on Control Application*, Dayton, Ohio, pp. 871–876.

Patton, R. J. and Chen, J. (1993a). Advances in fault diagnosis using analytical redundancy, *IEE Colloquium on "Plant Optimisation for Profit (Integrated Operations Management and Control)"*, pp. 6/1 – 6/12. IEE Colloquium Digest No. 1993/019.

Patton, R. J. and Chen, J. (1993b). A survey of robustness in quantitative model-based fault diagnosis, *Appl. Math. and Comp. Sci.* **3**(3): 399–416. ("Special Issue on Analytical and Artificial Intelligence Based Redundancy in Fault Diagnosis").

Patton, R. J. and Chen, J. (1994). A review of parity space approaches to fault diagnosis for aerospace systems, *J. of Guidance, Contr. & Dynamics* **17**(2): 278–285.

Patton, R. J. and Chen, J. (1996a). Neural networks in nonlinear dynamic systems fault diagnosis, *Engineering Simulation* **13**(6): 905–924. (Special Issue on "Engineering Diagnostics").

Patton, R. J. and Chen, J. (1996b). Robust fault detection and isolation (FDI) systems, *in* C. T. Leondes (ed.), *Dynamics and Control (Vol.74): Techniques in Discrete and Continuous Robust Systems*, Academic Press, pp. 171–224.

Patton, R. J. and Chen, J. (1997). Observer-based fault detection and isolation: robustness and applications, *Contr. Eng. Practice* **5**(5): 671–682.

Patton, R. J. and Chen, J. (1999). Uncertainty modelling and robust fault diagnosis for dynamic systems, *in* R. J. Patton, P. M. Frank and R. N. Clark (eds), *Advances in Fault Diagnosis of Dynamic Systems*, Springer-Verlag. to be published.

Patton, R. J. and Chen, J. (eds) (1998). *Proceedings of IFAC Symposium on Fault Detection, Supervision and Safety for Technical Processes – SAFE-PROSS'97*, Pergamon. ISBN 0-08-0423817.

Patton, R. J., Chen, J. and Liu, G. P. (1997). Robust fault detection of dynamic systems via genetic algorithms, *Proc. of IMechE Part I-J. of Sys. and Contr. Eng.* **211**(5): 357–364.

Patton, R. J., Chen, J. and Lopez-Toribio, C. J. (1998). Fuzzy observers for non-linear dynamic systems fault diagnosis. To appear in the *Proc. of the 37th IEEE Conf. on Decision and Control.*

Patton, R. J., Chen, J. and Millar, J. H. P. (1991). A robust disturbance decoupling approach to fault detection in process systems, *Proc. of the 30'th IEEE Conf. on Decision & Control*, Brighton, UK, pp. 1543–1548.

Patton, R. J., Chen, J. and Millar, J. H. P. (1992). Robust fault detection for a nuclear reactor system: A feasibility study, *Preprints of IFAC Int. Sympo. "On-line fault detection and supervision in the chemical process industries"*, Delaware, USA, pp. 120–125.

Patton, R. J., Chen, J., Millar, J. H. P. and Kiupel, N. (1991). Robust fault detection using eigenstructure assignment - a tutorial, *Proc. of 8'th Int. Conf. on Systems Engineering (ICSE'91)*, Coventry, UK, pp. 214–221.

Patton, R. J., Chen, J. and Nielsen, S. B. (1994). Model-based methods for fault diagnosis: Some guidelines, *Inst. M.C. Colloquium on "Quantitative & Qualitative Methods for Fault Diagnosis in Process Control"*, London. "also submitted to *Trans Inst MC*, 1994".

Patton, R. J., Chen, J. and Siew, T. M. (1994). Fault diagnosis in nonlinear dynamic systems via neural networks, *Proc. of the IEE Int. Conf.: Control' 94*, Peregrinus Press, Conf. Pub. No. 389, Warwick, UK, pp. 1346–1351.

Patton, R. J., Chen, J. and Zhang, H. Y. (1992). Modelling methods for improving robustness in fault diagnosis of jet engine system, *Proc. of the 31st IEEE Conf. on Control & Decision*, Tucson, Arizona, pp. 2330–2335.

Patton, R. J., Frank, P. M. and Clark, R. N. (eds) (1989). *Fault Diagnosis in Dynamic Systems, Theory and Application*, Control Engineering Series, Prentice Hall, New York.

Patton, R. J. and Hou, M. (1997). H_∞ estimation and robust fault detection, *Proc. of the 1997 European Contr. Conf.: ECC97 (CD-ROM)*, Brussels, Belgium.

Patton, R. J. and Kangethe, S. M. (1989). Robust fault diagnosis using eigenstructure assignment of observers, *in* R. J. Patton, P. M. Frank and R. N. Clark (eds), *Fault Diagnosis in Dynamic Systems, Theory and Application*, Prentice Hall, chapter 4, pp. 99–154.

Patton, R. J. and Liu, G. P. (1994). Robust control design via eigenstructure assignment, genetic algorithms and gradient-based optimization, *IEE Proc.-D: Contr. Theory & Appl.* **141**(3): 202–208.

Patton, R. J. and Willcox, S. W. (1987). Fault detection in dynamic systems using a robust output zeroing design method, *in* S. G. Tzafestas, M. G. Singh and G. Schmidt (eds), *System Fault Diagnostics, Reliability & Related Knowledge-based Approaches*, D. Reidel Press, Dordrecht, pp. 177–192 (Vol.1).

Patton, R. J., Willcox, S. W. and Winter, S. J. (1986). A parameter insensitive technique for aircraft sensor fault analysis, *Proc. of the AIAA Conf. on Guidance, Navigation & Control*, Williamsburg, Va., pp. 86–2029–CP. also published in a revised form on *J. of Guidance, Control & Dynamics*, Vol. 10, No.3, 359-367, 1987; translated into Russian and re-published in *J. Of Aero. and Space Techno.*, 82-91, August 1988.

Patton, R. J., Zhang, H. Y. and Chen, J. (1992). Modelling of uncertainties for robust fault diagnosis, *Proc. of the 31st IEEE Conf. on Decision & Control*, Tucson, Arizona, pp. 921–926.

Pau, L. F. (1975). *Failure Diagnosis and Performance Monitoring*, Contr. & Syst. Theory Series (Vol.11), Marcel Dekker, INC., New York.

Peng, Y. B., Youssouf, A., Arte, P. and Kinnaert, M. (1997). A complete procedure for residual generation and evaluation with application to a heat exchanger, *IEEE Trans. Contr. Sys. Techno.* **5**(6): 542–555.

Pfeufer, T. and Ayoubi, M. (1997). Application of a hybrid neuro-fuzzy system to the fault diagnosis of an automotive electromechanical actuator, *Fuzzy Sets and Systems* **89**(3): 351–360.

Pham, D. T. and Oh, S. J. (1992). A recurrent backpropagation neural network for dynamic system identification, *J. of Systems Eng.* **2**(4): 213–223.

Phatak, M. S. and Viswanadham, N. (1988). Actuator fault detection and isolation in linear systems, *Int. J. Sys. Sci.* **19**(12): 2593–2603.

Piercy, N. P. (1989). *A Redundancy Approach to Sensor Failure Detection with Application to Turbofan Engines*, PhD thesis, Univ. of Cambridge, Cambridge, UK.

Polycarpou, M. M. and Helmicki, A. J. (1995). Automated fault-detection and accommodation - a learning-systems approach, *IEEE Trans. Sys. Man & Cyber.* **25**(11): 1447–1458.

Polycarpou, M. M. and Vemuri, A. T. (1995). Learning methodology for failure-detection and accommodation, *IEEE Contr. Syst. Mag.* **15**(3): 16–24.

Potter, I. E. and Suman, M. C. (1977). Thresholdless redundancy management with arrays of skewed instruments, *Technical Report AGARDOGRAPH-224 (pp 15-11 to 15-25)*, AGARD. Integrity in Electronic Flight Control Systems.

Poulizeos, A. D. and Stravlakakis, G. S. (1994). *Real Time Fault Monitoring of Industrial Processes*, Int. Series on Micro-Based and Int. Sys. Eng., Vol.12, Series Editor Tzafestas, S., Kluwer Academic Press.

Qiu, Z. and Gertler, J. (1993). Robust FDI systems and H_∞-optimization: Disturbances and tall fault case, *Proc. of The 32nd IEEE Conf. on Decision & Control*, Texas, USA, pp. 1710–1715. A modified version also published in *Preprints of the IFAC Sympo. on Fault Detection, Supervision and Safety for Technical Processes: SAFEPROCESS'94*, Vol.1, 260-265, Espoo, Finland, June 13-16 1994.

Raghavan, S. and Hedrick, J. K. (1994). Observer design for a class of nonlinear systems, *Int. J. Contr.* **59**(2): 515–528.

Ragot, J., Maquin, D. and Kratz, F. (1993). Analytical redundancy for systems with unknown inputs – application to fault detection, *Control – Theory and Advanced Technology* **9**(3): 775–788.

Ray, A. and Luck, R. (1991). An introduction to sensor signal validation in redundant measurement systems, *IEEE Contr. Syst. Mag.* **11**(2): 44–49.

Roppenecker, G. (1986). On parametric state feedback design, *Int. J. Contr.* **43**(3): 793–804.

Ruokonen, T. (ed.) (1994). *Preprints of IFAC Symposium on Fault Detection, Supervision and Safety for Technical Processes – SAFEPROSS'94*, Espoo, Finland. ISBN 951-96042-6-X.

Sadrnia, M. A., Chen, J. and Patton, R. J. (1997a). Robust fault detection observer design using H_∞/μ techniques for uncertain flight control systems, *in* C. Bányász (ed.), *Preprints of the 2nd IFAC Symp. on Robust Contr. Design: RECOND97*, Budapest, Hungary, pp. 531–536.

Sadrnia, M. A., Chen, J. and Patton, R. J. (1997b). Robust H_∞/μ fault diagnosis observer design, *Proc. of the 1997 European Contr. Conf.: ECC97 (CD-ROM)*, Brussels, Belgium.

Sadrnia, M. A., Chen, J. and Patton, R. J. (1997c). Robust H_∞/μ observer-based residual generation for fault diagnosis, *Proc. of the IFAC Sympo. on Fault Detection, Supervision and Safety for Technical Processes: SAFE-PROCESS'97*, Pergamon 1998, Univ. of Hull, UK, pp. 155–162.

Saif, M. and Guan, Y. (1993). A new approach to robust fault-detection and identification, *IEEE Trans. Aero. & Electron. Syst.* **29**(3): 685–695.

Sauter, D., Rambeaux, F. and Hamelin, F. (1997). Robust fault diagnosis in a H_∞ setting, *Proc. of the IFAC Sympo. on Fault Detection, Supervision and Safety for Technical Processes: SAFEPROCESS'97*, Pergamon 1998, Univ. of Hull, UK, pp. 867–874.

Schneider, H. and Frank, P. M. (1994). Fuzzy logic based threshold adaptation for fault detection in robots, *Proc. of The Third IEEE Conf. on Control Applications*, Glasgow, Scotland, pp. 1127–1132.

Schneider, H. and Frank, P. M. (1996). Observer-based supervision and fault-detection in robots using nonlinear and fuzzy-logic residual evaluation, *IEEE Trans. Contr. Sys. Techno.* **4**(3): 274–282.

Schreier, G., Ragot, J., Patton, R. J. and Frank, P. M. (1997). Observer design for a class of non-linear systems, *Proc. of the IFAC Sympo. on Fault Detection, Supervision and Safety for Technical Processes: SAFEPROCESS'97*, Pergamon 1998, Univ. of Hull, UK, pp. 483–488.

Seliger, R. and Frank, P. M. (1991a). Fault diagnosis by disturbance decoupled nonlinear observers, *Proc. of the 30th IEEE Conf. on Decision & Control*, Brighton, UK, pp. 2248–225.

Seliger, R. and Frank, P. M. (1991b). Robust component fault detection and isolation in nonlinear dynamic systems using nonlinear unknown input observers, *Peprints of IFAC/IMACS Symp.: SAFEPROCESS'91*, Baden-Baden, pp. 313–318 (Vol.1).

Seliger, R. and Frank, P. M. (1993). Robust residual evaluation by threshold selection and a performance index for nonlinear observer-based fault diagnosis, *Proc. of Int. Conf. on Fault Diagnosis: TOOLDIAG'93*, Toulouse, pp. 496–504.

Seliger, R. and Frank, P. M. (1999). Robust observer-based fault diagnosis in non-linear uncertain systems, *in* R. J. Patton, P. M. Frank and R. N. Clark (eds), *Advances in Fault Diagnosis of Dynamic Systems*, Springer-Verlag. to be published.

Shaked, U. (1990). H_∞-minimum error state estimation of linear stationary processes, *IEEE Trans. Automat. Contr.* **AC-35**(5): 554–558.

Shaked, U. and Theodor, Y. (1992a). A frequency-domain approach to the problems of H_∞-minimum error state estimation and deconvolution, *IEEE Trans. On Signal Proc.* **40**(12): 3001–3011.

Shaked, U. and Theodor, Y. (1992b). H_∞-optimal estimation: an tutorial, *Proc. of the 31st Conf. on Decision & Control*, Tucson, AZ, USA, pp. 2278–2286.

Shen, L. C., Chang, S. K. and Hsu, P. L. (1998). Robust fault detection and isolation with unstructured uncertainty using eigenstructure assignment, *J. of Guidance, Contr. & Dynamics* **21**(1): 50–57.

Shen, Q. and Leitch, R. (1993). Fuzzy qualitative simulation, *IEEE Trans. on Sys., Man & Cybernetics* **SMC-23**(4): 1038–1061.

Shields, D. N. (1994). Robust fault detection for generalized state space systems, *Proc. of the IEE Int. Conf.: Control' 94*, Peregrinus Press, Conf. Pub. No. 389, Warwick, UK, pp. 1335–1349.

Shields, D. N. (1997). Observer design and detection for nonlinear descriptor systems, *Int. J. Contr.* **67**(2): 153–168.

Singh, M. G., Hindi, K. S., Schmidt, M. G. and Tzafestas, S. G. (eds) (1987). *Fault Detection and Reliability: Knowledge Based and Other Approaches*, Int. Series on Syst. & Contr., Pergamon Press, Oxford. ISBN 0-08-034922-6.

Slotine, J. J. E., Hedrick, J. K. and Misawa, E. A. (1987). On sliding observers for nonlinear systems, *J. Dyn. Sys., Meas. & Contr. - Trans. of the ASME* **109**(2): 245–252.

Smed, T. B., Carlsson, C. E., de Sonza, C. E. and Goodwin, G. C. (1988). Fault detection and diagnosis applied to gas turbines, *Technical Report EE8815*, Univ. of Newcastle, Australia.

Sobel, K. M. and Banda, S. S. (1989). Design of a modalized observer with eigenvalue sensitivity reduction, *J. of Guidance, Contr. & Dynamics* **12**(5): 762–764.

Sobel, K. M., Shapiro, E. Y. and Andry, A. N. (1994). Eigenstructure assignment, *Int. J. Contr.* **59**(1): 13–37.

Sohlberg, B. (1998). Monitoring and failure diagnosis of a steel strip process, *IEEE Trans. Contr. Sys. Techno.* **6**(2): 294–303.

Sorsa, T. and Koivo, H. N. (1993). Application of artificial neural networks in process fault-diagnosis, *Automatica* **29**(4): 843–849.

Sorsa, T., Koivo, H. N. and Koivisto, H. (1991). Neural networks in process fault diagnosis, *IEEE Trans. Systems, Man & Cybernetics* **21**(4): 815–825.

Speyer, J. L. and White, J. E. (1984). Shiryayev sequential probability ratio test for redundancy management, *J. of Guidance, Contr. & Dynamics* **7**(5): 588–595.

Staroswiecki, M., Cassar, J. P. and Cocquempot, V. (1993a). A general approach for multicriteria optimization of structured residuals, *Proc. of Int. Conf. on Fault Diagnosis: TOOLDIAG'93*, Toulouse, pp. 800–807.

Staroswiecki, M., Cassar, J. P. and Cocquempot, V. (1993b). Generation of optimal structured residuals in the parity space, *Preprints of The 12th IFAC World Congress*, Australia, pp. 299–305 (Vol.8).

Stein, J. L. (1993). Modelling and state estimator design issues for model-based monitoring systems, *J. Dyn. Sys., Meas. & Contr. - Trans. of the ASME* **115**(2B): 318–327.

Stengel, R. F. (1991). Intelligent failure-tolerant control, *IEEE Contr. Syst. Mag.* **11**(4): 14–23.

Stengel, R. F. (1993). Toward intelligent flight control, *IEEE Trans. on Sys., Man & Cybernetics* **23**(6): 1699–1717.

Stoustrup, J. and Grimble, M. J. (1997). Integrating control and fault diagnosis: a separation result, *Proc. of the IFAC Sympo. on Fault Detection, Supervision and Safety for Technical Processes: SAFEPROCESS'97*, Pergamon 1998, Univ. of Hull, UK, pp. 313–318.

Stoustrup, J., Grimble, M. J. and Niemann, H. (1997). Design of integrated systems for the control and detection of actuator/sensor faults, *Sensor Review* **17**(2): 138–149.

Stoustrup, J. and Niemann, H. (1998). Fault detection for nonlinear systems - a standard problem approach. To appear in the *Proc. of the 37th IEEE Conf. on Decision and Control.*

Strang, G. (1988). *Linear Algebra and its Applications*, 3rd edn, Harcourt Brace Jovanovich, San Diego.

Takagi, T. and Sugeno, M. (1985). Fuzzy identification of systems and its applications to modelling and control, *IEEE Trans. Sys. Man & Cyber.* **15**(1): 116–132.

Tanaka, K., Ikeda, T. and Wang, H. O. (1996). Robust stabilisation of a class of uncertain non-linear systems via fuzzy control: quadratic stabilisability, H_∞ control theory, and linear matrix inequalities, *IEEE Trans. Fuzzy Sys.* **4**(1): 1–13.

Tanaka, K., Ikeda, T. and Wang, H. O. (1998). Fuzzy regulators and fuzzy observers: relaxed stability conditions and LMI-based designs, *IEEE Trans. Fuzzy Sys.* **6**(2): 250–265.

Tanaka, K. and Sugeno, M. (1992). Stability analysis and design of fuzzy control systems, *Fuzzy Sets and Syst.* **45**(2): 135–156.

Tanaka, S. and Müller, P. C. (1990). Fault detection in linear discrete dynamic systems by a pattern-recongnition of a Generalized-Likelihood-Ratio, *J. Dyn. Sys., Meas. & Contr. - Trans. of the ASME* **112**(2): 276–282.

Tanaka, S. and Müller, P. C. (1993). Fault-detection in linear discrete dynamic-systems by a reduced order generalized-likelihood-ratio method, *Int. J. Sys. Sci.* **24**(4): 721–732.

Thau, F. E. (1973). Observing the state of non-linear dynamic systems, *Int. J. Contr.* **17**(3): 471–479.

Tsai, T. M. and Chou, H. P. (1993). Sensor fault-detection with the single sensor parity relation, *Nuclear Science and Engineering* **114**(2): 141–148.

Tufts, D. W., Kumaresan, R. and Kirsteins, I. (1982). Data adaptive signal estimation by singular value decomposition of a data matrix, *Proc. IEEE* **70**(6): 684–685.

Tzafestas, S. G. (1989). System fault diagnosis using the knowledge-based methodology, *in* R. J. Patton, P. M. Frank and R. N. Clark (eds), *Fault Diagnosis in Dynamic Systems, Theory and Application*, Prentice Hall, chapter 15, pp. 509–572.

Tzafestas, S. G. and Watanabe, K. (1990). Modern approaches to system/sensor fault detection and diagnosis, *Journal A* **31**(4): 42–57.

Ulieru, M. (1993). A fuzzy logic based computer assisted fault diagnosis system, *Proc. of Int. Conf. on Fault Diagnosis: TOOLDIAG'93*, Toulouse, France, pp. 689–699 (Vol.2).

Ulieru, M. and Isermann, R. (1993). Design of fuzzy-logic based diagnostic model for technical process, *Fuzzy Set and Systems* **58**(3): 249–271.

van Schrick, D. (1991). Investigations of reliability for instrument fault detection state-estimator schemes, *European J. of Diagnosis and Safety in Automation (Revue européenne Diagnostic et sûreté de fonctionnement)* **1**(1): 63–78.

van Schrick, D. (1993). *Zustandsschätzerschemen zur Fehlerkenung, deren Zuverlässigkeit und Anwendung auf den spurgeführten Omnibus*, PhD thesis, University of Wuppertal, Germany.

van Schrick, D. (1994a). A comparison of IFD schemes: A decision aid for designers, *Proc. of The Third IEEE Conf. on Control Applications*, Glasgow, Scotland, pp. 889–894.

van Schrick, D. (1994b). FDI residual generators - a comparison, *Preprints of the IFAC Sympo. on Fault Detection, Supervision and Safety for Technical Processes: SAFEPROCESS'94*, Espoo, Finland, pp. 545–550 (Vol.2).

van Schrick, D. (1997). Remarks on terminology in the field of supervision, fault detection and diagnosis, *Proc. of the IFAC Sympo. on Fault Detection, Supervision and Safety for Technical Processes: SAFEPROCESS'97*, Pergamon 1998, Univ. of Hull, UK, pp. 959–964.

Vardulakis, A. (1991). *Linear Multivariable Control Algebriac Analysis and Synthesis Methods*, John Wiley.

Vemuri, A. T. and Polycarpou, M. M. (1997a). Neural-network-based robust fault diagnosis in robotic systems, *IEEE Trans. On Neural Networks* **8**(6): 1410–1420.

Vemuri, A. T. and Polycarpou, M. M. (1997b). Robust nonlinear fault diagnosis in input-output systems, *Int. J. Contr.* **68**(2): 343–360.

Vemuri, A. T., Polycarpou, M. M. and Diakourtis, S. A. (1998). Neural network based fault detection in robotic manipulators, *IEEE Trans. On Robotics and Automation* **14**(2): 342–348.

Vidyasagar, M. (1985). *Control Systems Synthesis: A Factorization Approach*, North Holland System and Control Series (Vol.8), MIT Press, Cambridge, MA.

Vidyasagar, M. (1993). *Nonlinear Systems Analysis*, 2nd edn, Prentice Hall, New Jersey.

Villaneuva, H., Merrington, G., Ninness, B. M. and Goodwin, G. C. (1991). Application of robust fault detection methods to F404 gas turbine engines, *Peprint of IFAC/IMACS Symp.: SAFEPROCESS'91*, Baden-Baden, pp. 223–228 (Vol.2).

Viswanadham, N. and Minto, K. D. (1988). Robust observer design with application to fault detection, *Proc. the 1988 Amer. Control Conf.*, Atlanta, GA, pp. 1393–1399 (Vol.3).

Viswanadham, N., Sarma, V. V. S. and Singh, M. G. (1987). *Reliability of Compueter and Control Systems*, North Holland System and Control Series (Vol.8), North-Holland, Amsterdam.

Viswanadham, N. and Srichander, R. (1987). Fault detection using unknown input observers, *Control – Theory and Advanced Technology* **3**(2): 91–101.

Viswanadham, N., Taylor, J. H. and Luce, E. C. (1987). A frequency-domain approach to failure detection and isolation with application to GE-21 turbine engine control systems, *Control – Theory and Advanced Technology* **3**(1): 45–72.

Wald, A. (1947). *Sequential Analysis*, Wiley, New York.

Walker, B. (1989). Fault detection threshold determination using markov theory, *in* R. J. Patton, P. M. Frank and R. N. Clark (eds), *Fault Diagnosis in Dynamic Systems, Theory and Application*, Prentice Hall, chapter 14, pp. 477–508.

Walker, B. K. (1983). Recent developments in fault diagnosis and accommodation, *Proc. of 1983 AIAA Guidance & Control Conf.*, Gatlingburg Ten., pp. 83–2358–CP.

Walker, B. K. and Gai, E. (1979). Fault detection threshold determination techniques using Markov theory, *J. of Guidance, Contr. & Dynamics* **2**(4): 313–319.

Wang, H. and Daley, S. (1993). A fault-detection method for unknown systems with unknown input and its application to hydraulic-turbine monitoring, *Int. J. Contr.* **57**(2): 247–260.

Wang, H., Kropholler, H. and Daley, S. (1993). Robust observer based FDI and its application to the monitoring of a distillation column, *Trans Inst MC* **15**(5): 221–227.

Wang, S. H., Davison, E. J. and Dorato, P. (1975). Observing the states of systems with unmeasurable disturbance, *IEEE Trans. Automat. Contr.* **AC-20**: 716–717.

Wang, Y. Y. and Wu, N. E. (1993). An approach to configuration of robust control systems for robust failure detection, *Proc. of the 32nd IEEE Conf. on Decision and Contr.*, Texas, pp. 1704–1709.

Watanabe, K. and Himmelblau, D. M. (1982). Instrument fault detection in systems with uncertainties, *Int. J. Sys. Sci.* **13**(2): 137–158.

Watanabe, K., Matsuura, I., Abe, M., Kubota, M. and Himmelblau, D. M. (1989). Incipient fault diagnosis of chemical processes via artificial neural networks, *AIChE J.* **35**(11): 1803–1812.

Weiss, J. L. and Hsu, J. Y. (1985). Design and evaluation of a failure detection and isolation algorithm for restructurable control systems, *Technical Report NASA-CR-178213*, NASA.

White, B. A. (1991). Eigenstructure assignment by output-feedback, *Int. J. Contr.* **53**(6): 1413–1429.

White, J. E. and Speyer, J. L. (1987). Detection filter design: spectral theory and algorithm, *IEEE Trans. Automat. Contr.* **AC-32**(7): 593–603.

Willis, M. J., Massimo, C. D., Montague, G. A., Tham, M. T. and Morris, A. J. (1991). Artificial neural networks in process engineering, *IEE Proc.-D: Contr. Theory & Appl.* **138**(3): 256–266.

Willsky, A. S. (1976). A survey of design methods for failure detection in dynamic systems, *Automatica* **12**(6): 601–611.

Willsky, A. S., Deyst, J. J. and Crawford, B. S. (1974). Adaptive filtering and self-test methods for failure detection and compensation, *Proc. of The 1974 Joint Amer. Contr. Conf.*, Austin, pp. 637–645.

Willsky, A. S., Deyst, J. J. and Crawford, B. S. (1975). Two self-test methods applied to internal system problem, *J. Spacecrafts and Rockets* **12**(7): 434–437.

Willsky, A. S. and Jones, H. L. (1974). A generalized likelihood approach to state estimation in linear systems subjected to abruppt changes, *Proc. of The 1974 IEEE Conf. on Control and Decision*, Arizona.

Willsky, A. S. and Jones, H. L. (1976). A generalized likelihood ratio approach to the detection and estimation of jumps in linear systems, *IEEE Trans. Automat. Contr.* **AC-21**: 108–121.

Wise, B. M. and Gallagher, N. B. (1996). The process chemometrics approach to process monitoring and fault- detection, *J. of Process Control* **6**(6): 329–348.

Wu, N. E. (1992). Failure sensitizing reconfigurable control design, *Proc. of the 31st IEEE Conf. on Control & Decision*, Tucson, AZ, pp. 44–49.

Wu, N. E. and Wang, Y. Y. (1993). Robust failure detection with parity check on filtered measurements, *Proc. of the 32nd IEEE Conf. on Decision and Contr.*, Texas, pp. 255–256.

Wu, N. E. and Wang, Y. Y. (1995). Robust failure-detection with parity check on filtered measurements, *IEEE Trans. Aero. & Electron. Syst.* **31**(1): 489–491.

Wünnenberg, J. (1990). *Observer-based Fault Detection in Dynamic Systems*, PhD thesis, Univ. of Duisburg, Germany.

Wünnenberg, J. and Frank, P. M. (1987). Sensor fault using detection via robust observers, *in* S. G. Tzafestas, M. G. Singh and G. Schmidt (eds), *System Fault Diagnostics, Reliability & Related Knowledge-based Approaches*, D. Reidel Press, Dordrecht, pp. 147–160.

Wünnenberg, J. and Frank, P. M. (1988). Model-based residual generation for dynamic systems with unknown inputs, *Proc. 12th IMACS World Congress on Scientific Computation*, Paris, pp. 435–437 (Vol.2).

Wünnenberg, J. and Frank, P. M. (1990). Robust observer-based detection for linear and non-linear systems with application to robots, *Proc. of IMACS Annals on Computing & Applied Mathematics* $MIM - S^2$*:90*, Brussels.

Xia, X. and Zeitz, M. (1997). On nonlinear continuous observers, *Int. J. Contr.* **66**(6): 943–954.

Yaesh, I. and Shaked, U. (1992). Game-theory approach to optimal linear state estimation and its relation to the minimum H_∞-norm estimation, *IEEE Trans. Automat. Contr.* **37**(6): 828–831.

Yang, F. and Richard, W. W. (1988). Observers for linear systems with unknown inputs, *IEEE Trans. Automat. Contr.* **AC-33**(7): 677–681.

Yang, H. L. and Saif, M. (1997). State observation, failure detection and isolation (FDI) in bilinear systems, *Int. J. Contr.* **67**(6): 901–920.

Yao, Y. X., Schaefers, J. and Darouach, M. (1994). Robust fault detection based on factorization approach, *Preprints of the IFAC Sympo. on Fault Detection, Supervision and Safety for Technical Processes: SAFEPROCESS'94*, Espoo, Finland, pp. 525–530 (Vol.2).

Yu, D. L. and Shields, D. N. (1994). A fault detection methods for a nonlinear system and its application to a hydraulic test rig, *Preprints of the IFAC Sympo. on Fault Detection, Supervision and Safety for Technical Processes: SAFEPROCESS'94*, Espoo, Finland, pp. 305–310 (Vol.1).

Yu, D. L. and Shields, D. N. (1996). A bilinear fault detection observer, *Automatica* **32**(11): 1597–1602.

Yu, D. L. and Shields, D. N. (1997). A bilinear fault detection filter, *Int. J. Contr.* **68**(3): 417–430.

Yu, D. L., Shields, D. N. and Daley, S. (1996). A bilinear fault-detection observer and its application to a hydraulic drive system, *Int. J. Contr.* **64**(6): 1023–1047.

Yu, D. L., Shields, D. N. and Mahtani, J. L. (1994a). Fault detection for bilinear systems with application to a hydraulic system, *Proc. of The Third IEEE Conf. on Control Applications*, Glasgow, Scotland, pp. 1379–1384.

Yu, D. L., Shields, D. N. and Mahtani, J. L. (1994b). A nonlinear fault detection method for a hydraulic system, *Proc. of the IEE Int. Conf.: Control' 94*, Peregrinus Press, Conf. Pub. No. 389, Warwick, UK, pp. 1318–1322.

Zakian, V. (1979). New formulation for the method of inequalities, *Proc. IEE* **126**(6): 579–584.

Zakian, V. and Al-Naib, U. (1973). Design of dynamical and control systems by the method of inequalities, *IEE Proc.-D: Contr. Theory & Appl.* **120**(11): 1421–1427.

Zhang, J. (1991). *Expert Systems in On-line Process Control and Fault Diagnosis*, PhD thesis, City Univ., London.

Zhang, J., Martin, E. B. and Morris, A. J. (1996). Fault-detection and diagnosis using multivariate statistical techniques, *Chemical Engineering Research & Design* **74**(A1): 89–96.

Zhang, J., Martin, E. B. and Morris, A. J. (1997). Process monitoring using non-linear statistical techniques, *Chemical Eng. J.* **67**(3): 181–189.

Zhang, J. and Morris, J. (1996). Process modeling and fault-diagnosis using fuzzy neural networks, *Fuzzy Sets and Systems* **79**(1): 127–140.

Zhang, Q., Slater, G. L. and Allemang, R. J. (1990). Suppression of undesired inputs of linear systems by eigenspace assignment, *J. of Guidance, Contr. & Dynamics* **13**(2): 330–336.

Zhou, K., Doyle, J. C. and Golver, K. (1996). *Robust and Optimal Control*, Prentice Hall, New Jersey.

Zhou, K. and Khargonekar, P. P. (1988). An algebraic riccati equation approach to H_∞ optimization, *Syst. Contr. Lett.* **11**(2): 85–91.

Zhuang, Z. and Frank, P. M. (1997). Qualitative observer and its application to fault detection and isolation systems, *Proc. IMechE Part I-J. of Sys. and Contr. Eng.* **211**(4): 253–262.

Zolghadri, A. (1996). An algorithm for real-time failure-detection in Kalman filters, *IEEE Trans. Automat. Contr.* **41**(10): 1537–1539.

Index

About the Authors

Dr Jie Chen received BEng and MSc degrees in Control Systems Engineering from Beijing University of Aeronautics and Astronautics, China, in 1984 and 1987 respectively, and DPhil degree in Electronic Engineering from University of York, UK, in 1995. He joined the Department of Mechanical Engineering of Brunel University as a Lecturer of Aeronautical Engineering in February 1998. Before that, he worked in the University of Hull, UK as a Lecturer of Control Systems Engineering between July 1995 to January 1998. From March 1990 to September 1994, he worked as a Research Associate, in the University of York, UK, while he pursued his DPhil degree. From October 1994 to June 1995, he spent a short period in the University of Strathclyde, UK as Post-Doctoral Research Fellow. He has worked in the field of fault diagnosis and fault-tolerant control over eight years and has published over 50 papers in international journals and conference proceedings on the subject. He is a member of IFAC Technical Committee: *SAFEPROCESS*. He was awarded (jointly with R. J. Patton) the 1997 IEE Kelvin Premium for a paper published in IEE Proceedings-D. His current research interests are: model-based fault diagnosis and applications to non-linear systems, robust and fault-tolerant control, neuro-fuzzy techniques for control and fault diagnosis.

Professor Ron J. Patton graduated in Electrical & Electronic Engineering ('72). His post-graduate training includes two years' experience with the BBC Research Department ('72/74), followed by an MEng degree in Control Systems Engineering ('75) and PhD research into Mathematical Modelling of Biological Control Systems ('75/77), University of Sheffield. After one year as SERC Research Assistant, Sheffield City Polytechnic ('77/78) he was appointed lecturer in Control Systems and Electronics at Sheffield City Polytechnic ('78/81) and, subsequently as Lecturer, Department of Electronics, University of York ('81), promoted to Senior Lecturer in 1987. Since January 1995, he has been with Department of Electronic Engineering, the University of Hull as a Professor of Control and Intelligent Systems Engineering. Prof. Patton has a long track record of fundamental contributions to model-based fault diagnosis and control systems design via eigenstructure assignment. He has served on a number of professional committees, e.g., as Chairman of the IFAC Technical Committee: Fault Detection, Safety and Supervision of Technical Processes (*SAFEPROCESS*), Chairman of IEE C8 PG Control Systems Theory and Design. He has also served as consultant for NATO, and several European organizations and research consortia. Ron Patton is a co-proposer and co-organizer of the European Science Foundation (ESF) programme on Complex Control Systems (COSY), and has been the chairman of the theme group on fault-tolerant control. He has published over 240 papers on fault diagnosis and robust control systems in leading journals, conference proceedings and book articles. He was awarded (jointly with J. Chen) the 1997 IEE Kelvin Premium for a paper published in IEE Proceedings-D. He is the co-editor of two authoritative books on fault diagnosis.

Printed in the United States
127181LV00002B/100/A

9 780792 384113